The ICE Conditions of Contract: Seventh Edition

Brian Eggleston
CEng, FICE, FIStructE, FCIArb

Blackwell Science

© Brian Eggleston, 1993, 2001

Blackwell Science Ltd
Editorial Offices:
Osney Mead, Oxford OX2 0EL
25 John Street, London WC1N 2BS
23 Ainslie Place, Edinburgh EH3 6AJ
350 Main Street, Malden
    MA 02148 5018, USA
54 University Street, Carlton
    Victoria 3053, Australia
10, rue Casimir Delavigne
    75006 Paris, France

Other Editorial Offices:

Blackwell Wissenschafts-Verlag GmbH
Kurfürstendamm 57
10707 Berlin, Germany

Blackwell Science KK
MG Kodenmacho Building
7–10 Kodenmacho Nihombashi
Chuo-ku, Tokyo 104, Japan

Iowa State University Press
A Blackwell Science Company
2121 S. State Avenue
Ames, Iowa 50014-8300, USA

The right of the Author to be identified as the
Author of this Work has been asserted in
accordance with the Copyright, Designs and
Patents Act 1988.

First edition of *The ICE Conditions Sixth Edition*
published 1993
Reprinted 1994
Second edition of *The ICE Conditions Sixth Edition*
published 2001 under the title *The ICE Conditions
Seventh Edition*

Set in 10/12 Palatino
by DP Photosetting, Aylesbury, Bucks
Printed and bound in Great Britain by
MPG Books Ltd, Bodmin, Cornwall

The Blackwell Science logo is a
trade mark of Blackwell Science Ltd,
registered at the United Kingdom
Trade Marks Registry

DISTRIBUTORS
    Marston Book Services Ltd
    PO Box 269
    Abingdon
    Oxon OX14 4YN
    (*Orders:* Tel: 01235 465500
            Fax: 01235 465555)

USA
    Blackwell Science, Inc.
    Commerce Place
    350 Main Street
    Malden, MA 02148 5018
    (*Orders:* Tel: 800 759 6102
            781 388 8250
            Fax: 781 388 8255)

Canada
    Login Brothers Book Company
    324 Saulteaux Crescent
    Winnipeg, Manitoba R3J 3T2
    (*Orders:* Tel: 204 837-2987
            Fax: 204 837-3116)

Australia
    Blackwell Science Pty Ltd
    54 University Street
    Carlton, Victoria 3053
    (*Orders:* Tel: 03 9347 0300
            Fax: 03 9347 5001)

A catalogue record for this title
is available from the British Library

ISBN 0-632-05196-5

Library of Congress
Cataloging-in-Publication Data
is available

For further information on
Blackwell Science, visit our website:
www.blackwell-science.com

Contents

Preface

At the Institution of Civil Engineers' launch of the Seventh edition of ICE Conditions the questions put to the presentation panel from the floor were more concerned with the logic of publishing a Seventh edition than with the contents. One strongly expressed view was that with the Institution's New Engineering Contract (now called NEC Engineering and Construction Contract) steadily gaining ground and recognition there was no point in continuing the publication of traditional ICE Conditions. It was suggested they were a relic of another age and after 50 years or so of use it was time for change.

My own view is that not only is there room for both sets of conditions in the industry but there is need for both. It is a matter of horses for courses as to which type of contract is best for any particular project. The New Engineering Contract requires project management skills, high staffing levels and a prominent place on office desks. Traditional ICE contracts have been satisfactorily run for decades and in thousands by civil engineers practising their ordinarily professional skills, frequently with mud on their boots and with the Conditions rarely on the office desk. The strength of the Conditions is in their consistency and continuity and by any standards they must be regarded as one of the most successful standard forms ever published.

I look forward to reviewing the Eighth edition in my retirement.

Brian Eggleston
April 2001

Notes on the text

Format

The format of most chapters is an introductory section followed by one or more sections with general information on the subject matter of the chapter and thereafter detailed analysis of the relevant clauses of the Conditions.

Phraseology

Rather than repeat throughout the book the full title of the Conditions – the ICE Conditions of Contract, Measurement Version, 7th edition – I use abbreviations where appropriate.

Generally, when I refer to the Conditions I mean the Seventh edition and when I refer to ICE Conditions I mean the Seventh edition and its predecessors. References to the New Engineering Contract are intended as references to the NEC Engineering and Construction Contract.

Capitals

As a matter of style capital letters have been used sparingly. The result, I hope, is easier reading. However, it does mean that defined contractual terms such as the employer and the contractor appear in lower case except in quotations.

Lists

Bullet points are generally used for lists, but where the lists are of sub-clauses from the Conditions, letters are used.

Text of the Conditions

For commercial reasons it has not been possible to include the text of the Conditions in this book. Readers may find it helpful to have to hand a copy of the Conditions.

Chapter 1
Introduction

1.1 Publication

The First edition of ICE Conditions of Contract was published in 1945; the Second in 1950; the Third in 1951; the Fourth in 1955; the Fifth in 1973; the Sixth in 1991; and the Seventh in September 1999. The Seventh edition is published by Thomas Telford Publishing on behalf of the Institution of Civil Engineers, the Association of Consulting Engineers and the Civil Engineering Contractors Association.

By one measure of reckoning the Seventh edition appeared about 10 years earlier than might have been expected. The Fourth edition lasted 18 years; the Fifth edition lasted 18 years; and had the Sixth edition done likewise the Seventh edition was not due until 2009.

To the extent that publication of the Seventh edition Conditions can be viewed as premature, the explanation seems to lie in a desire on the part of the sponsoring bodies to see the amendments to the Sixth edition brought about by the Housing Grants, Construction and Regeneration Act 1996 properly incorporated into a newly printed version of the Conditions. But for those amendments the changes from the Sixth to the Seventh edition would not in themselves have justified the release of a new edition.

1.2 Policy

The Seventh edition retains the key characteristics of traditional ICE Conditions:

- valuation by remeasurement
- engineer responsible for design (unless expressly stated otherwise)
- engineer as the impartial certifier and valuer
- engineer's decision as the first stage of dispute resolution (now renamed as decisions on matters of dissatisfaction).

Whether the Seventh edition retains as firmly as previous editions the primacy of the contractor's tendered rates and prices in the valuation process is questionable given the introduction of quotations for variations.

Some further uncertainty on long term intentions for ICE Conditions might be taken from the title of the Seventh edition as published in 1999 – ICE Conditions of Contract, Measurement Version. This could hint at the possibility in due course of a lump sum version but it may be no more than part of a scheme to have distinguishing titles for each of the standard forms in the ICE set of Conditions – e.g. ICE Design and Construct Conditions of Contract; ICE Conditions of Contract for Minor Works; ICE Conditions of Contract, Measurement Version.

Drafting aims

The text of the Seventh edition suggests that the primary aims of the drafting committee were:

- to incorporate amendments made to the Sixth edition since its publication in 1991
- to incorporate certain ideals from Sir Michael Latham's 1994 report 'Constructing the Team' – e.g. pre-pricing of variations
- to modernise certain provisions which with the passage of time have come to look old-fashioned – e.g. those relating to title, to war and to frustration
- to amend certain provisions of the Sixth edition which have attracted justifiable criticism – e.g. clause 11 on the provision of information by the employer
- to rectify conspicuous omissions from the text of earlier editions of ICE Conditions – e.g. extension of time for employer's default
- to correct small errors and faults detected in the text of the Sixth edition – e.g. liability for costs of tests etc.
- to modernise certain terms – e.g. 'carrying out' for 'execution'.

There is little in the Seventh edition other than in relation to performance bonds to suggest that the drafting committee felt it necessary to make changes in recognition of or to counter the impact of recent legal decisions. Some such change might have been welcome, for instance to clause 12 to take account of the decision in *Humber Oils* v. *Harbour & General* (1991) on the meaning of 'physical conditions'.

Overall, the policy of the Seventh edition is perhaps best reflected in the following extracts from the introductory section of the Guidance Notes to the Conditions:

> 'Like all previous editions of the ICE Conditions of Contract, the Seventh Edition is based on the traditional pattern of Engineer-designed, Contractor-built Works with valuation by admeasurement. The traditional role of the Engineer in advising his client, designing the Works, supervising construction, certifying payment and deciding what are now called matters of dissatisfaction is fully maintained.
>
> The Contract is drafted bearing in mind the benefits of team working and current procurement initiatives. If its procedures are followed the parties to the Contract will be provided with an "early warning" of cir-

cumstances that may give rise to additional costs or delay in a stage or completion of the Contract.

To enable the minimisation of additional cost and/or delay as well as potential for dispute all possible steps should be taken to avoid changes to pre-planned Works information. Design work should be completed before tender information is sent out and then not be subject to change. Changes in requirements after contracts have been let will normally lead to substantial increases in cost and undermine the principle of mutual co-operation.'

The mention in the Guidance Notes of 'early warning' procedures should not be taken as heralding the arrival in the Seventh edition of formalised early warning procedures similar to those in the New Engineering Contract (NEC Engineering and Construction Contract). It seems to mean no more than that if the notice requirements of the Conditions are strictly followed then there will be early warning of additional cost and delay situations.

1.3 *Changes from the Sixth edition*

Comparison of the Seventh edition with bound versions of the Sixth edition shows a level of change between the Sixth and Seventh editions which overstates the true position. Account needs to be taken of amendments to the Sixth edition since its publication in 1991. Some of these were incorporated in the 1995 and 1997 reprints but the important 1998 amendments arising from the Housing Grants, Construction and Regeneration Act 1996 remained in loose leaf form and it is only in the Seventh edition that they appear in a bound set of the Conditions.

The 1998 amendments relate principally to payments and to dispute resolution procedures and the background and impact of these amendments are explained in Chapters 20 and 28. They may look like changes from the Sixth to the Seventh edition but as amendments to the Conditions pre-dating publication of the Seventh edition they are not true changes.

The true changes can be seen, in the broadest sense, as intended or unintended – with further sub-division to significant and insignificant.

Intended changes of significance

- clause 3 – new clause 3(2) introduced to counter application of the Contracts (Rights of Third Parties) Act 1999
- clause 4 – clause 4(2) reworded to provide additional controls on the appointment of subcontractors
- clause 11 – clauses 11(1) and 11(3) reworded changing the obligations of the parties on the provision and interpretation of site information
- clause 12 – clause 12(6) reworded to expand and clarify the engineer's duties in respect of delay and extra cost claims for unforeseen conditions
- clause 26 – new clause 26(4) introduced stating the contractor's entitle-

ments when the engineer issues instructions or the employer is in default in respect of statutes/regulations etc.

- clause 36 – new clause 36(4) introduced to the effect that the cost of tests falls on the employer unless necessitated by the contractor's default
- clause 41 – clause 41(1)(b) reworded so that where the engineer notifies the start date it is no earlier than 14 days after the award of contract
- clause 42 – clauses 42(2) and 42(3) reworded, the major effect being that unless the contract prescribes otherwise the employer is required to give possession of the whole of the site at commencement
- clause 44 – new clause 44(1)(e) introduced stating any delay, impediment, prevention or default of the employer as ground for extension of time
- clause 52 – new clauses 52(1) and 52(2) introduced on the pre-pricing of variations and the valuation of variations by quotation
- clause 54 – clause 54(1) reworded deleting the employer's deemed ownership of the contractor's equipment, temporary works etc.
- clause 59 – new clause 59(4)(f) introduced stating contractor's entitlement to extension of time for nominated subcontractor's default
- clause 63 – Seventh edition clause 63 is a composite and simplified replacement for clauses 64 (frustration) and 65 (war) of the Sixth edition
- clause 64 – new clause stating the contractor's right to terminate his own employment under the contract in the event of certain specified defaults of the employer
- clause 68 – new clause 68(3) states that notices to be given to the engineer shall be delivered as the engineer may direct
- clause 69 – new clause 69(1)(b) provides for adjustment of the contract price in respect of landfill tax fluctuations.

Unintended changes of potential significance

One of the recognised difficulties (or hazards) of amending documents is that changes in one part can have unintended consequences in respect of other parts. Only time will tell the full extent to which ICE Seventh edition Conditions have acquired traps and surprises from the apparently modest changes made from the Sixth edition. Attempts are made in later chapters of this book to identify potential problems in a clause by clause analysis. Not everything will have been found and it is speculative to assess the importance of those that have been found. The following are given here simply by way of examples:

- approvals/consents/acceptances – the Seventh edition retains in clauses 1(1)(f) and 1(1)(g) the rule that specifications and drawings approved in writing by the engineer from time to time are classed as being within the 'Contract' (as defined). Sensibly many of the references in other clauses of the Conditions, which in the Sixth edition were to approvals, are changed in the Seventh edition to 'consents' and 'acceptances' to avoid contractor produced documents unwittingly being given contractually binding status. However the changes may have gone too far in that under clause 7(6), drawings, specifications and the like relating to permanent works designed by the contractor are now submitted only for 'acceptance' and yet these are documents which should be binding on the contractor.

- sufficiency of tender – by clause 11(3) of the Sixth edition the contractor was deemed to have based his tender on information made available by the employer and his own inspection and examination. Under the re-worded clause 11(3) of the Seventh edition the contractor is deemed to have based his tender on his own inspection and examination and on 'all information whether obtainable by him or made available by the Employer'. This is potentially of much wider application and it could impact significantly on the contractor's entitlements under clause 12 in respect of unforeseen conditions.
- suspension for weather conditions – under clause 40 of the Sixth edition the engineer was entitled to take delay caused by suspension due to weather conditions into account when assessing the contractor's entitle-ment to extension of time. The wording of clause 40 is re-arranged in the Seventh edition and, accidentally it is thought, the engineer is now precluded from taking into account suspension due to weather con-ditions.
- delay caused by nominated subcontractor's default – the new clause 59(4)(f) of the Seventh edition giving the contractor entitlement to exten-sion of time for nominated subcontractor's default may be intended to apply only to defaults involving termination of the nominated subcontract but the term 'default' is not defined and may have wider and ordinary application.

1.4 Major areas of interest

Consideration of how the Seventh edition will operate in practice leads to two major areas of interest:

- whether the new provisions in clauses 52(1) and 52(2) for the valuation of variations by quotation are workable as drafted and, if workable, whether they will lead to reduction of global claims for delay and disruption as well as to reduction of disputes on the valuation of variations
- whether the dispute resolution procedures in clause 66 are compliant with the Housing Grants, Construction and Regeneration Act 1996 require-ments on adjudication and whether the procedures, taken as a whole, are an improvement on the procedures in ICE Conditions which pre-dated operation of the Act or whether they have lost some of the best features of those procedures.

Both these topics are considered in detail in later chapters but in summary the reasons for interest are as follows.

Valuation of variations

The introduction into the Seventh edition of a quotation system for the valuation of variations comes as no surprise given the recommendations in Sir Michael Latham's 1994 report *Constructing the Team* on the pre-pricing of variations. Traditionally, however, responsibility for the valuation of variations under ICE Conditions has fallen on the engineer under a scheme using the contractor's tendered rates and prices as the basis of valuation. Departure from this practice raises questions as to the extent to which a quotation system is appropriate for variations which are necessary (as opposed to optional) and the extent to which the contractor should be permitted in his quotations to depart from the rates and prices in his tender and which still in the Seventh edition form the basis of the contract.

The wording of the new clauses 52(1) and 52(2) is disappointingly obscure on many points of detail and it clearly needs some improvement. And it is interesting in the way it tries to bring into quotations the costs and extent of delay (if not the costs of disruption) and in that it imposes a duty on the engineer to negotiate with the contractor.

Dispute resolution procedures

When the Housing Grants, Construction and Regeneration Act 1996 came into force in 1998 and gave the parties in the UK (on what are described in the Act as 'construction contracts') a statutory right to adjudication, the drafting committee of ICE Conditions had a choice: to leave intact and unchanged well tried and generally well understood dispute resolution procedures which had been developed in ICE Conditions over the years as a contractual scheme accepted by the parties – with the statutory scheme for adjudication standing alongside as a parallel or alternative procedure; or alternatively to amend the contractual scheme in an endeavour to encompass the statutory right to adjudication and thereby avoid application of the statutory scheme.

The drafting committee chose the latter option whilst at the same time trying to retain the engineer's involvement in the dispute resolution process by the introduction of a new step in the procedure – referral to the engineer of 'matters of dissatisfaction' instead of as previously matters of 'dispute'.

Opinions are divided as to whether the new procedure is compliant with the Act and until the point is settled by a decision of the courts users of the Conditions are left with uncertainty as to whether adjudication under the statutory scheme is permissible and are left with a contractual scheme which, by its complexity, is a poor substitute for the contractual scheme it replaced.

Chapter 2
Definitions, interpretation and notices

2.1 Introduction

The Seventh edition maintains a longstanding drafting rule of ICE Conditions that certain terms of significance should commence with capitals and that such terms should, with a few exceptions, be listed as defined terms in clause 1 of the Conditions. The benefit of this is that terms in the Conditions commencing with capitals can generally be taken to be defined terms.

A minor drawback of the way the rule operates is that it fails to distinguish between descriptive terms, which require definition to give them precise meaning, and identification terms, which are fixed by factual data. This has the effect that some of the definitions are disappointingly circular. It is not an uncommon fault since many standard forms use the same rule, but the approach of the New Engineering Contract which gives defined terms capitals but puts identification terms in italics is a thoughtful alternative.

There are few changes of significance in either clause 1 (definitions and interpretation) or clause 68 (notices) between the Sixth and Seventh editions. Clause 1(6) is expanded to allow communications in writing to be sent by 'other means' (e.g. email). But this change first appeared in the March 1998 amendment to the Sixth edition. Clause 68(3) is a new sub-clause correcting a problem in the old clause 68 by allowing the engineer to direct how any notice to be given by him, or to him, should be delivered. Minor improvements are made to clause 1(1)(m) – nominated subcontractors and clause 1(1)(r) certificate of substantial completion, but for the rest the changes are principally word replacements, e.g. 'carrying out' for 'execution'.

Application of definitions

Clause 1(1) commences with a statement to the effect that in all the contract documents the defined terms have the meaning assigned to them except where the context requires otherwise.

A point worth noting on this is that the proper construction of any contract requires consideration of all the documents forming the contract. The Conditions will usually be taken as the key document for determining the obligations, rights and liabilities of the parties but in the Seventh edition, as in earlier editions, neither the Conditions nor any other documents are given precedence. There is a need, therefore, for those drafting any of the contract documents to be aware that one of the documents, the Conditions, endeavours or purports to standardise certain definitions for all the documents.

2.2. *Identification definitions*

Clauses 1(1(a), 1(1)(b), (1)(1)(c) and 1(1)(d) are the identification definitions for the employer, the contractor, the engineer and the engineer's representative respectively.

For the proper working of the contract all these need to be identified with precision but the necessity for absolute certainty on the identities of the employer and the contractor cannot be over-stated. As the parties to the contract they carry the legal obligations and liabilities of the contract and it is they who may eventually become subject to adjudicators' decisions, arbitrators' awards or court judgements. Nevertheless, it is surprising how often discrepant names and titles appear in contract documents and how often the first task in legal proceedings is to determine by enquiry exactly who are the parties to the contract – a process which can lead to unexpected and sometimes unwelcome conclusions.

It is possible for the identity of all the participants – the employer, the contractor, the engineer and the engineer's representative – to change during the life of the contract, but there is an essential difference between the position as regards the parties (the employer and the contractor) and the position as regards the engineer and his representative. The identity of the parties can only change by some form of legal succession or assignment, but the identity of the engineer can be changed by direction of the employer and similarly the identify of the engineer's representative can be changed by direction of the engineer. A point worth noting where one or both of the parties is a limited company is that the true identity of a company lies in its registration number and not in its name – which can be changed at will by a simple application to the Registrar of Companies.

Employer

Clause 1(1)(a) states that the employer means the person or persons, firm, company or other body named in the appendix to the form of tender. It adds that the employer includes the employer's personal representatives, successors and permitted assignees.

Where the formalities of making the contract go no further than acceptance of the contractor's tender, the name of the employer inserted in the appendix to the form of tender will normally close any question as to who is the employer. However, if the formalities extend to completion of a form of agreement executed as a deed and there is any discrepancy between the name of the employer in the appendix and that in the deed then the latter may well have the higher legal standing.

Interpretation of the concluding part of clause 1(1)(a) in its reference to the employer's personal representatives, successors and permitted assignees is not without its difficulties. The intention with regard to successors and assignees is clear enough – they acquire legal rights and obligations according to the circumstances of their appointment and in some cases will stand in full for the original named employer. However, that may not be the intention with regard to the employer's personal representatives. The normal situation is that such representatives may bind the employer by giving instructions and the like but only rarely will it be intended that they should acquire the employer's legal obligations and liabilities.

The probability is that the clause intends no more than that a personal representative can act for the employer. That would put the representative in a similar position to an agent where the general rule is that providing the existence or identity of the principal is disclosed the agent does not acquire personal liability. But the general rule has its exceptions and persons acting as agents need to proceed with caution as shown by the case of *Sika Contracts Ltd* v. *Gill* (1978). Mr Gill, a chartered civil engineer, was held personally liable for payment on a contract where he was acting both as agent for the employer and in his professional capacity. His main mistake was to have accepted a quotation from Sika on his own headed notepaper.

Contractor

Clause 1(1)(b) states that the contractor means the person or persons, firm or company to whom the contract has been awarded by the employer. It includes the contractor's personal representatives, successors and permitted assignees.

The award of the contract is most likely to be by a letter of acceptance of the contractor's tender and where the standard ICE form of tender is used such a letter expressly creates a binding contract. In such situations, providing that the same name for the contractor follows through into any later form of agreement, the identity of the contractor at least at commencement should be beyond doubt. As to change that takes place thereafter it is a feature of contracting that companies come and go. Frequently this is accommodated by the appointment of a new contractor by a process known as novation. There can however be difficult questions to resolve on identity when claims are pursued after a company failure by others maintaining they have by some means or other acquired the legal right to stand in the shoes of the original contractor.

Engineer

Clause 1(1)(c) defines the engineer as the person, firm or company appointed by the employer to act as the engineer for the purposes of the contract and who is either:

- named in the appendix to the form of tender, or
- appointed from time to time and notified in writing as such to the contractor.

The Seventh edition retains by clause 2(2)(a) the rule introduced in the Sixth edition that where the appointed engineer is a firm or company, a single named chartered engineer who will assume the full responsibilities of the engineer under the contract has to be notified to the contractor. This allows the employer the benefit of appointing a firm or company of appropriate status for the project, complete with its corresponding level of professional indemnity insurance, whilst at the same time ensuring that contractual decisions affecting the contractor are subject to individual judgement.

The Seventh edition also retains in clause 1(1)(c) the employer's power to re-appoint – without which the employer might be in some difficulty if the first appointed engineer was to die or retire or, if a firm, go out of business. Note, however, that the wording also gives the employer the power to replace his appointed engineer at will – a power which in some standard forms can only be exercised with the contractor's consent. The employer is not obliged to give reasons for re-appointing.

The power of the employer to re-appoint also creates a duty as an implied term if the appointed engineer is no longer able to continue. In *Croudace Ltd v. London Borough of Lambeth* (1986) it was held that Lambeth was in breach of contract in failing to nominate a successor to an architect who had retired, thereby leaving no one legally able to issue certificates.

Engineer's representative

Clause 1(1)(d) defines the engineer's representative as a person notified as such from time to time by the engineer under clause 2(3)(a). The requirement for the notification to be in writing is found in clause 2(3)(a).

Both clause 1(1)(d) and 2(3)(a) are written in terms which suggest there is to be only one engineer's representative – and that representative is to be a person and not a firm. If this does not suit the employer's preferred arrangements for the contract it is best that the matter is addressed openly in an amendment.

2.3 *Contract documents*

Clauses 1(1)(e), 1(1)(f) and 1(1)(g) define which documents comprise the contract and what is meant by the terms specification and drawings.

Contract

Clause 1(1)(e) defines the contract as including:

- conditions of contract
- specification
- drawings
- bill of quantities
- form of tender
- written acceptance
- form of agreement (if completed).

The definition raises questions as to how, if at all, documents which do not fit readily into any of the listed categories can or should be given contractual status. Common examples of such documents are pre-tender letters, tender clarification letters, letters of intent, tender programmes and method statements.

In the IChemE forms of contract and in certain civil engineering forms of subcontract the question is resolved by provisions that only listed documents have contractual effect. ICE Conditions of Contract have traditionally taken what appears to be a less prescriptive approach but this may be something of an illusion. ICE Conditions, Seventh edition included, do not leave open the question of which documents form the contract – as do some standard forms. The listing of documents by category in the Conditions is itself restrictive. And this restriction is followed through into the standard form of agreement where the same categories of documents as in the Conditions are listed as comprising the contract documents.

One approach to overcoming the problem of additional documents in the ICE Conditions is to bind copies of all documents intended to be contract documents into one or more of the listed categories. Another is to include within the letter of acceptance and the form of agreement, if completed, a schedule of the particular documents which are to form the contract.

Precedence of documents

Many standard forms expressly provide a precise order of precedence for the various categories of documents forming the contract. ICE Conditions do not. The Seventh edition, following the policy of its predecessors, states in clause 5 that the several documents forming the contract are to be taken as mutually explanatory of one another. Comment on the implications of this is given in Chapter 5, Sections 5.2 and 5.3.

Conditions of contract

It is common practice in civil engineering that conditions of contract are incorporated by reference.

One effect of this practice is that amendments to standard ICE Conditions are often made by statements in a separate document rather than in the printed set of conditions. In the building industry this rarely happens because the standard conditions usually preclude extraneous amendments; any changes have to be made in the pages of the standard conditions and initialled by both parties. There is much to be said for this because it imposes discipline on amending, the amendments are conspicuous, and amendments are literally within the standard conditions.

Where the standard ICE Conditions are amended in a separate document, references to the conditions of contract in the form of tender, form of agreement and other formal documents should be suitably annotated.

The dangers of the ICE practice of amending in separate documents was illustrated in the case of *Harbour & General Works* v. *Environment Agency* (1999). The contractor did not notice an amendment to the standard Sixth edition Conditions made in the August 1993 corrigenda and as a result found himself out of time in serving first, a notice of conciliation, and later, a notice of arbitration.

Special conditions

Clause 72 of the Conditions provides a starting place for the inclusion in the contract of any special conditions. These are intended to be additional to, and not amendments to, the standard conditions in clauses 1–71. Frequently they cover such matters as price fluctuations, corruption, restrictions on publicity and the like.

The amendment of standard conditions by way of special conditions is not recommended. Less still is the practice, still occasionally found, of trying to amend the standard conditions by notes on drawings and in specifications.

Specification

Clause 1(1)(f) states that the 'Specification' means the specification referred to in the form of tender and any modification thereof or additions thereto as may from time to time be furnished or approved in writing by the engineer.

The form of tender simply refers to the 'Specification' and to identify a particular specification as the applicable specification for the contract it is necessary to look elsewhere for the details.

At first sight it may appear that the definition in clause 1(1)(f) is circular but it is not so much circular as intentionally open. It is drafted to ensure that where obligations relating to the specification are found in the Conditions of Contract, those obligations extend not just to an identified and stipulated specification but to additions and modifications of the specification made thereafter by the engineer.

Various clauses of the Conditions empower the engineer to make additions and modifications to the specification, most notably clause 7(1) – further drawings, specifications and instructions; clause 12(4) – action by the engineer on adverse physical conditions; clause 13(3) – instructions; and clause 51(1) – variations. The contract provides for time and price adjust-

ments in respect of changes made by the engineer to the original specification – a point which emphasises that notwithstanding the generality of the
definition in clause 1(1)(f) it is necessary to be able to identify with certainty
the original stipulated specification on which the contractor based his
tender.

Drawings

Clause 1(1)(g) states that 'Drawings' means the drawings referred to in the
'Specification' and any modifications of such drawings approved in writing
by the engineer and such other drawings as may from time to time be
furnished by or approved in writing by the engineer.

It is not clear why clause 1(1)(g) takes the approach of defining 'Drawings'
by reference to the 'Specification', rather than by reference to the form of
tender as clause 1(1)(f) does for the 'Specification'. The change introduces the
possibility of serious omission since contract drawings are frequently
identified by lists rather than in specifications.

Approved in writing by the engineer

Both clauses 1(1)(f) and 1(1)(g) conclude with the phrase 'or approved in
writing by the Engineer'. The effect is that any specification or drawing
approved in writing by the engineer acquires status within the meaning of
the corresponding defined term. The implications of this can be considerable
when it comes to assessing the contractor's entitlements to additional time
and money.

The engineer should be careful in the use of the term 'approval', particularly in respect of proposals, specifications and drawings put forward by
the contractor. It can inadvertently create variations where none is intended.

The point has been recognised in the drafting of the Seventh edition and
clauses which in the Sixth edition referred to 'approval by the Engineer' now
generally refer to 'acceptance by the Engineer'.

2.4 Financial definitions

Clauses 1(1)(h), 1(1)(i) and 1(1)(j) define respectively, the bill of quantities,
the tender total and the contract price.

Bill of quantities

Clause 1(1)(h) states that the bill of quantities means the priced and completed bill of quantities.

The purpose of this definition is to make clear that references elsewhere in the Conditions to the bill of quantities are to the priced and completed bill and cannot be taken as referring to an unpriced bill.

Tender total

Clause 1(1)(i) states that the tender total means the total of the bill of quantities at the date of the award of the contract or, in the absence of a bill of quantities, the agreed estimated total value of the works at that date.

The only contractual significance of the tender total is in fixing the amount of performance security (clause 10), and in fixing limits on retention (clauses 60(5) and 60(6), and the appendix to the form of tender).

The tender total has no direct relationship with the contract price – something which frequently surprises employers more familiar with lump sum contracts than re-measurement contracts.

The reference in the clause to the agreed estimated total value of the works is presumably to the situation where the Seventh edition is used for schedule of rates contracts – which is not uncommon – or for lump sum contracts, in which case various amendments need to be made to the valuation clauses.

Contract price

Clause 1(1)(j) defines the contract price as the sum to be ascertained and paid in accordance with the provisions in the Conditions for the construction and completion of the works in accordance with the contract.

The term 'contract price' has little usage in the Conditions since it represents a figure which is indeterminate until the works are completed. Unlike the position in many standard forms the contractor does not undertake to complete the works for the contract price. In the form of tender the contractor undertakes to complete the works 'for such sum as may be ascertained in accordance with the Conditions of Contract'. This reflects the uncertainties of re-measurement and the frequency in the Conditions of clauses entitling the contractor to additional payments.

Note, however, that in the form of agreement the employer does covenant to pay the contract price.

It is an interesting point as to whether claims paid as damages for breach of contract can properly be included in the contract price. There are implications for both retention and value added tax (which is arguably not payable on an award for damages) and it is questionable whether there is anything in the Conditions (other than in clause 66) which empowers the engineer to value claims for breach which are not expressed in the Conditions as contractual entitlements. Nevertheless engineers do frequently take on the burden of valuing damages claims, treating them in much the same way as contractual claims. It may be convenient and sometimes sensible practice but it should be treated with caution.

2.5 *Prime cost items and provisional sums*

Clauses 1(1)(k) and 1(1)(l) provide the meaning, for the purposes of the Seventh edition, of the much used construction billing terms, prime cost items and provisional sums. Neither definition stands completely on its own and clauses 58(1) and 58(2), which deal with the use of provisional sums and prime cost items, need to be considered for a practical understanding of the terms.

Prime cost items

Clause 1(1)(k) simply explains that prime cost item means an item which contains a sum referred to as prime cost which is to be used for the carrying out of work or the supply of goods, materials or services.

Clause 58(2) expands the meaning by stating that the engineer may use prime cost items either under the contractual rules for nominated subcontracting or, with the consent of the contractor, under arrangements whereby the contractor himself does the work and/or provides the goods, materials or services.

Provisional sums

Clause 1(1)(l) defines a provisional sum as a sum designated as such, and included in the contract as a specific contingency for the carrying out of work or the supply of goods, materials or services and which may be used in whole or in part at the direction and discretion of the engineer.

Clause 58(1) permits the engineer to use provisional sums to order that the work, goods and services be provided by the contractor himself, in which case the valuation rules for variations apply, or that work, goods and services be provided by a nominated subcontractor, in which case the valuation rules of clause 59 apply.

For further comment on prime cost items and provisional sums see Chapter 24.

2.6 *Nominated subcontractors*

Clause 1(1)(m) defines a nominated subcontractor as a person, firm (or others) nominated in accordance with the contract to be employed by the contractor for carrying out work or supplying goods, materials or services for which a prime cost item or a provisional sum has been included in the contract.

The wording of clause 1(1)(m) in the Seventh edition is slightly changed from that in the Sixth edition but this seems to be in the interests of brevity rather than a policy change.

An important point which comes out of clause 1(1)(m) is that there is no provision in the contract for the status of nominated subcontractor to be conferred except where there is an applicable prime cost item or provisional sum included in the contract.

In Chapter 14 the position is considered on suppliers and specialist sub-contractors who are named in the contract (commonly in the specification) but whose works, goods and services are not covered by a prime cost item or provisional sum.

2.7 *Works definitions*

Clauses 1(1)(n), 1(1)(o) and 1(1)(p) define respectively the meaning of the terms 'Permanent Works', 'Temporary Works' and the 'Works'.

Clause 1(1)(n) states that 'Permanent Works' means the permanent works to be constructed and completed in accordance with the contract.

Clause 1(1)(o) states that 'Temporary Works' means all temporary works of every kind required in or about the construction and completion of the works.

Clause 1(1)(p) states that the 'Works' means the permanent works together with the temporary works.

The simplicity and circularity of the definitions prevents them from being of any assistance in identifying in practical terms the distinction between permanent and temporary works. They serve principally to allow distinctions to be made in certain clauses of the Conditions for matters such as design responsibility and liability for loss or damage.

In the case of *Norwest Holst Construction Ltd* v. *Renfrewshire Council* (1996) which concerned a case under ICE Conditions, Fifth edition, where the contractor was given design responsibility for certain piling, one of the questions the court had to decide was whether contractor's design came within the scope of permanent works or temporary works as defined in the contract. The court took the view that the contractor's obligation as set out in clauses 8(1) and 13(1) was to 'construct and complete the Works' and that its obligation to design part of the works was additional to and not part of that obligation. From that the court concluded that contractor's design was not within the contractual definitions of permanent works or temporary works.

For further comment on this case and how the court went on to decide how clause 13 deals with works designed by the contractor which are impossible to construct, see Chapter 9, Section 9.3.

2.8 *Commencement and completion definitions*

Clause 1(1)(q) defines by reference to clause 41(1) the term 'Works Commencement Date'. Clause 1(1)(r) states by reference to clause 48 what is meant by 'Certificate of Substantial Completion'.

The purpose of bringing the definitions forward into the first section of the Conditions is simply to endow the terms with formal meaning in order to bring certainty to usage of the terms in other clauses of the Conditions.

Commencement

Clause 1(1)(q) states only that the works commencement date is as defined in clause 41(1).

Clause 41(1) is in fact not so much a definition as a mechanism for fixing the works commencement date for any particular contract. It offers three possibilities – a specified date, a date notified by the engineer, or an agreed date. The meaning of the term 'Works Commencement Date' is perhaps best derived from clause 43 which indicates that time starts to run in respect of completion obligations from that date.

The works commencement date is also relevant to other obligations and liabilities of the parties – particularly those relating to care of the works and insurances.

Completion

Clause 1(1)(r) states that a 'Certificate of Substantial Completion' means a certificate issued under clause 48.

In the Sixth edition the wording was slightly different in that it referred to a certificate issued under clause 48(2)(a), 48(3) or 48(4). There is no change between the Sixth and Seventh editions in the wording of clause 48 (or any of its sub-clauses) and the change in clause 1(1)(r) seems to be no more than correction of a drafting error in the Sixth edition which failed to take account of the reference to a certificate of substantial completion in parts of clause 48 other than those listed.

Clause 48 does not actually define or offer any guidance on what is meant by substantial completion. That is a difficult subject which is considered later in this book in Chapter 18. What clause 48 does is to set out the circumstances by which a certificate can be issued by the engineer confirming that the contractor has fulfilled certain obligations in respect of completion of the whole or parts of the works.

2.9 Defects definitions

Clauses 1(1)(s) and 1(1)(t) state respectively what is meant by the 'Defects Correction period' and the 'Defects Correction Certificate'.

Defects correction period

Clause 1(1)(s) describes the 'Defects Correction Period' as the period stated in the appendix to the form of tender calculated from the date on which the contractor becomes entitled to a certificate of substantial completion for the works or any section or part.

The period, usually 52 weeks, is fixed by the employer in part 1 of the appendix, and all that the definition in clause 1(1)(s) adds is information on when the period starts to run and confirmation that there will be more than one defects correction period if certificates of substantial completion are issued for parts or sections of the works.

The formal definition of the term 'Defects Correction Period' amounts to little more therefore than identification of a period of time and stipulation as to when it commences. The general meaning of the term is to be found in clause 49 from which it can be deduced that the defects correction period is a period after completion when the contractor has an obligation and entitlement to complete outstanding works and to remedy defects.

Defects correction certificate

Clause 1(1)(t) states only that the 'Defects Correction Certificate' is as defined in clause 61(1).

Clause 61(1) does not define 'Defects Correction Certificate' in so many words but it does indicate that it is a certificate issued by the engineer after the end of the defects correction period, or the end of the last such period if more than one certificate of substantial completion has been issued. It states the date on which the contractor has completed his obligations to construct and complete the works to the engineer's satisfaction.

Under ICE Conditions, Seventh edition and earlier editions, the policy has always been to avoid wording which could be construed as conferring absolute finality on certificates in respect of the obligations and liabilities of the parties. For the avoidance of doubt clause 61(2) states expressly that the issue of the defects correction certificate shall not relieve either party of its liabilities.

2.10 *Sections and parts of the works*

Clause 1(1)(u) defines what is meant by a 'Section'. It is a part of the works separately identified in the appendix to the form of tender.

In fact the appendix identifies only the times for completion of sections (if there are any) and it is completion which is of contractual significance – not the physical boundaries of any sections. These need to be identified elsewhere.

Parts

The Conditions do not define the much used phrase 'part of the Works'. But that is not to say that a part of the works has no contractual standing. A part of the works which is finished early or used by the employer may attract its own certificate of substantial completion and its own defects correction period subject to compliance with the provisions in clause 48(3) or 48(4). That in turn can have an impact on the employer's entitlement to liquidated damages through the proportioning down mechanisms in clauses 47(1)(b) and 47(2)(b).

The essential difference between a section of the works and a part of the works is that the contractor has an obligation to complete a section within the stipulated time or face liquidated damages, whereas there is no contractual sanction for failure to finish a part any earlier than the whole.

Endeavours to impose partial completion obligations on the contractor without using the sectional completion provisions frequently end in dispute and recrimination.

2.11 The site

Clause 1(1)(v) states that 'Site' means:

- the lands and other places on, under, and in, or through, which the works are to be constructed
- any other lands or places provided by the employer for the purposes of the contract
- such other places as may be designated in the contract
- such other places as may subsequently be agreed by the engineer as forming part of the site.

The definition, which is essentially the same as in the Sixth edition, has a number of problems of interpretation. One, which is probably more of a talking point than a problem of practical application, is how the definition applies to underground works such as tunnels. Is the land above a tunnel part of the site and if so is the employer obliged to give possession of that land under clause 42(2)? The answer may depend on the details of the work and the controls, if any, to be exercised at ground level.

A more commonplace problem is how the site as defined, which is not obviously restricted to land provided by the employer, is compatible with the obligation in clause 42(2) for the employer to give possession of the whole of the site. And arising from this is the question of whether the employer is liable under clause 26(2) for the rates of parts of the site which are not in his ownership.

Taking clause 1(1)(v) on its own, the definition of the site cannot be construed as restricted to land provided by the employer – under the final part of the clause it could, for example, include land acquired by the contractor as the site compound or for storage purposes.

2.12 Contractor's equipment

Clause 1(1)(w) states what is meant by 'Contractor's Equipment'. In short it is anything required to construct and complete the works excluding materials and things intended to form part of the permanent works.

From its own wording and the wording of clauses 63(3) and 64(2) which deal with removal from site of contractor's equipment, the term would seem to include materials used in temporary works but there are other clauses, 54(1) and 65(2), which seem to distinguish between equipment and materials.

2.13 Interpretative rules

Clauses 1(2), 1(3) and 1(4) set down some basic particularised rules of interpretation for the contract – albeit that the rules from clauses 1(3) and 1(4) are only applicable to interpretation of the Conditions of Contract.

Singular and plural

Clause 1(2) states that words importing the singular also include the plural and vice versa when the context requires.

Similar clauses are found in many contracts and they lead to simpler text and an element of flexibility. In this clause, however, the key words are at the end in the phrase 'where the context requires'. This provides a useful test for whether words in the singular should be given plural meaning and it limits the application of the clause to situations where the test is met.

Headings and marginal notes

Clause 1(3) states that the headings and marginal notes in the Conditions shall not be deemed to be part thereof or taken into consideration in the interpretation or construction of the contract.

From a practical point of view headings and marginal notes are a beneficial aspect of the Conditions. Headings make for easier reading and understanding of the text and marginal notes provide useful labels for the various clauses. From a legal point of view it may be said that a contract should be read as a whole for its proper construction and treating it in sections under block headings can lead to misinterpretation; and in similar vein that the brevity of marginal notes makes them vulnerable to misinterpretation.

There is a line of thought which says it is inappropriate to have a clause in a contract which renders part of the text inoperative and that could be said to apply to clause 2(3), but on balance it is better to have inoperative headings and marginal notes than none at all.

Clause references

Clause 1(4) states that references to clauses in the Conditions are references only to clauses in the Conditions and not to those in other contract documents. This would normally be taken as self evident and if there is any necessity behind clause 1(4) it is perhaps that ICE Conditions, Seventh edition included, do make generous use of cross-referencing between clauses and documents.

2.14 Cost

Clause 1(5) is a definition of major importance. It concerns the meaning of 'cost' – a word which appears frequently throughout the Conditions and is the basis for valuation of the majority of the contractor's claims (variations excepted). The definition states that 'cost' when used in the Conditions means all expenditure properly incurred, whether on or off the site, including overhead, finance and other charges properly allocable thereto but excluding any allowance for profit.

It is not uncommon in civil engineering claims for cost to be equated with loss and expense – a term held in *Wraight Ltd* v. *P.H. & T. (Holdings) Ltd* (1968) to be the equivalent to damages for breach of contract. There is, however, an obvious distinction between 'cost' and 'loss and expense'. Cost and loss are potentially different concepts.

The exclusion of losses from cost claims can severely diminish the amount recoverable. In many delay and disruption claims the sums included for lost overheads, lost opportunities and the like can be the major items. To avoid this difficulty claims are sometimes presented as claims for damages rather than as claims under the contract for express entitlements based on cost. There are risks for the contractor in this in being able to prove breach of contract and in overcoming the argument that where there is a clear contractual entitlement the common law remedy of damages for breach is excluded. However the legal position on this latter point is not as clear as it might be since there is one school of thought which says that legal rights can only be excluded by express words and another which says that clearly stated alternative provisions have equivalent effect to express words.

'Expenditure properly incurred'

Various points arise out of the phrase 'expenditure properly incurred' which describes 'cost' in clause 1(5). One, as discussed above, is whether cost can extend to loss. The answer may be that it depends on the type of loss. Thus, loss of profit would rarely qualify as 'expenditure properly incurred' whereas the financial loss on a particular contract due to delaying circumstances or the like might be matched not by costs on that contract, but by costs on another contract or elsewhere within the contractor's organisation. Such a situation can occur, for example, when the contractor's company owned plant is kept on site longer than expected by adverse ground conditions which could not be foreseen and as a result the contractor is obliged to hire in replacement plant on the contract to which the delayed plant was to be moved. In such circumstances it is sometimes argued that the contractor is not entitled to recover under contract A costs incurred under contract B. However, such a rule does not follow naturally from the wording of clause 1(5), where there is no restriction to costs incurred under the contract (as opposed to costs incurred more widely by the contractor) and there is express entitlement to recovery of costs incurred off the site.

A matter of commonplace relevance to the above is precisely what entitlements the contractor has to recover as cost, its own or sister company plant charges. Plant in the ownership of sister companies with separate legal identity can usually be treated as being in the same category as externally hired plant but otherwise the cost to the contractor of its own plant is no more than depreciation and operating costs. Contractors often claim they are entitled to recover the equivalent of reasonable hire charges in respect of their own plant but, excepting the possible circumstances of loss of opportunity to use elsewhere as discussed above, there is no legal basis for substituting actual cost with notional hire charges. In the case of *Alfred McAlpine Homes North Ltd* v. *Property and Land Contractors Ltd* (1995) it was held:

> 'in ascertaining direct loss or expense under clause 26 of the JCT conditions in respect of plant owned by the contractor the actual loss or expense incurred by the contractor must be ascertained and not any hypothetical loss or expense that might have been incurred whether by way of assumed or typical hire charges or otherwise.'

Another commonplace point arising out of the phrase 'expenditure properly incurred' is whether the contractor is entitled to be paid the relevant cost when liability is incurred or only when expenditure (by way of payment out) can be proved. As a general rule entitlement to recover cost incurred would not normally be subject to proof of payment but in clause 1(5) the word 'expenditure' hints at a different policy. However, given that under the clause cost can also mean 'expenditure ... to be incurred' it is not certain what is the policy. Nevertheless as a matter of practice, payment of cost under ICE Conditions is usually made on proof of liability rather than on proof of payment.

An interesting legal point which comes out of clause 1(5) is whether the phrase 'all expenditure properly incurred' can be taken to avoid the rules for remoteness of damage developed from the case of *Hadley* v. *Baxendale* (1854). In other words if the contractor incurs a cost, however remote, is that cost recoverable under the wording of clause 1(5) even though it would not be recoverable as damages? The phrase 'properly incurred' may have some restrictive effect but it is not wholly conclusive.

Expenditure – 'to be incurred'

On one interpretation of clause 1(5), the provision for cost to be 'expenditure ... to be incurred' is easily understood. It allows payment to be made on proof of liability.

An alternative interpretation is that the provision allows the contractor to look ahead when making an application for payment and to include in the application for future expected costs. Under the Sixth edition that seemed a possible interpretation because of inconsistencies in the wording of the various entitlement to payment clauses, some of which restricted payment to costs incurred whilst others did not. Those inconsistencies have been largely removed in the Seventh edition and it is now more difficult to sustain the alternative interpretation.

Cost – 'on or off the site'

The inclusion in clause 1(5) of the phrase 'on or off the site' may simply be intended to put beyond doubt the contractor's entitlement to recover head office overheads and the like as an element of cost. The phrase is hardly necessary to establish that the contractor can recover off-site construction costs. These would readily fall within the definition of cost without the phrase.

The interesting aspect of the phrase is whether it expands the definition of cost to the point that the contractor can legitimately claim consequential costs.

Cost – amounts recoverable under insurances

There is little legal authority on the question of whether amounts recoverable as cost should take into account amounts recovered by the contractor under insurances. In the case of *Amec Civil Engineering and Alfred McAlpine Construction* v. *Cheshire County Council* (1999) it was said that there was no authority 'touching directly on the relevance of insurance proceeds to the determination of a fair and reasonable sum'.

In that case, which concerned a contract under ICE Fifth edition Conditions, the employer and the joint venture contractors entered into an agreement for acceleration of the works on general terms that the contractors would be paid their costs for using their best endeavours to complete the works by a particular date. The contractors accelerated and achieved substantial completion as required but the costs of acceleration greatly exceeded the original estimates and the parties fell into dispute. The parties agreed to litigate rather than to arbitrate their dispute and one of the points which came before the court was whether, in determining fair and reasonable recompense for the contractors, account should be taken of several hundred thousands of pounds which the contractors had recovered for insurance events including flooding and the like. The court held:

- the contractors were not required to account to the Council for sums received under insurance policies

- the insurances were primarily for the benefit of the contractors and only incidentally for the employer/Council
- recovery by the contractors from the Council would not result in double recovery as the joint venture would have to account to the insurers for payments received
- 'costs' in the formula used by the parties did not take account of sums recovered from insurers and to take them into account would vary the proper method of evaluation of quantum meruit.

It would, perhaps, be going too far to say that the case is authority for the proposition that 'cost' as defined in ICE Conditions excludes amounts recovered under insurances since the judgment concentrates on the quantum meruit aspect of the contractors' entitlement rather than on cost as defined. Nevertheless it throws some light on a legally obscure but fairly commonplace problem.

Overhead, finance and other charges

The contractor's entitlement to recover overheads and financing charges is examined in detail in Chapter 21. The point considered here is how the phrase in clause 1(5) 'other charges properly allocatable thereto' is to be construed. The question, which is one of some significance when it comes to proof of cost, is whether the words 'allocatable thereto' apply to 'other charges' or additionally to the overhead and finance elements of cost.

If it is permissible to 'allocate' overhead charges to cost, particularly head office overheads, then usually the contractor will be better placed recovering such charges on a percentage addition to cost basis rather than proving on the basis of expenditure that additional head office overheads have been incurred as a result of the claimable event.

On the other hand if the words 'other charges properly allocatable' simply add to the elements of provable cost, then it would seem that both overhead and finance charges are subject to the same rules for qualification as 'expenditure properly incurred' as other cost elements.

No allowance for profit

The general approach of the Conditions to profit is that it is allowable only for the additional work element of cost based claims. Most of the claim clauses include the sentence (in connection with cost) 'Profit shall be added thereto in respect of any additional permanent or temporary work'. Clause 12 (adverse physical conditions) is the exception in that it allows for profit on all costs.

The exclusion of profit in the definition of cost in clause 1(5) matches this general approach.

2.15 Communications

Clause 1(6) states that communications which are required under the contract to be in writing may be handwritten, typed or printed and may be sent by hand, post, telex, cable, facsimile or other means resulting in a permanent record.

The 'other means' part of the clause was first introduced in the March 1998 amendment to the Sixth edition. It is presumably intended to cover electronic transmissions such as email.

The communications which the Conditions require to be in writing are mostly notices or notifications (if a distinction is intended between the two) and to this extent clause 1(6) supplements clause 68 (notices) which also requires all notices to be served in writing. The difference between the clauses is that whereas clause 1(6) is concerned with the method of communication, clause 68 is concerned more with places of delivery.

2.16 Notices

Clause 68(1) requires any notice to be given to the contractor under the terms of the contract to be in writing and to be served at the contractor's principal place of business, or if the contractor is a company, at its registered office.

As this clause stood under the Sixth edition it was onerous if not unworkable. It appeared to catch routine notices given by the engineer to the contractor which, for good management, would be better served on site or at the contractor's local address. That fault has been rectified in the Seventh edition by the new clause 68(3) which states that any notice to be given by or to the engineer under the contract shall be 'delivered' as the engineer may direct.

The use of the word 'delivered' in the new clause 68(3) is interesting in so far that it differs from the word 'served' used in clauses 68(1) and 68(2). Arguably it could extend the engineer's powers of direction to include the means of communication in clause 1(6) as well as the destination.

Clause 68(2), which deals with notices to be given to the employer, mirrors clause 68(1) except that the place of service is the employer's last known address (as opposed to the contractor's principal place of business).

In the case of *Anglian Water Authority* v. *RDL Contracting Ltd* (1988) it was held, in relation to the near identical clause 68(2) of the ICE Fifth edition, on the facts that service to the divisional address of the employer was effective as service to 'the employer's last known address'. Nevertheless the case highlighted the potential uncertainty of the phrase. Employers wishing to have contractual notices sent to a particular address should consider amending clause 68(2).

Chapter 3
The engineer and engineer's representative

3.1 Introduction

The Seventh edition retains the traditional title of 'Engineer' for the contract administrator against the trend elsewhere for more modern alternatives – project manager, employer's representative and the like. This is in keeping with the policy of minimising change through successive editions of the ICE Conditions and the maintenance in the Seventh edition of the same long standing core duties – designer, supervisor, contract administrator and certifier – which over the years have been recognised as comprising the role of the engineer.

The changes from the Sixth to the Seventh edition in the clauses dealing with the engineer, the engineer's representative and assistants, are more cosmetic than significant shifts in policy. A new clause 2(1)(d) appears stating that consents and approvals given by the engineer do not relieve the contractor of any of his obligations. The old clause 2(7) entitled 'reference on dissatisfaction' is relocated to clause 66. The corrections to minor drafting errors in the Sixth edition first addressed in a corrigenda sheet issued in August 1993 are included.

Legal duties of engineers

By his contract of engagement with the employer, the engineer has legal duties to the employer in contract and in tort for all aspects of the obligations he undertakes. The majority of legal actions against engineers used to relate to design failures or the consequences of poor or late design information in exposing the employer to claims for delay and extra cost. Actions relating to inadequate supervision of the works, usually a side effect of actions by the employer against the contractor for faulty construction, probably ranked next in the engineer's risk list. Actions by the employer against the engineer in his role as contract administrator or certifier were comparatively uncommon.

A new order now seems to be in place. Actions by employers against engineers, architects, project managers and the like for maladministration or erroneous certification routinely follow successful claims against employers by the contractor. The engineer who over-certifies or under-certifies time or money is in the firing line. The need for care and competence in administering contracts has never been greater.

Various factors are behind this: modern trends in litigation and attitudes towards recovery of loss; increasing use of consultants as opposed to directly employed engineers; and budgetary pressures on employers. But, most recently, the factor which seems to be the driving force is the introduction to construction contracts of adjudication through the Housing Grants, Construction and Regeneration Act 1996. The act, by its quick and cheap mode of dispute resolution, has created a surge of claims and disputes at all levels of the industry and engineers are caught with the rest.

Engineer/contractor relationship

Fortunately for engineers adjudication under the Act has no application to the relationship between engineers and contractors under ICE Conditions. The Act applies to construction contracts made in writing; that draws in contracts of engagement between engineers and employers along with main contracts and subcontracts, but rarely will there be any contract, less still a contract in writing, between the engineer and the contractor.

However, that leaves open the general question – can the contractor sue an engineer for negligence, or put another way, does the engineer owe a duty of care to the contractor on a non-contractual basis?

In *Pacific Associates* v. *Baxter* (1988) the Court of Appeal held that the engineer owed no duty of care to the contractor in respect of his certificates. This was to some extent a surprising decision because the prior legal trend through a series of cases suggested that there could be such a duty. In *Sutcliffe* v. *Thackrah* (1974) it had been decided that an architect when certifying was not acting in an arbitral capacity and was not immune from suit by the employer. In *Arenson* v. *Arenson* (1977) and in *Salliss* v. *Calil* (1987) there was clear support for the proposition that a contractor enjoyed similar rights in this respect to an employer.

The general trend in English law towards exposure to actions in negligence continues unabated with a host of recent decisions, leaving the decision in *Pacific Associates* looking vulnerable to the suggestion that it is particular on its facts. It was not followed in the case of *Edgeworth Construction* v. *Lea & Associates* (1993) where the Canadian Supreme Court found that a road building contractor could sue the consulting engineers who had prepared the bills of quantities for the employer for economic loss sustained as a result of negligent errors in the bills.

Engineers in teams

In civil engineering contracts, particularly those under ICE Conditions, the engineer's firm is more often than not the sole professional firm engaged by the employer. In building contracts it is more commonplace for the employer to engage a team of professionals, sometimes led by a firm of project managers.

Questions arose in the case of *Chesham Properties* v. *Bucknall Austin Project Management Services* (1996) on the duties of the various professionals involved to warn the employer of deficiencies in their own or other professionals' performance. The judgment went mainly against the project management firm – it was found to have a duty to warn against failings of the architect, the structural engineer and the quantity surveyor; with the exception that the architect was found to have a duty to warn against failings of the quantity surveyor, the architect, the structural engineer and the quantity surveyor avoided liability.

Engineers to contracts who find themselves in the role of project managers should be mindful of this development in the law.

Engineer's duty on design

The general rule is that the design responsibility of engineers, architects and the like cannot be delegated without the express agreement of the employer. This follows the decision in *Moresk Cleaners Ltd* v. *Thomas Henwood Hicks* (1966) where it was held that an architect had no power and no implied authority to delegate part of his design duty to the building contractor.

A warning for firms engaged as designers can be illustrated by the case of *Richard Roberts Holdings* v. *Douglas Smith Stimpson Partnership* (1988). In that case the architects designed an effluent tank but were later instructed to consider cheaper alternatives. The architects approved a quotation from a specialist firm whose lining was then installed. The lining failed. The court held that although the architects were not knowledgeable about linings they had a duty to their client to make investigations and/or to advise the client to seek specialist advice. The architects were held to be negligent and in breach of contract notwithstanding, as a side issue, that they had charged no additional fee for dealing with the alternative lining.

Engineer/third party relationship

Engineers should also be aware of another development in the law relating to their liability to third parties. In the case of *Baxall Securities* v. *Sheard Walshaw Partnership* (2000) it was held that an architect owed a duty of care to subsequent occupiers of a property in respect of latent defects in a building arising from the architect's design – the problem in that particular case being inadequate roof drainage.

3.2 *Duties and authority of the engineer*

Clauses 2(1)(a) to 2(1)(d) deal in general terms with the duties and authority of the engineer. This is a growth area of the ICE Conditions with successive editions endeavouring to keep pace with the trend of the times for the extent of duties and powers to be specified. Clauses 2(1)(a) to 2(1)(c) first appeared in the Sixth edition and they are repeated word for word in the Seventh edition. Clause 2(1)(d) is a new clause which serves as a warning to the contractor that approvals and consents do not give relief from contractual obligations.

Obligations of the engineer

Clause 2(1)(a) states that the engineer shall carry out the duties specified in or necessarily to be implied from the contract.

At face value this appears to be no more than a directive to the engineer to do his job properly. And, since the engineer will by one means or other be engaged by the employer, it is questionable what function such a directive has in the contract between the employer and the contractor.

If the explanation is that the engineer owes duties to both the employer and the contractor in properly administering the contract and supervising the works it is arguable that the engineer is not as protected from action for breach of duty by the contractor as the *Pacific Associates* case mentioned in section 3.1 above apparently suggests.

An alternative explanation is that the employer warrants to the contractor that the engineer will perform properly and that failure by the engineer to do so will be a breach for which the employer accepts liability.

In respect of certain duties of the engineer the Conditions do expressly provide for the employer to recompense the contractor for delay and cost caused by late performance of the engineer – for example under clause 7(4)(a) delay in issue of drawings, and under clause 14(8) delay in consent to proposed methods of construction. However, apart from what appears in clause 2(1)(a) there is no general provision for recompense to cover situations which can arise under other clauses of the Conditions, for example, late attending for work ready to be covered up under clause 38(1). There is therefore some logic in assuming that as between the employer and the contractor the purpose of clause 2(1)(a) is to allocate liability for the performance of the engineer to the employer.

Extent of engineer's authority

Clause 2(1)(b) states firstly that the engineer may exercise the authority specified in or necessarily to be implied from the contract.

It would be odd if things were otherwise and the purpose of the provision is not immediately obvious unless it is to complement clause 2(1)(a). There is a drafting convention that duties are indicated by use of the word 'shall' and the exercise of authority or discretion is indicated by use of the word 'may'. Taken together, clause 2(1)(a) and the opening provision of clause 2(1)(b) cover the exercise of both duties and discretion and conveniently illustrate the use of the convention in the Conditions. The drafting of the Conditions would be improved if the two were in the same clause.

Specific approval

The second part of clause 2(1)(b) states that where the engineer is required under the terms of his appointment to obtain the specific approval of the employer before exercising any authority, particulars of the requirement are to be set out in the appendix to the form of tender. This provision first appeared in ICE Conditions in the Sixth edition. It gave belated recognition to the potential conflict between clauses of the contract which apparently endow the engineer with unrestricted power and the reality of the engineer's authority as derived from his contract of appointment with the employer. For example, clause 51(1) requires the engineer to order any variation necessary for completion of the works without, in the clause, any obvious regard to the cost; whereas the employer for reasons of prudence and financial management may well fix limits on the value of variations which can be ordered by the engineer without prior approval.

Something clearly had to be included in the Sixth edition to address the conflict, not least because, for the first time, the Sixth edition expressly stated the engineer's duty to act impartially – a duty repeated in the Seventh edition at clause 2(7). The contractor's entitlement to knowledge of those aspects of the engineer's contract with the employer with a bearing on the engineer's powers was duly incorporated into the Sixth edition at clause 2(1)(b) and the provision remains unchanged in the Seventh edition.

If practice under the Seventh edition follows that under the Sixth, it is likely that entries in the appendix under Clause 2(1)(b) will be mainly related to financial limits on variations and little else. Entries restricting the engineer's powers to grant extensions of time and to accept alternative designs from the contractor have been seen but only rarely.

In the event that no entry is made in the appendix even when the engineer is bound by approval restrictions the contractor could have a case for damages for breach of contract or misrepresentation if resulting loss could be proved.

The final part of clause 2(1)(b) states that approval shall be deemed to have been given by the employer for authority exercised by the engineer particulars of which have been provided in the appendix. The contractor is not obliged therefore to seek evidence from the engineer that he has obtained specific approval from the employer before exercising his authority. One consequence is that the employer cannot avoid liability to the contractor arising from the engineer's action on the grounds that it was an unauthorised action.

No authority to amend the contract

Clause 2(1)(c) provides that except as expressly stated in the contract the engineer has no authority to amend the contract nor to relieve the contractor of any of his obligations.

The Sixth edition used the phrase 'no authority to amend the Terms and

Conditions of the Contract' whereas the Seventh edition uses the phrase 'no authority to amend the Contract'. The Seventh edition clause is potentially of wider scope than the Sixth edition clause to the extent that the contract may contain more than terms and conditions. But arguably it is less clear in its meaning since lack of power to amend 'the contract' is not as precise a restriction as lack of power to amend 'the terms and conditions'.

The clause touches on a frequently arising question of when is the engineer, as a named individual, not the engineer for the purposes of the contract. The answer would seem to be when the engineer is acting outside the authority conferred by the contract. If that is the correct answer clause 2(1)(c) is probably redundant on strict analysis. However it is the case that the activities of the named engineer are not always readily distinguishable between activities in the 'engineer' role and activities as employer's agent or in some other capacity. The consequences of this can be troublesome. For instance, if the person named as engineer tells the contractor that the employer will not be deducting liquidated damages for late completion, that will not bind the employer if done in the role of engineer but it may bind the employer by estoppel if done in the role of employer's agent.

If clause 2(1)(c) does nothing else it does at least draw attention to the problem.

Giving of consent and approvals

Clause 2(1)(d) is a new clause which provides that the giving of any consent or approval by or on behalf of the engineer shall not relieve the contractor of any of his obligations nor of his duty to ensure correctness and accuracy.

Notwithstanding this new clause, the Seventh edition retains two similar clauses from the Sixth edition – clause 7(7) relating to permanent works designed by the contractor and clause 14(9) relating to the contractor's programme and methods of construction.

If the new clause adds anything, other than by serving as a catch-all provision, it is perhaps that by referring to a duty of the contractor 'to ensure the correctness or accuracy of the matter or thing which is the subject of the consent or approval', it may be going further than the reasonable skill and care test for design responsibility stated in clause 8(2) and some way towards an absolute test such as fitness for purpose.

3.3 Engineer to be named and chartered

The Seventh edition retains in clause 2(2)(a) the policy adopted in the Sixth edition that where the engineer as defined in clause 1(1)(c) is not a named and chartered individual then such an individual shall be named within 7 days of the award of the contract and before the works commencement date. This allows the employer the benefit of having a firm backed by adequate resources and professional indemnity cover as the defined engineer whilst recognising the need for an identified individual to act as the engineer.

The requirement in clause 2(2)(a) for the engineer to be a named individual is in some organisations less of a problem than the requirement that the individual should also be a chartered engineer. The standing orders of some public sector organisations require a particular post holder or departmental head to take the responsibilities of the engineer. Such a person may not be chartered. If the problem can be foreseen the Conditions can be amended; if the problem is unforeseen and the point is made during the contract that there is in place no engineer compliant with clause 2(2)(a) the employer could be in serious difficulties. It would be unwise for the employer to continue without a contract compliant engineer in the face of opposition from the contractor.

Replacement of named individual

Clause 2(2)(b) permits the engineer (as defined) to replace a named individual acting as engineer by written notice to the contractor.

This is similar to the provision in clause 1(1)(c) whereby the employer can replace at will the defined engineer by notice in writing. Neither clause requires the consent of the contractor to the new appointment nor debars the substitution of originally named independent firms/persons with firms/persons attached to the employer's organisation. Some standard forms do debar such changes unless made with the contractor's consent.

There is nothing in either clause 2(2)(a) or 2(2)(b) requiring the defined engineer to notify the employer of the person named to act as engineer. Presumably this is left to be dealt with under the defined engineer's contract with the employer.

3.4 Engineer's representative

ICE Conditions are drafted on the assumption that the engineer will not himself undertake watching and supervising the works. The expectation is that site duties will be carried out by others – resident engineers, assistant resident engineers, clerks of works and the like. Principal amongst these is the person who carries the title 'Engineer's Representative'.

The Conditions do not expressly require the appointment of an engineer's representative but it is implicit in clause 2(3)(a) that there will be such an appointment. The clause states that the engineer's representative shall be responsible to the engineer, who shall notify his appointment to the contractor in writing.

Some engineers appoint two engineer's representatives, one site based, the other office based, reflecting the distribution of the engineer's duties. The Conditions do not prohibit this but by the requirement for the engineer's representative to watch and supervise the works it is clearly not intended. It is, in any event, unnecessary because if the engineer wishes to confer administrative or valuation duties on office staff, it can be done without conferring a title.

Duties and responsibilities

Clause 2(3)(b) commences by stating that the engineer's representative shall watch and supervise the construction and completion of the works. The clause goes on to say that he shall have no authority to relieve the contractor of any of his duties or obligations nor, except as expressly provided, order work involving delay or extra payment or make any variation of the works.

It is doubtful if the duty to watch and supervise, taken by itself, gives the engineer's representative any authority to instruct the contractor on construction of the works. The power of instruction is retained throughout the Conditions by the engineer unless there is delegation. Even clause 39, which deals with the removal of unsatisfactory work and materials during the progress of the works, is expressed in terms of the powers of the engineer. The only exception is in clause 19 in respect of safety and security.

The consequences of this are firstly, that the engineer's representative without delegated power is little more than the eyes and ears of the engineer and, secondly, the contractor cannot rely on things said and done by the engineer's representative in claims for additional payments in the absence of delegated powers.

There is one minor drafting change in clause 2(3)(b) between the Sixth and the Seventh editions. The words of the Sixth edition qualifying the provision that the engineer's representative has no authority to order work involving delay, extra payment or variations 'nor except as expressly provided hereunder', are changed in the Seventh edition to 'nor except as expressly provided for in sub-clause (4) of this clause'. Sub-clause (4) relates to delegated powers and thus the Seventh edition closes a potential loophole in the Sixth edition where the phrase 'hereunder' could be taken to refer to any of the clauses in the Conditions following clause 2(3)(b).

Reference on dissatisfaction

A further change between the Sixth and Seventh editions in connection with the engineer's representative is that the Seventh edition transfers the old clause 2(7) of the Sixth edition, reference on dissatisfaction, to the expanded clause 66 – 'avoidance and settlement of disputes'.

3.5 *Delegation by the engineer*

Clause 2(4) permits the engineer to delegate any of his duties or authorities except decisions to be taken or certificates to be issued in respect of:

- clause 12(6) – delay and extra cost for unforeseen conditions
- clause 44   – extension of time for completion
- clause 46(3) – accelerated completion
- clause 48   – certificates of substantial completion
- clause 60(4) – final account certificate
- clause 61   – defects correction certificate
- clause 65   – termination for contractor's default
- clause 66   – avoidance and settlement of disputes.

Delegation under clause 2(4) may be 'from to time' and to 'the Engineer's Representative or any other person responsible to the Engineer'. It can at any time be revoked.

Clause 2(4)(a) requires delegation to be in writing and confirms that it does not take effect until a copy of the notice has been delivered to the contractor. From the wording of this clause it is clear that written notice is to be given to the delegate – a procedural point which raises questions as to whether the common practice of confirming delegation simply by notification letters to the contractor is strictly effective.

Clause 2(4)(b) states that delegation continues in force until such time as the engineer notifies the contractor in writing that it has been revoked. This is something which is frequently overlooked when there are staff changes and management reorganisations and it can lead to difficult situations if it is discovered late in the day that the person with delegated authority has long been absent from the contractual scene.

Nature of delegation

When complications do occur in the operation of delegated powers a question which can arise is whether the engineer, having delegated a power, still has that power himself or whether it resides solely in the person to whom it has been delegated until formally revoked. Opinions are divided on this and there is little case law on the point. The answer may be that it is necessary to look at the circumstances in which the delegation takes place to determine what is intended. In the ICE Conditions, relevant circumstances could be that delegation is 'from time to time'; that it can be revoked 'at any time'; and that to have two persons capable of simultaneously exercising the same power in respect of matters which relate to the contractor's entitlements could lead to uncertainty and injustice.

Even if it is the case under ICE Conditions that the engineer retains the powers he has delegated, that does not diminish the effectiveness of actions taken by persons with delegated powers. They stand as actions taken by the engineer.

Restrictions on delegation

Clause 2(4) does not expressly restrict delegation of the engineer's powers to one person nor does it expressly restrict delegation of a particular power to a single person. It is not uncommon for engineers to spread delegated powers between various persons and there can be good organisational reasons for this. However the practice occasionally adopted of giving various persons the same delegated power is probably outside the intention of the clause and is a recipe for trouble.

The restrictions in clause 2(4)(c) debarring delegation under eight particular clauses, all of which impact significantly on the contractor's rights, are not always treated with the respect they deserve. Sometimes enthusiastic subordinates of the engineer take it upon themselves to make decisions and issue certificates outside the scope of their authority and sometimes engineers distance themselves too far from the contracts under their charge and allow others to undertake duties which should not be delegated.

The consequences of such prohibited actions are dangerously unpredictable. Much depends on whether the contractor challenges their effectiveness. But they can be potentially costly to the employer if, for example, maladministration of the provisions for extensions of time in clause 44 deprives the employer of his right to liquidated damages, or maladministration of the default procedures in clause 65 leaves the employer liable for wrongful termination of the contractor's employment.

Taking advice

It is of course permissible for the engineer to rely on others for advice and information in the performance of his duties and this does not breach the contractual rules on delegation providing the engineer makes the ultimate decision. Confirmation of this can be taken from the case of *Perini* v. *Commonwealth of Australia* (1969) where the judge said:

> 'I cannot accept all the arguments submitted by learned counsel for the plaintiff that the Director is bound to investigate every dependent fact himself; this conclusion would, I think, be to ignore the realities of the situation. I am of opinion, though, that by this agreement and by his mandate he may act upon the findings and opinions of other persons, be they subordinates or independent persons such as architects or meteorological observers; he may also consider and pay attention to the recommendations of subordinates with respect to the very application he is considering. I do agree though that the actual decision must be one which flows from the volition of his own mind and I am of the opinion that it is quite irrelevant that that decision is expressed by the placing of his initials upon the recommendation of a subordinate officer'.

The decision maker

Engineers should be careful in taking advice that the style of their correspondence/certificates/communications does not create the wrong impression. A case in point is *Anglian Water Authority* v. *RDL Contracting* (1988) where a letter conveying an engineer's decision under clause 66 was headed – 'This matter is being handled by Mr Baxter'. Mr Baxter was the project manager not the engineer. A question arose as to whether the letter was a valid engineer's decision. The judge held that it was – influenced it seems by the fact that the engineer personally signed the letter. The judge said:

> 'In the commercial world many decisions are made by people such as Mr Rouse, who append their signature to letters drafted by others. It would require compelling evidence to establish in such circumstances that the decision was not that of the signatory. The facts that Mr Baxter was the Project Engineer and had taken an active part previously in the contract had no probative value'.

The decision might well have been different if the engineer had allowed a subordinate to sign in his (the engineer's) name or, worse still, in the sub-ordinate's name.

3.6 Assistants

Clause 2(5)(a) permits the engineer or the engineer's representative to appoint any number of persons to assist the employer's representative in carrying out his duties. The clause refers specifically to duties under sub-clause (3)(b) and (4) – which are watching and supervising, (3)(b), and delegated duties, (4). The contractor is to be notified of the names, duties and scope of authority of such persons.

The clause is clearly directed at site based staff such as assistant resident engineers, clerks of works and technicians with the intention that the contractor should be informed of the roles of all persons with an interest in his performance. The clause does not expressly require the notification to the contractor to be in writing but good practice demands that it should be.

Powers of assistants

The Seventh edition follows the Sixth edition in dealing with the powers of assistants. Compared with the earlier Fifth edition where powers were limited to assisting the employer's representative in watching and supervising, they can be surprisingly wide powers. Clause 2(5)(a) of the Seventh includes not only assistance in watching and supervising but also assistance in respect of any delegated powers.

Clause 2(5)(b) seeks to impose some checks by stating that assistants have no authority to issue instructions except as necessary to enable them to carry out their duties and to secure the acceptance of materials and workmanship as being in accordance with the contract. This, however, confuses rather than clarifies the situation. It implies that assistants have powers in relation to the acceptance of materials and workmanship which are additional to their notified duties. It is by no means clear how these additional powers can be derived from the contract and there is a hint in clause 2(5)(b) that the draftsmen have recognised the anomaly. The Sixth edition used the phrase 'to secure their acceptance of materials and workmanship'. In the Seventh edition this is changed to 'the acceptance'. A small step perhaps to curbing exuberance of authority.

The final part of clause 2(5)(b) states that any instruction given by an assistant shall, where appropriate, be in writing and be deemed to have been given by the engineer's representative. On instructions in writing this is a less strict requirement than applies elsewhere in the Conditions. For instance clause 2(6) requires instructions given by the engineer and the engineer's representative to be in writing. On the deemed to have been given by the engineer's representative point, the clause raises the disturbing prospect that an assistant, exercising powers to assist in delegated duties can, even by an oral instruction, intentionally or otherwise operate the delegated authority of the engineer's representative.

It must be questionable given the scope for misapplication and mis-understanding on the powers of assistants that the engineer should share with the engineer's representative the powers of appointment/notification of duties conferred by clause 2(5)(a). This is serious business over which the engineer should retain control unless relinquished by a considered and deliberate policy of delegation.

Dissatisfaction with assistant's instructions

Clause 2(5(c) states that if the contractor is dissatisfied by reason of any instruction of any assistant he shall be entitled to refer the matter to the engineer's representative who shall thereupon confirm, reverse or vary the instruction.

Having regard to other provisions in the contract relating to matters of dissatisfaction, notably clause 66, the probable intention of clause 2(5)(c) is that it should operate at a practical level in enabling the contractor to challenge an instruction before carrying it out.

The wording of the clause is weak however in comparison with the wording of similar clauses in other standard forms in that it says the contractor is 'entitled to refer' not that the contractor 'shall refer'. This weakness leaves the door open to the contractor complying with an ill-advised instruction from an assistant and then relying on that instruction to seek payment for its foreseeable consequences.

Where the contractor does operate clause 2(5)(c) by referring a matter of dissatisfaction to the engineer's representative, it is more than likely that in some circumstances there will be delay and extra cost consequences – if only whilst the engineer's representative considers the matter. The clause itself has nothing to say on such consequences and it is possible that they can be dealt with under clause 13 (instructions) or clause 51 (variations) – although both arguably require the engineer's representative to have delegated powers in respect of these clauses.

3.7 *Instructions*

Clause 2(6) of the Seventh edition contains some minor amendments to the text of the Sixth edition made in the 1993 corrigenda. Most significantly these replace the phrase 'Engineer's Representative' with 'any person' to ensure that the application of the clause is not unintentionally restricted.

Instructions to be in writing

Clause 2(6)(a) requires instructions given by the engineer or any person exercising delegated duties to be in writing. It concludes with the proviso that if it is considered necessary to give an oral instruction the contractor shall comply.

The purpose of the proviso is probably to prevent the contractor from insisting on an instruction in writing in situations requiring immediate action. It is questionable if it takes away the contractor's right to a written instruction in less urgent circumstances. The test is in the words 'considered necessary'.

Although it is probably the intention of the clause that all instructions shall be in writing it does not actually say so and, if anything, it establishes a three tier rule:

- all instructions given by the engineer to be in writing
- instructions given by the engineer's representative or assistants under delegated powers to be in writing
- other instructions given by the engineer's representative or assistants not required to be in writing.

Confirmation of oral instructions

Clause 2(6)(b) addresses the proviso on oral instructions in clause 2(6)(a) by requiring all oral instructions within the scope of that clause to be confirmed in writing as soon as possible in the circumstances. The proviso is also recognised in clause 51(2) thereby bringing the rules on variations into line with those on instructions.

Clause 2(6)(b) additionally states that if the contractor confirms in writing an oral instruction and this is not contradicted forthwith in writing by the engineer or the engineer's representative, it is deemed to be an instruction given by the engineer.

Failure by the engineer or the engineer's representative to confirm an oral instruction in writing could amount to a breach entitling the contractor to damages but the better remedy is given by the clause in enabling the contractor to take the initiative in establishing the status of the instruction.

Disputes do arise on confirmations from the contractor which are contradicted in writing and it cannot be assumed that the contradiction by itself is always sufficient to dispose of an argument on whether a contractor has been given an oral instruction.

Specification of authority

Clause 2(6)(c) entitles the contractor to request, and obliges the engineer or any person exercising delegated authority to specify, on what basis an instruction is given.

As a matter of good practice all instructions should state the basis on which they are given – not only to establish a proper record but also to indicate that the giver of the instruction knows what he is doing. Frequently it is the case that when the contractor has to ask for the basis of authority, that is the prelude to a dispute of lengthy magnitude and a sign of poor project management.

In the event that the engineer is unable to identify the authority for an instruction the best course is that it be reconsidered.

3.8 Impartiality

Clause 2(7) states that the engineer, the engineer's representative and any person exercising delegated duties and authorities shall act impartially within the terms of the contract having regard to all the circumstances.

The wording of clause 2(7) differs significantly from that in the corresponding clause in the Sixth edition – clause 2(8). There it was only the engineer who was required to act impartially and even that was curiously qualified by the phrase 'except in connection with matters requiring the specific approval of the Employer'. The duty of the engineer's representative to act impartially will be noted with interest by contractors looking for improved relationships with the engineer's staff.

Meaning of 'impartially'

The word 'impartially' can lead to some difficulty since in the narrow sense of affording both parties a hearing it can be distinguished from fairness, as this extract from the judgment of Viscount Dilhorne in the case of *Sutcliffe* v. *Thackrah* (1974) illustrates:

'Here the architect is required to issue interim certificates for the "total value of the work properly executed" and in valuing that work he has to use his professional skill and knowledge. He is not employed to be unfair to the builder. He is not required to determine a dispute between his employer and the builder. As I see it, there is no question of his having to act impartially between them. He must, if he exercises his professional skill and knowledge as it should be exercised, assess the total value of the works properly executed at what he honestly believes to be its true value. If he does that he is acting fairly'.

The wider meaning of impartially can be gathered from the *Shorter Oxford Dictionary* which defines impartial as: 'not favouring one more than the other, unprejudiced, unbiased, fair, just, equitable'. And this is more in accord with the usual interpretation of the phrase 'duty to act impartially'.

Lord Reid in his judgment in the *Sutcliffe* v. *Thackrah* case expressed such a duty in these words:

'It has often been said, I think rightly, that the architect has two different types of function to perform. In many matters he is bound to act on his client's instructions, whether he agrees with them or not; but in many other matters requiring professional skill he must form and act on his own opinion.

Many matters may arise in the course of the execution of a building contract where a decision has to be made which will affect the amount of money which the contractor gets. Under the RIBA contract many such decisions have to be made by the architect and the parties agree to accept his decisions. For example, he decides whether the contractor should be reimbursed for loss under clause 11 (variation), clause 24 (disturbance) or clause 34 (antiquities); whether he should be allowed extra time (clause 23); or where work ought reasonably to have been completed (clause 22). And, perhaps most important, he has to decide whether work is defective. These decisions will be reflected in the amounts contained in certificates issued by the architect.

The building owner and the contractor make their contract on the understanding that in all such matters the architect will act in a fair and unbiased manner and it must therefore be implicit in the owner's contract with the architect that he shall not only exercise due care and skill but also reach such decisions fairly holding the balance between his client and the contractor.'

The application of those principles to civil engineering contracts was confirmed in the New Zealand case of *Canterbury Pipe Lines* v. *Christchurch Drainage Board* (1979) where it was said, referring to *Sutcliffe* v. *Thackrah* and other cases:

'In our opinion it should be held in the light of these authorities that in certifying or acting under Clause 13 here the engineer, though not bound to act judicially in the ordinary sense, was bound to act fairly and impartially'.

More recently in the English case of *John Barker Construction* v. *London Portman Hotel* (1996) it was held that an architect was obliged to act fairly in granting extensions of time. And in *Balfour Beatty* v. *Docklands Light Railway* (1996) it was accepted that under an amended version of the ICE Fifth edition where the employer replaced the engineer as the certifier, the employer was nevertheless bound to act fairly and reasonably.

Purpose of clause 2(7)

Considering clause 2(7) in the light of the legal authorities, its general purpose is perhaps no more than to state the common law duty of certifiers to act fairly. If in doing so it brings home the message to the engineer and his staff that failure to act fairly invalidates certificates it serves a useful purpose.

However, to the extent that on its wording the clause requires the engineer and his staff to act impartially in matters generally, it goes beyond the common law duty and arguably too far. The engineer and his staff can legitimately have duties to the employer where impartiality would not be appropriate.

It may be that the phrase 'having regard to all the circumstances' is intended to regulate the scope of the clause but on one interpretation it is more concerned with how impartiality is undertaken rather than with the extent to which impartiality is required.

Chapter 4
Assignment and subcontracting

4.1 Introduction

Clauses 3 and 4 of the Seventh edition deal respectively with assignment and subcontracting. Both clauses are significantly changed from the Sixth edition.

Clause 3 is expanded to include a new provision designed to exclude third parties from acquiring rights under the contract by application of the Contracts (Rights of Third Parties) Act 1999.

Clause 4 is expanded to re-introduce some control by the engineer over the appointment of subcontractors. The requirement for consent of the engineer to subcontracting parts of the works was a long standing feature of ICE Conditions until swept away in the Sixth edition. That was done to improve the commercial freedom of contractors but it was not popular with employers, many of whom reverted to the old order, amending the Conditions by restoring clause 4 from the Fifth edition in place of clause 4 from the Sixth edition.

4.2 Assignment and subcontracting generally

Assignment is the transfer of legal rights. It can include transfer of title, interest in land, property or contractual rights. Assignment operates as an exemption to the doctrine of privity – a doctrine to the effect that only the parties to a contract acquire rights and obligations under the contract.

Privity of contract

Until comparatively recently assignment provided one of the few escapes from the application of the rule of privity of contract. It has long been recognised as a rule of potential injustice with a string of cases *Crow* v. *Rogers* (1724), *Price* v. *Easton* (1833), *Tweddle* v. *Atkinson* (1861), and *Beswick* v. *Beswick* (1968) – all revealing the difficulties facing an intended beneficiary in enforcing a benefit when not a party to the contract establishing the benefit. But, as Lord Haldane put it in the case of *Dunlop Pneumatic Tyre Co Ltd* v. *Selfridge & Co Ltd* (1915):

> 'In the law of England certain principles are fundamental. One is that only a person who is a party to a contract can sue on it.'

In that case Dunlop was attempting to sue a retailer for breach of a retail price agreement but Dunlop's contract was with the wholesaler who was not himself in breach.

In construction contracts the effects of privity of contract are readily apparent. An employer cannot sue (at least in contract) a subcontractor for damages arising from faulty work. A subcontractor cannot sue an employer for work done but not paid for by the main contractor. And, taking it further by way of example, a pension company buying as an investment a building from a property developer and far removed contractually from the contractor, the designers, subcontractors and suppliers, is not well placed to sue for building defects.

The construction industry has devised various methods of overcoming some of the difficulties caused by the rule of privity – collateral contracts and collateral warranties, of which more is said later in this chapter – being amongst the most effective. Assignment remains as the only exception to privity expressly stated in ICE Conditions and many other standard forms but by itself it does not wholly address the problems of the construction industry because as a general rule English law permits only the assignment of rights and not obligations – that is, benefits but not burdens.

Recent legal developments

Two important recent developments have eased strict application of the rule of privity. The first is a statutory measure – the passing into law of the Contracts (Rights of Third Parties) Act 1999. The second is a trend in judgments on common law actions for damages towards recognising the rights of third parties.

The Contracts (Rights of Third Parties) Act 1999 is a general legal reform which applies to most contracts, including construction contracts and related professional service contracts. Under the Act, which came into force in May 2000, parties to contracts can confer directly enforceable rights on third parties.

Examination of the common law trend is beyond the scope of this book but it seems reasonably clear from House of Lords' decisions in *St Martins Corporation Ltd* v. *Sir Robert McAlpine & Sons* (1997) and *Alfred McAlpine Construction Ltd* v. *Panatown Ltd* (2000) that the rule that a person cannot recover damages for breach of contract when he himself has suffered no loss is in the process of being partly abolished.

Novation

Novation is a tripartite agreement whereby a contract is rescinded in consideration of a new contract being entered into, on the same or similar terms as the old contract, by one of the original parties and a third party. This can occur when one of the original parties changes its legal status or goes into receivership.

Novation has the effect of discharging the parties from the obligations of the original contract and in this respect it differs from assignment. Thus the displaced party in a novation is free from further obligation whereas, for example, when a lease to a property is assigned the original leaseholder remains liable to the landlord for payments until the lease expires.

Subcontracting

Subcontracting has been described as an arrangement to secure vicarious performance of contractual obligations. Lord Greene, in the case of *Davies* v. *Collins* (1945) described it as follows:

> 'In many contracts all that is stipulated for is that the work shall be done and the actual hand to do it need not be that of the contracting party himself; the other party will be bound to accept performance carried out by somebody else. The contracting party, of course, is the only party who remains liable. He cannot assign his liability to a sub-contractor, but his liability in those cases is to see that the work is done, and if it is not properly done he is liable. It is quite a mistake to regard that as an assignment of contract; it is not.'

Subcontractors fall into three main categories:

- domestic subcontractors
- named or specified subcontractors
- nominated subcontractors.

Domestic subcontractors are selected and appointed by the main contractor to suit his own arrangements. The main contractor usually remains fully responsible for the performance of domestic subcontractors as though he had undertaken the work himself.

Named, or specified subcontractors, are selected by the employer or the engineer but usually the terms of business are left to the contractor. Subject to certain exceptions full responsibility for their performance will normally rest with the main contractor.

Nominated subcontractors are either specified in the contract or their use is instructed by the engineer under prime cost or provisional sums. Standard forms of contract differ as to how they allocate responsibility for their performance.

4.3 *Novation, assignment and subcontracting distinguished*

Many cases have come before the courts where the distinction between novation, assignment and subcontracting has been less than immediately obvious. No better analysis is available than that given by Lord Justice Staughton in the Court of Appeal decision in *Linden Gardens Trust Ltd* v. *Lenesta Sludge Disposals Ltd* (1992) as illustrated by the following extracts:

'(a) Novation
This is the process by which a contract between A and B is transformed into a contract between A and C. It can only be achieved by agreement between all three of them, A, B and C. Unless there is such an agreement, and therefore a novation, neither A nor B can rid himself of any obligation which he owes to the other under the contract. This is commonly expressed in the proposition that the burden of a contract cannot be assigned unilaterally. If A is entitled to look to B for payment under the contract, he cannot be compelled to look to C instead, unless there is a novation. Otherwise B remains liable, even if he has assigned his rights under the contract to C.

Similarly, the nature and content of the contractual obligations cannot be altered unilaterally. If a tailor (A) has contracted to make a suit for B, he cannot by assignment be placed under an obligation to make a suit for C, whose dimensions may be quite different. It may be that C by assignment would become entitled to enforce the contract – although specific performance seems somewhat implausible – or to claim damages for his breach. But it would still be a contract to make a suit that fitted B, and B would still be liable to A for the price.

(b) Assignment
This consists in the transfer from B to C of the benefit of one or more obligations that A owes to B. These may be obligations to pay money, or to perform other contractual promises, or to pay damages for a breach of contract, subject of course to the common law prohibition on the assignment of a bare cause of action.

But the nature and content of the obligation, as I have said, may not be changed by an assignment. It is this concept which lies, in my view, behind the doctrine that personal contracts are not assignable.

Thus if A agrees to serve B as chauffeur, gardener or valet, his obligations cannot by an assignment make him liable to serve C, who may have different tastes in cars, or plants, or the care of his clothes.

There is no reason in principle why a party who has earned his fee for performing a personal contract should not assign the right to receive it; nor, so far as I can see, why B for whom the tailor has completed a suit should not assign to C the right to receive it.

(c) Subcontracting
I turn now to the topic of subcontracting, or what has been called in this and other cases vicarious performance. In many types of contract it is immaterial whether a party performs his obligations personally, or by somebody else. Thus a contract to sell soya beans, by shipping them from a United States port and tendering the bill of lading to the buyer, can be and frequently is performed by the seller tendering a bill of lading for soya

beans that somebody else has shipped. On the other hand a contract to sing the part of Hans Sachs at Covent Garden Opera House will not be fulfilled by procuring someone else to do so. That is not because the burden of a contract may not be assigned unilaterally; in each case the original contractor would still be liable if the obligation were not performed or were performed badly. It is because some contractual obligations are personal; they must be performed by the party who has contracted to perform them and nobody else.'

4.4 Collateral contracts and warranties

The construction industry more than most has felt the effect of the rules of privity of contract. The financial consequences of defective construction work can frequently be beyond the resources of the contractor – a situation in part of under-capitalisation and low profitability. Employers and others with long term interests in construction projects can find themselves exposed to risks which it is not commercially prudent to withstand. Not surprisingly there has long been a search for methods of protection.

Early in the 1980s it appeared that actions for negligence might become an effective remedy against defaults of firms or persons where there was no direct contract. This was fuelled by a series of landmark decisions on duties of care – including the House of Lords' decision in *Junior Books Ltd* v. *The Veitchi Co Ltd* (1982) where it was held that the relationship between a nominated subcontractor and the employer was such as to give rise to a duty of care in the performance of the subcontractor's work.

Later decisions, however, reversed the trend towards actions in tort by severely restricting the recovery of purely economic loss – most notably, *D&F Estates Ltd* v. *Church Commissioners for England* (1988) and *Department of the Environment* v. *Thomas Bates & Jones Ltd* (1990).

This threw attention back to earlier remedies, collateral contracts and collateral warranties.

Collateral contracts

A collateral contract, if its existence can be established, forms a useful device for overcoming the rule of privity and bridging a gap in a contractual claim. Such a contract is usually founded on a warranty given to or given by a party who is not a party to the principal contract. As such it is not far removed from a guarantee.

Various attempts have been made to argue that nomination creates a collateral contract between the employer and the subcontractor on the basis that in consideration for being nominated the subcontractor warrants performance to the employer. As a general proposition the argument fails but there are at least two construction cases on record in respect of named suppliers, *Shanklin Pier Ltd* v. *Detel Products Ltd* (1951) and *Greater London*

Council v. *Ryarsh Brick Co. Ltd* (1986), where the courts have accepted the existence of collateral contracts because of the particular circumstances.

In the *Shanklin Pier* case the contractor was instructed to use paint supplied by Detel on the basis of a representation made to the employer by Detel that the paint would last for seven years. When the paint failed the employer, Shanklin Pier, successfully sued the supplier, Detel, on the basis that consideration lay in specifying the paint in return for the promise of its durability.

Collateral warranties

The term 'collateral warranty' as used in the construction industry usually relates to a formally executed agreement designed to create a contractual relationship where otherwise none would exist – as, for example, between employers and subcontractors, developers and designers, and the like. In commercial construction projects such warranties are commonplace. They enable actions to be brought in contract for defective performance by a range of participants by widening liability beyond the basic employer/contractor relationship.

4.5 Assignment in the Seventh edition

Clause 3(1) of the Seventh edition provides that neither the employer nor the contractor shall assign the contract or any part thereof or any benefit or interest without the prior written consent of the other party, which shall not be unreasonably withheld. This is identical to clause 3 of the Sixth edition.

Similar clauses to clause 3(1) are found in most standard forms for construction contracts. They are not without difficulties in their application and they have troubled the courts on many occasions. The problem seems to be a conflict between what is sometimes thought to be legal right to assign benefits and the scope of the contractual restriction.

Restriction on assignment

The legal effectiveness of restrictions is not without question. As long ago as 1908 in the case of *Tom Shaw & Co* v. *Moss Empires Ltd* Mr Justice Darling said of a prohibition on assignment:

'it could no more operate to invalidate the assignment than it could interfere with the laws of gravitation'.

And Lord Justice Staughton in the *Linden Gardens* case referred to at section 4.3 above said:

'If it were free from authority, I should be inclined to think that a prohibition on assignment in a contract was ineffective. Seeing that assignment cannot increase the obligation or alter it in any way, but only change the person who is to benefit from it, there are no very powerful reasons for allowing it to be prohibited.'

The authority referred to was principally the decision in the civil engineering case of *Helstan Securities Ltd* v. *Hertfordshire County Council* (1978). A contractor under the ICE Fourth edition had assigned to Helstan amounts due for work undertaken for the employer, Hertfordshire. It was held that Helstan could not recover since the contract provided that the contractor should not 'assign the contract or any part thereof or any benefit or interest therein or thereunder without the written consent of the employer'.

In the *Linden Gardens* case the Court of Appeal resolved the dilemma by holding that a prohibition simply on assigning the contract did not prohibit the assignment of benefits under the contract. The House of Lords took a more robust view and held that:

'an assignment of contractual rights in breach of a prohibition against such assignment is ineffective to vest the contractual rights in the assignee'.

Lord Brown-Wilkinson in addressing the issue of whether a prohibition on assignment is void as being contrary to public policy made these points:

'It was submitted that it is normally unlawful as being contrary to public policy to seek to render property inalienable. Since contractual rights are a species of property, it is said that a prohibition against assigning such rights is void as being illegal.

In the face of authority, the House is being invited to change the law by holding that such a prohibition is void as contrary to public policy. For myself I can see no good reason for so doing. Nothing was urged in argument as showing that such a prohibition was contrary to the public interest beyond the fact that such prohibition renders the chose in action inalienable.

A party to a building contract, as I have sought to explain, can have a genuine commercial interest in seeking to ensure that he is in contractual relations only with a person who he has selected as the other party to the contract. In the circumstances, I can see no policy reason why a contractual prohibition on assignment of contractual rights should be held contrary to public policy.

The existing authorities establish that an attempted assignment of contractual rights in breach of a contractual prohibition is ineffective to transfer such contractual rights. I regard the law as being satisfactorily settled in that sense. If the law were otherwise, it would defeat the legitimate commercial reason for inserting the contractual prohibition viz, to ensure that the original parties to the contract are not brought into direct contractual relations with third parties.'

Assignment of monies due

Some standard forms of contract seek to avoid dispute on whether the assignment of monies due is caught by express restrictions on assignment by adding into the assignment clause a phrase such as 'save that the Contractor may without such consent assign either absolutely or by way of charge any money which is or may become due to him under the contract'.

ICE Conditions, Seventh edition included, have not traditionally included such wording as standard but it is a common amendment.

Consent to be not unnecessarily withheld

The concluding words of clause 3(1) 'which consent shall not unreasonably be withheld' inevitably attract the question – what if it is?

The answer to that would seem to be that an action for damages for breach of contract might follow. As to what would be 'unreasonable', that would be a matter of fact to be determined in the circumstances of the case.

Assignment without consent

So far as assignment under the Seventh edition is concerned and the question of what is the position if either of the parties does assign without consent, note should be taken of the default provisions of the contract in clauses 64(1) and 65(1).

If the employer is in breach, clause 64(1) permits the contractor to terminate his own employment and if the contractor is in breach, clause 65(1) permits the employer to terminate the contractor's employment.

4.6 *Third party rights*

The new clause 3(2) of the Seventh edition states that nothing in the contract shall confer or purport to confer on any third party any benefit or right to enforce any term of the contract.

The purpose of this clause is quite simply to exclude the application of the Contracts (Rights of Third Parties) Act 1999.

This is a perfectly legitimate approach designed to avoid some of the problems which could arise if third parties could rely on the contract, not least, the possibility of:

- contractors suing the employer's professional team (and the reverse)
- employers suing subcontractors (and the reverse)
- subcontractors suing one another
- future owners of the works suing firms and individuals involved in the construction process.

It could be said that application of the Act would avoid the need for collateral warranties but that potential benefit is probably not sufficient to outweigh the uncertainties of contractual rights passing to third parties.

4.7 Subcontracting in the Seventh edition

In summary the Seventh edition regulates subcontracting as follows:

- clause 4(1) – the contractor shall not subcontract the whole of the works without the consent of the employer
- clause 4(2) – the contractor may subcontract parts of the works unless expressly stated otherwise in the appendix to the form of tender
- clause 4(2) – the contractor shall notify the engineer of the extent of the work to be subcontracted with names of the subcontractors
- clause 4(2) – the engineer can, with good reason, object to proposed subcontractors
- clause 4(3) – labour only subcontracting does not require notification
- clause 4(4) – the contractor remains liable for all work subcontracted and the acts/defaults of subcontractors
- clause 4(5) – the engineer can require the removal from site of subcontractors for misconduct and the like.

The whole of the works

Clause 4(1) states that the contractor shall not subcontract the whole of the works without the prior written consent of the employer.

This clause does not include the words 'which consent shall not unnecessarily be without' which qualify rights to assign in clause 3(1). It is therefore in the employer's absolute discretion as to whether to consent to subcontracting of the whole of the works. Note that the decision rests with the employer and not the engineer.

It is not absolutely clear whether the phrase 'the Whole of the Works' is intended to apply to the whole as a single package or the whole as a collection of packages. Arguably it applies to both but it probably does not prevent the contractor subcontracting the whole of the construction operations providing management control is retained.

The Seventh edition does not expressly state a remedy for breach by the contractor of clause 4(1) but it would seem to be caught by clause 65(1)(j) of the default provisions, i.e. 'fundamentally in breach of his obligations under the Contract'.

Nevertheless, the absence of an express remedy in the Conditions does leave open the question which sometimes arises more generally in construction contracts – is the employer bound to pay for work which is subcontracted without consent?

The legal position on this was reviewed in the case of *Southway Group Ltd* v. *Wolff and Wolff* (1991). The matter in appeal concerned whether the contractor was entitled to perform his obligations vicariously and whether the purchaser was obliged to accept vicarious performance as good performance under the contract. It was held that there was an element of personal confidence in the selection of the contractor and that he had no right to subcontract. Accordingly the purchaser was not obliged to accept performance by others.

The learned authors *of Building Law Reports* commenting on the *Southway* case say this in 57 BLR at page 35:

> 'It might be thought that these principles [referring to the judgment] would not have much application to the modern construction industry. Yet choices are frequently made on the basis of reputation, e.g. of an architect to design a building or an extension to an existing building or a contractor or a subcontractor on the basis of specialist skill or generally dependable personal performance. It may even be present when competitive tenders are obtained since those selected for tender may have been selected because of the confidence placed in them. In these instances the degree to which subcontracts may be permissible in law, in the absence of contrary contract provisions, may be severely limited.'

Subcontracting parts of the works

Clause 4(2) of the Seventh edition allows for subcontracting parts of the works subject to certain controls. The first sentence of the clause states that except where otherwise provided in the appendix to the form of tender, the contractor may subcontract any parts of the works or their design.

At first sight this suggests that there may be parts of the works which the contractor will not be permitted to subcontract. However, the section of the appendix relevant to clause 4(2) states 'Parts or Sections of the Works which shall not be subcontracted without the Engineer's prior written approval'. It might appear, therefore, that the first part of clause 4(2) is intended to be read as 'except when otherwise provided in the appendix the contractor may subcontract any parts of the works without obtaining the engineer's prior approval'. However, note has to be taken of the remainder of clause 4(2) which requires the contractor to provide details of all subcontractors and gives the engineer rights to object to particular subcontractors.

In effect this reverses, at least for parts of the works identified in the appendix, the policy of subcontracting without control adopted in the Sixth edition and re-imposes to some extent the controls which existed in the Fifth edition. Many employers and engineers will welcome the change. Contractors may be concerned however at the ease by which the clause could be converted to a ban on subcontracting parts of the works by the simple deletion of the words in the appendix 'without the Engineer's prior written approval'. The opening sentence of clause 4(2) would then operate to allow only subcontracting of parts of the works not listed in the appendix.

Details of subcontractors

The second sentence of clause 4(2) states that the extent of the work to be subcontracted and the names and addresses of subcontractors must be notified to the engineer in writing as soon as practicable and in any event not later than 14 days prior to the subcontractor's entry to the site or, in the case of design, on appointment.

There are two distinct matters here – the work to be subcontracted and the particulars of the subcontractor. This raises the question of whether any approval required by virtue of listing in the appendix is to be an approval simply related to work content or whether it can also be related to the choice of subcontractor. Having regard to the 'objection' provisions which follow later in clause 4(2) the answer would seem to be that the engineer should treat applications to subcontract parts of the works and applications to use particular subcontractors as separate issues.

If this is correct then it can be taken that the second sentence of clause 4(2) applies to all subcontractors and not simply to those involved with work stipulated as requiring approval prior to subcontracting.

Objections to subcontractors

The final part of clause 4(2) provides that if the engineer, for good reason, objects to a proposed subcontractor within 7 days of receipt of details, the subcontractor shall not be employed on or in connection with the works. Any objection must be accompanied by reasons in writing.

The engineer had no such right of objection in the Sixth edition – similar details had to be provided but apparently for information or record purposes only.

This new part of clause 4(2) is unlikely to be welcomed by contractors and it may not be welcomed by all engineers. Contractors tender on the basis of subcontract prices and the financial consequences of rejection of a proposed subcontractor can be significant. Engineers may foresee arguments with contractors on the rationality of their reasons for objection and pressure from employers to explain when things have gone wrong why certain subcontractors were allowed on site in the first instance.

Engineers should be careful in giving reasons for objections, mindful that amongst other things their reasons will almost certainly be conveyed back to the rejected subcontractor. The legal implications could be considerable. If an aggrieved contractor is able to establish that the engineer's reasons are not 'good reasons' that could leave the employer liable to claims for damages, delay or extra cost. And a subcontractor, in possession of reasons in writing for its rejection, might itself seek legal redress against the engineer if incautious or inaccurate statements are made.

Labour-only subcontractors

Clause 4(3) provides that the employment of labour-only subcontractors does not require notification to the engineer under clause 4(2).

Identical or similar clauses were to be found in earlier editions of ICE Conditions and generally it is understood that the contractor has wide freedom to employ casual labour, agency supplied labour and firms undertaking work on a labour-only basis.

The difficulties which do arise with this type of clause are frequently whether a subcontractor is genuinely labour-only or whether, as with drainage subcontractors and the like, by use of owned plant or the provision of materials for temporary works a subcontractor ceases to be 'labour-only'. The strict approach can be that anything more than the use of hand tools and personal equipment is going too far but a more relaxed approach is frequently tolerated.

Contractor liable for subcontractor's defaults

Clause 4(4) provides that the contractor shall be liable under the contract for all subcontracted work and for the costs, defects and neglects of any subcontractors.

This clause confirms the ordinary legal position that the contractor cannot avoid liability under the main contract by pleading the fault of a subcontractor. The words may even be sufficient to make the contractor liable to the employer under the contract for negligence of a subcontractor – if, for example, the subcontractor damaged the employer's property.

It is doubtful, however, if the clause burdens the contractor with the subcontractor's negligence on a wider basis. In law a subcontractor is generally regarded as an independent contractor responsible for his own negligence. In *D&F Estates Ltd* v. *The Church Commissions for England* (1988) in an action in negligence by a third party it was held that the contractor discharged his duty of care to third parties by appointing a subcontractor he believed to be reasonably competent and any duty to supervise the subcontractor's work was solely in contract.

It is also doubtful if the clause offers any protection to the subcontractor against claims of negligence by the employer or others. It is difficult to see how a subcontractor sued for negligence can raise as his defence the terms of a contract to which he is not a party. In *Southern Water Authority* v. *Lewis and Duvivier* (1984) it was held in respect of a similar defence involving the old model form MF'A', that a subcontractor could not obtain the benefit of an exclusion of liability clause applying to the main contractor.

Clause 4(4) of the Seventh edition contains one small wording change from clause 4(4) of the Sixth edition. Under the Sixth edition the clause referred to the contractor's liability for 'all work sub-contracted under this clause'. In the Seventh edition that becomes 'all work sub-contracted by him'. It was possible to read the Sixth edition clause as excluding work subcontracted under the nomination provisions of clause 59. There was some logic in this because under clause 59 the contractor does not have full responsibility for the performance of nominated subcontractors in all circumstances. That remains the position in the Seventh edition and there is therefore a potential point of conflict between clause 4(4) and clause 59. It may not amount to much but as the following two examples show, the courts are capable of springing big surprises on minor oddities of wording.

'Beyond the contractor's control'

This phrase was given unexpectedly wide meaning by the House of Lords in the case of *Scott Lithgow Ltd* v. *Secretary of State for Defence* (1989). It was ruled that the contractor was entitled to payment under a provision which included, amongst other things, as grounds for extra payment the phrase 'or any other cause beyond the contractor's control'. Lord Keith said:

> 'Failures by such suppliers or sub-contractors, in breach of their contractual obligations to Scott Lithgow, are not matters which, according to the ordinary use of language, can be regarded as within Scott Lithgow's control.'

The case concerned nominated specialist suppliers and subcontractors but it may nevertheless have wider applications. Accordingly many standard forms of contract which include the phrase 'beyond the Contractor's control' in relation to payment, extension of time or other matters have been amended.

The JCT Minor Works Agreement, for example, which uses the phrase in its extension of time clause, now has an extra sentence:

> 'Reasons within the control of the Contractor include any default of the Contractor or of others employed or engaged by or under him or in connection with the Works and of any supplier of goods or materials for the Works'.

Restricted interpretation of 'contractor'

On general principles one would not expect the meaning of 'contractor' when used in the provisions of a standard form, to exclude 'sub-contractors'. However in *Jarvis Ltd v Rockdale Housing Association Ltd* (1986) the Court of Appeal held that 'contractor' in the phrase 'unless caused by some negligence or default of the contractor' in the determination provisions of JCT 80 applied only to the contractor as main contractor and did not include nominated subcontractors.

These two extracts from the judgment of Lord Justice Bingham explain the logic of the decision:

'I cannot accept the employer's submission that "the Contractor" in 28.1.3.4. is to be understood as including subcontractors and their servants and agents. Both in this clause and elsewhere in the contract the draftsman has made express reference to subcontractors, their servants and agents. The absence of any such reference here is not explained by the context, or can it be dismissed as of no significance. Even an experienced draftsman could be relied upon to avoid such an error.

"The Contractor" can in my judgment be naturally and sensibly understood as referring to, in this case, John Jarvis Ltd, its servants and agents, through whom alone it can, as a corporation act.'

It is clearly most important therefore that contract draftsmen express their intentions with both clarity and consistency. One essential rule is 'never change your wording unless you intend to change your meaning'.

Removal of subcontractors

Clause 4(5) empowers the engineer, after due warning in writing, to require the contractor to remove from the works any subcontractor for misconduct, incompetence, negligence or breach of health and safety rules.

The clause seems to be aimed particularly at firms rather than individuals. Clause 16 deals with the removal of persons.

Clause 4(5) of the Seventh edition differs slightly from the corresponding clause in the Sixth edition by including the words 'or their design' in the phrase 'remove from the Works or their design'. This extends the engineer's powers of removal to designers employed by the contractor – an interesting if not potentially contentious development.

Failure by the contractor to comply with a removal order could have serious consequences. It is not expressly stated in clause 65 as a default justifying termination but it could come within the scope of the default 'persistently or fundamentally' in breach of obligations under the contract. On a lesser scale it might be treated as grounds for suspension of work under clause 40.

It should be noted however that a valid order for removal only follows a 'due warning' in writing. The Conditions do not specify any time limits between the warning and the removal order but the correct approach is to give the contractor sufficient and appropriate time to notify the defaulting subcontractor of the need to mend his ways.

Chapter 5
Contract documents

5.1 Introduction

Clauses 5 to 7 are placed in the Conditions under the heading 'Contract Documents'. It is perhaps as well that clause 1(3) provides that the headings in the Conditions shall not be deemed to be part of the contract or be taken into account in interpretation of the contract because the purpose of clauses 5 to 7 is not to identify or regulate the documents forming the contract. To the extent that this is done at all in the Conditions, it is done in clause 1(1)(e).

Clauses 5 to 7 do no more than state (in clause 5) that the contract documents are to be taken as mutually explanatory; provide (in clause 6) for the supply of documents and copyright; and detail (in clause 7) the position on certain documents produced by the engineer or the contractor during the progress of the works.

The changes in clauses 5 to 7 between the Sixth and the Seventh editions are, on the face of it, comparatively minor. Clause 7(6)(a) uses slightly different terminology to describe how permanent works designed by the contractor shall satisfy the engineer. And throughout clause 7 the phrase which in the Sixth edition appears as 'approved by the Engineer' appears in the Seventh edition as 'accepted by the Engineer'. This latter change may have unintended consequences.

Documents accepted/approved

By virtue of clauses 1(1)(e), 1(1)(f) and 1(1)(g), specifications or drawings 'approved in writing' by the engineer form part of the contract. In section 2.3 of Chapter 2 it is suggested that the change in clause 7 from 'approved' to 'accepted' may have been in recognition of the potential problems created by the possibility that anything put forward by the contractor and 'approved' by the engineer could gain contractual status. That was a genuine problem which needed to be addressed. However, the position now in the Seventh edition is that the engineer is only required to 'accept' documents for permanent works designed by the contractor. 'Acceptance' of these documents does not put them within the scope of contract documents as defined in clauses 1(1)(e), 1(1)(f) and 1(1)(g). This may not have been the intention of the draftsmen and it seems more likely that the change was intended to prevent the possibility of the contractor's temporary works details inadvertently gaining contractual status by being approved.

There is no clear or general rule in the Conditions to the effect that documents which are 'accepted' by the engineer bind the contractor in like manner to documents which are 'approved'. Nor is it likely that such a rule can be implied. Clause 7(2) expressly states that the contractor is bound by 'further documents' requested by the engineer but the general rule as stated in clause 7(7) is simply that 'acceptance' by the engineer shall not relieve the contractor of his responsibilities under the contract.

Note, in connection with clause 7(7), that whilst the text of the clause refers to 'acceptance', the marginal note refers to 'approval'. This seems to be a printing error, but one which is conveniently rendered inconsequential by clause 1(3) – marginal notes not to be taken into account.

5.2 Construction of contracts

The extent to which clause 5 of the Conditions is concerned with the legal construction of the contract is questionable, as explained in section 5.3 below. However, it is worth noting here some general rules of construction developed by the courts over the years. Briefly, the more important rules are:

- intention to be found from the contract itself. The courts will not go outside the written documents and substitute the presumed intention of the parties
- words to be given their ordinary or plain meaning. In trade or technical contracts the customary meaning applies
- words to be construed to make a contract valid rather than invalid
- the intention of the parties to be derived by construing a contract as a whole
- effect to be given to all terms – treating some terms as redundant to be avoided
- unless the documents expressly provide otherwise, particular conditions prevail over standard conditions
- *expressio unius* – express inclusion of a certain thing excludes others of a similar nature
- *ejusdem generis* -when words of a particular class are followed by general words, the general words are taken to apply to things of the same class
- *contra proferentem* – where there is ambiguity in a document the words are to be construed against the party who put forward the document.

The general principles applied by the courts in the construction of documents were summarised by Lord Hoffman in the House of Lords ruling in *Investors Compensation Scheme* v. *West Bromwich Building Society* (1998) as follows:

'(1) Interpretation is the ascertainment of the meaning which the document would convey to a reasonable person having all the background knowledge which would reasonably have been available to the parties in the situation in which they were at the time of the contract

(2) The background was famously referred to by Lord Wilberforce as the "matrix of fact". But this phrase is if anything an understated description of what the background may include. Subject to the requirement that it should have been reasonably available to the parties and to the exception to be mentioned next, it includes absolutely anything which would have affected the way in which the language of the document would have been understood by a reasonable man

(3) The law excludes from the admissible background the previous negotiations of the parties and their declarations of subjective intent. They are admissible only in an action for rectification. The law makes this distinction for reasons of practical policy and, in this respect only, legal interpretation differs from the way we would interpret utterances in ordinary life. The boundaries of this exception are in some respects unclear. But this is not the occasion on which to explore them

(4) The meaning which a document (or any other utterance) would convey to a reasonable man is not the same thing as the meaning of its words. The meaning of words is a matter of dictionaries and grammars; the meaning of the document is what the parties using those words against the relevant background would reasonably have been understood to mean. The background may not merely enable the reasonable man to choose between the possible meanings of words which are ambiguous but even (as occasionally happens in ordinary life) to conclude that the parties must, for whatever reason, have used the wrong words or syntax: see *Mannai Investments Co Ltd* v. *Eagle Star Life Assurance Co Ltd* (1997) AC 749

(5) The "rule" that words should be given their "natural and ordinary meaning" reflects the common sense proposition that we do not easily accept that people have made linguistic mistakes, particularly in formal documents. On the other hand, if one would nevertheless conclude from the background that something must have gone wrong with the language, the law does not require judges to attribute to the parties an intention which they plainly could not have had. Lord Diplock made this point more vigorously when he said in *Antaios Companta Naviera SA* v. *Salen Rederierna AB* (1985) AC 191.201:

> "if detailed semantic and syntactical analysis of words in a commercial contract is going to lead to a conclusion that flouts business commonsense, it must be made to yield to business commonsense".'

5.3 *Explanation and adjustment of documents*

Clause 5 provides that the documents forming the contract are to be taken as mutually explanatory of one another and that ambiguities and discrepancies are to be explained and adjusted by the engineer who shall issue instructions in writing which are to be regarded as instructions issued in accordance with clause 13.

Although clause 5 has a long-standing place in ICE Conditions (it is unchanged since publication of the Fifth edition in 1973 and there was a similar clause in the Fourth edition, published 1955) it remains a clause which still attracts disputes as to its purpose and application. The question is whether it is intended to be explanatory of the legal position on documents or simply directory as to the engineer's obligations in the event of uncertainty on what the contractor is to construct. Put another way, the question is whether the clause empowers the engineer to make decisions on legal interpretation of the contract or whether it empowers him only to give instructions of practical effect.

There are fundamental differences between the two tasks – one of which is that if the engineer decides on legal interpretation, that is not a decision which binds the contractor, whereas if the engineer gives instructions of practical effect they do bind the contractor by virtue of clause 13.

On its wording there can be little doubt that the main purpose of clause 5 is to oblige the engineer to give instructions when there is uncertainty in the documents as to what the contractor is required to do. However, in referring to the documents forming the contract and in stating that the contract documents are to be taken as mutually explanatory of one another, the clause may intrude on legal matters.

Documents forming the contract

The documents forming the contract will for any particular contract be a matter of fact depending on the circumstances. Clause 1(1)(e) defines the 'contract' in terms of certain categories of documents and it is not entirely clear what is the position if there are documents which one or both parties contend should be in the contract but which do not fall readily into any of the stated categories.

It is unlikely that a standard form general definition prevails over the particulars in such a situation and the engineer may then become involved in a dispute on what documents form the contract.

Documents mutually explanatory

Under standard ICE Conditions there is no stated order of precedence of documents. However, from the rules of construction and from the particulars in some cases there will be an order of precedence – if no more than that particulars prevail over the general.

The statement in clause 5 that the several documents forming the contract are to taken as mutually explanatory of one another is not, it is suggested, intended to reduce all documents to the same level for all purposes. Taken in the context of the clause the statement can be seen to have two effects: one, it allows the engineer freedom to make a decision on what instruction should be given for the works to be constructed to his satisfaction; the other, it restricts the contractor's entitlement to claim that reliance was placed on some of the documents to the exclusion of others.

Ambiguities or discrepancies

Clause 5 is frequently quoted along with the *contra proferentum* rule in claims relating to ambiguity in the contract documents. Such claims often overlook the point that for the *contra proferentum* rule to apply there needs to be genuine ambiguity in the documents and not just error or omission.

But in any event the ambiguities or discrepancies referred to in clause 5 are matters to be explained and adjusted by the engineer and to the extent there are cost or delay consequences the proper basis for any claim is more likely to be compliance with instructions and the remedies provided in clause 13(3) than simple reliance on clause 5.

Engineer to explain and adjust

Clause 5 requires the engineer to explain and adjust ambiguities or discrepancies and thereupon to issue instructions. The burden is clearly on the engineer rather than the contractor to operate the clause, although in practice it is usually the contractor who starts the process.

It is apparent from the reference in clause 5 to clause 13 that the task of the engineer is to explain and adjust the documents so that in practical terms the works can be constructed to his satisfaction. This may involve amending drawings, specifications, bills of quantities and the like.

If there are ambiguities or discrepancies in the documents of legal rather than practical effect – for example, whether or not the contract provides for liquidated damages – and the engineer purports to give an instruction or decision under clause 5 dealing with the matter, it is questionable whether either would have any standing. Clause 2(1)(c) confirms that the engineer has no authority to amend the contract and it is most unlikely that clause 5 can be read as giving power to amend under the exception in clause 2(1)(c) which reads 'Except as expressly stated in the Contract'. If there are disputes between the parties on the terms of the contract, the involvement of the engineer should be confined to the disputes resolution procedures in clause 66.

Instructions

Clause 5 concludes with the requirement that the engineer shall issue to the contractor appropriate instructions in writing which shall be regarded as instructions in accordance with clause 13.

The fact that the outcome of the engineer's involvement in clause 5 is the issuance of instructions to the contractor (no mention of employer) is further indication, if it be needed, that the engineer's role is not to resolve legal disputes.

The stipulation that instructions are to be regarded as instructions under clause 13 entitles the contractor to claim for extra cost which could not reasonably have been foreseen at the time of tender and/or valuation of a variation if appropriate.

The opening words of clause 13(3), 'If in pursuance of clause 5 … the Engineer shall issue instructions', appear to be superfluous but they do dispose of the proposition, at least in connection with instructions given under clause 5, that entitlement to payment under clause 13(3) is conditional upon the instructions being given to overcome impossibility. Further comment on this aspect of clause 13 is given in Chapter 9.

5.4 Supply and copyright of documents

Clause 6(1) deals with the supply to the contractor of documents produced by or on behalf of the employer.

Upon award of the contract, the contractor is to be provided, free of charge with:

(a) four copies of the Conditions, the specification and unpriced bill of quantities, and
(b) such number and type of copies of drawings listed in the specification as stipulated in the appendix to the form of tender.

This issue of documents can probably be taken as meaning the issue of additional sets of the documents provided at tender stage. Nevertheless cautious contractors usually have a careful checking system to ensure that any changes between the tender documents and this later issue are identified. Where there are changes with time and cost implications the contractor will normally be entitled to have such changes treated as variations.

Two other points that contractors need to have in mind when taking receipt of additional sets of documents (and distributing them for site use) are:

- whether the sets are complete – there may be documents relating to the tender which are not covered by those listed in clause 6(1)(a) but which need to be in the possession of both the contractor's and the engineer's project management teams, and
- whether the drawings issued on award are definitely intended for construction use – from the wording of clauses 6(1) and 7(1) this is clearly the intention but in practice it is not always the case.

Documents produced by the contractor

Clause 6(2) deals with documents produced by the contractor under any responsibilities he may have for design of the permanent works.

The clause provides that the contractor shall upon approval by the engineer under clause 7(6), supply free of charge to the engineer, four copies of all documents submitted to the engineer and shall supply, at the employer's expense, such further copies as the engineer may request.

By a drafting oversight or typing error, clause 6(2) refers to 'approval' by the engineer under clause 7(6) when what is required by that clause is 'acceptance' by the engineer.

Copyright

Clause 6(3) concerns the copyright of documents as between the parties. It confirms that the copyright of documents supplied by the employer or the engineer to the contractor does not pass to the contractor but it permits the contractor to make copies at his own expense for the purposes of the contract. Similarly, it confirms that copyright of documents supplied by the contractor under clause 7(6) remains with the contractor but the employer and the engineer have power to reproduce and use 'for the purpose of completing operating, maintaining and adjusting the Works'.

In comparison with the terms of other standard forms these are very simple copyright provisions. They say nothing about the rights of third parties and they do not require indemnities against breaches of copyright. It is not unusual to see extensive supplementary copyright clauses added to ICE Conditions and, where the contractor has significant responsibility for design of the permanent works, it is advisable.

5.5 *Further drawings, specifications and instructions*

Clauses 7(1) to 7(4) relate to further information necessary for the construction of the works.

Clause 7(1) requires the engineer to supply the contractor from time to time during the progress of the works with such modified or further drawings, specifications and instructions as, in the engineer's opinion, are necessary for the proper and adequate construction and completion of the works and binds the contractor to complying with the same.

The clause concludes with the statement that if any such drawings, specifications or instructions require any variation to any part of the works, a variation shall be deemed to have been issued pursuant to clause 51.

Perhaps the first thing to note about clause 7(1) is that the engineer's duty is not expressed as being dependent on an application for further information by the contractor. The test for supply is what is necessary 'in the Engineer's opinion'. The clause can be seen, therefore, as containing both a power and a duty to supply further information.

The extent to which the engineer can use his power to supply further information during the progress of the works is not without dispute. There is a line of argument which says that the engineer has a duty to supply at commencement all information then known to be necessary for construction of the works and that any later supply is breach of contract

unless relating to variations and/or resolution of ambiguities and discrepancies.

The argument is sometimes run on the basis of interpretation of the contract, in particular clauses 6(1), 7(1) and 51(1) and sometimes on the basis
of an implied term that the contractor is entitled to receive all the information known to be necessary for construction of the works at commencement. The counter to the argument, whichever way it is presented, is that
clause 7(1) provides the remedy for any valid complaint by the contractor
in its concluding part. That part provides that when further information
requires a variation, a variation is deemed to have been issued. That is to
say the employer carries the risk in respect of all information issued by
the engineer after the award of contract, that such information is potentially a variation.

As to whether such further information actually constitutes a variation the
answer in any particular case will depend upon comparison between the
information on the scope and detail of the works given to the contractor prior
to award of the contract (and incorporated into the contract) and the information provided later.

An interesting point in the concluding part of clause 7(1) is that if further
information requires a variation then a variation 'shall be deemed to have
been issued'. This suggests that the issue of revised or additional drawings
or specifications stands as an order in writing for the purposes of clause 51(2)
(ordered variations to be in writing). It is not open to the engineer, or
employer, to resist a valid claim on the grounds that no formal written order
was given.

In so far as the contractor's complaint on further information is the timing
of its delivery and the claim is for late delivery that would seem to be best
accommodated under the provisions of clause 7(4)(a) – delay in issue.

Contractor to provide further documents

Clause 7(2) states that where clause 7(6) applies (permanent works designed
by the contractor) the engineer may require the contractor to supply such
further documents as, in the opinion of the engineer, are necessary for the
proper and adequate construction, completion and maintenance of the
works and that, when accepted, the contractor shall be bound by the same.

Clause 7(2) does not itself deal with the acceptance process and that is
retained in clause 7(6). There is, however, an interesting change in wording
between the two clauses. Under clause 7(6) the contractor is required to
submit for acceptance such documents as are necessary 'to satisfy the
Engineer that the Contractor's design generally complies with the requirements of the Contract', whereas under clause 7(2) the contractor is required
to supply such further documents 'as shall in the Engineer's opinion be
necessary for the purpose of the proper and adequate construction completion and maintenance of the Works'. In short a request for further
information under clause 7(2) is apparently not restricted to rectifying any

perceived deficiencies in the contractor's submission under clause 7(6) and it can validly be a request for information going beyond that necessary to show that the contractor's design generally complies with the requirements of the contract.

One explanation for this may be that under clause 7(7) the engineer is responsible for the integration and co-ordination of the contractor's design with the rest of the works. Sufficient information may not be forthcoming under clause 7(6) but it should be available under clause 7(2).

Failure by the contractor to comply with a request for further information could have quite different practical and legal consequences from failure by the engineer to comply with a request for further information by the contractor. In the event of the latter the progress of the works might be disrupted and the contractor might have recompense under a contractual claim or as damages for breach of contract. In the event of the former it would be for the engineer to decide whether the contractor's default could be dealt with under clause 7(4)(b) as a reduction to a claim entitlement or whether it was sufficiently serious to invoke clause 40 (suspension) or clause 65 (contractor's default).

Notice by contractor

Clause 7(3) states that the contractor shall give adequate notice in writing to the engineer of any further drawing or specification required for the construction and completion of the works or otherwise under the contract.

The key questions arising from this clause are what is meant by 'adequate notice' and what is the consequence of the contractor's failure to give adequate notice or any notice. Does failure relieve the engineer of his obligation to provide the information necessary for construction of the works?

What is 'adequate notice' will be a matter of fact to be decided in the circumstances of each case. However some guidance can be gained from the decisions in *London Borough of Merton* v. *Leach* (1985) where, in a case under a JCT building contract, the contractor had set out in his programme the dates by which all instructions (information) were required. It was held that such an application might be made at commencement for all the instructions the contractor could foresee would be required provided the dates specified for the delivery of the instructions were not unreasonably distant from or too close to the relevant date.

The decision in *Merton* v. *Leach* highlights the point that requirements for further information may arise from different causes. It may be that the engineer has not completed his design at commencement of the works and further drawings and specifications are to be issued during the progress of the works, or it may be that unforeseen conditions or other complications create the need for instructions and further information.

In the first situation, failure by the contractor to give adequate notice might diminish the contractor's prospects of successfully claiming for delay and disruption for late information, but it would not relieve the engineer of his obligation to provide the information in time for the contractor to complete the works within the time for completion. So much can be deduced from decisions of the courts (including *Merton* v. *Leach*), on implied terms. There is reluctance to find an implied term that an employer (through his engineer) must take all reasonable steps to enable a contractor to execute the works in a regular and orderly manner but generally no difficulty in finding an implied term that the employer must not prevent the contractor from fulfilling his obligations under the contract.

5.6 Delay in issue of further information

Clause 7(4)(a) provides a remedy for the contractor when there is failure by the engineer to issue further information at a time reasonable in all the circumstances. Clause 7(4)(b) allows the engineer to take into account in assessing the contractor's entitlement to additional cost and/or time whether his failure to issue information was caused by the contractor's own failure to submit required information.

Delay in issue – engineer's failure

The provisions of clause 7(4)(a) can be summarised as follows:

- if by reason of any failure or inability of the engineer to issue at a time reasonable in all circumstances, drawings, specifications or instructions requested by the contractor
  and
- considered necessary by the engineer in accordance with clause 7(1)
  and
- if the contractor suffers delay or incurs additional cost
  then
- the engineer shall determine any extension of time due under clause 44,
  and
- subject to clause 53, the contractor shall be paid the amount of such cost as is reasonable.

The scope of clause 7(4)(a) is limited on its wording to the late issue of information requested by the contractor. For information not so requested the contractor must look elsewhere for a remedy: possibly clause 13(3) – delay and cost arising from instructions which could not reasonably have been foreseen; or possibly clause 52 – valuation of ordered variations. The alternative is a claim for damages for breach of contract – a claim which under the Sixth edition sometimes led to the contention in respect of additional time that as clause 44 failed to provide for such delay, time was 'at large' and accordingly the provisions for liquidated damages were ineffective. Under the Seventh edition, that contention would have difficulty in overcoming the amended wording of clause 44 which does now include any delay or default by the employer as a ground for extension of time.

A time reasonable in all the circumstances

As repeated elsewhere in this book, questions such as what is a time reasonable in all the circumstances have a stock answer – it is a matter of fact to be determined in the particular circumstances. That is not to say, however, that decisions of the courts do not offer guidance on the application of legal principles.

For example, in *Neodox Ltd* v. *Swinton and Pendlebury Borough Council* (1958) the court was asked to decide whether there was an implied term in the contract that the engineer would give all details and instructions necessary for the execution of the works in sufficient time for the contractor to execute and complete the works in an economic and expeditious manner and/or in sufficient time to prevent the contractor from being delayed. In holding that it was impossible to imply such a term in the manner framed, Mr Justice Diplock said:

> 'It is clear from these clauses which I have read that to give business efficacy to the contract, details and instructions necessary for the execution of the works must be given by the engineer from time to time in the course of the contract and must be given in reasonable time. In giving such instructions, the engineer is acting as agent for his principals, the Corporation, and if he fails to give such instructions within a reasonable time, the Corporation are liable in damages for breach of contract.
>
> What is a reasonable time does not depend solely upon the convenience and financial interests of the claimants. No doubt it is to their interest to have every detail cut and dried on the day the contract is signed, but the contract does not contemplate that. It contemplates further details and instructions being provided, and the engineer is to have a time to provide them which is reasonable having regard to the point of view of him and his staff and the point of view of the Corporation, as well as the point of view of the contractors.
>
> In determining what is a reasonable time as respects any particular details and instructions, factors which must obviously be borne in mind are such matters as the order in which the engineer has determined the works shall be carried out (as he is entitled to do under clause 2 of the specification), whether requests for particular details or instructions have been made by the contractors, whether the instructions relate to a variation of the contract which the engineer is entitled to make from time to time during the execution of the contract, or whether they relate to part of the original works, and also the time, including any extension of time, within which the contractors are contractually bound to complete the works.
>
> In mentioning these matters, I want to make it perfectly clear that they

are not intended to be exhaustive or anything like it. What is a reasonable time is a question of fact having regard to all the circumstances of the case, and the case stated does not disclose sufficient details of the circumstances relating to any particular details or instructions to make it possible for me to indicate what would be all the relevant factors in determining what was a reasonable time within which such details and instructions should have been given. What I have mentioned are merely some examples of factors which may or may not be relevant to any particular details or instructions given which the arbitrator has considered.'

In *A. McAlpine & Son* v. *Transvaal Provincial Administration* (1974), a South African case, a motorway contractor asked the court to define an implied term on the time for supplying information and giving instructions on variations as either:

- a time convenient and profitable to himself
- a time not causing loss and expense
- a time so that the works could be executed efficiently and economically.

The court declined on the grounds that under the contract, variations could be ordered at any time, irrespective of the progress of the works, and that drawings and instructions should be given within a reasonable time after the obligation arose.

More recently in the case of *Glenlion Construction Ltd* v. *The Guinness Trust* (1987) the question arose under a JCT building contract of whether if the contractor had programmed to finish earlier than the due completion date, the employer (through his architect) was obliged to supply information so as to enable the contractor to carry out the works in accordance with the programme and to complete by the programmed date. Under the express terms of the contract the contractor was entitled to recover loss and expense only if the architect failed to supply information in time for completion by the due date but, in the event, the contractor finished earlier than the due date but later than the programmed date. This led the contractor to rely on an implied term and the question put to the judge was as follows:

'...Whether there was an implied term of the contract between the applicant and the respondent that, if and insofar as the programme showed a completion date before the date for completion the employer by himself, his servants or agents should so perform the said agreement as to enable the contractor to carry out the works in accordance with the programme and to complete the works on the said completion date.'

The judge decided:

'The answer to the question must be "No". It is not suggested by Glenlion that they were both entitled and obliged to finish by the earlier completion date. If there is such an implied term it imposed an obligation on the Trust but none on Glenlion.

It is not immediately apparent why it is reasonable or equitable that a unilateral absolute obligation should be placed on an employer.'

It should not be assumed from the decision in *Glenlion* that claims for late information based on shortened programmes will invariably fail. Under some standard forms such as the New Engineering Contract and the Institution of Chemical Engineers' Red and Green Books the employer's obligations are expressly related to the contractor's programme and such claims may well succeed. The distinction is between claims based on implied terms and claims based on express terms.

The position under ICE Conditions (Seventh edition and earlier) is open to argument. For further information requested by the contractor the express obligation on the engineer under clause 7(4)(a) is to supply information at a reasonable time in all the circumstances – and acceptance of a shortened programme by the engineer might be a relevant circumstance. For information not so requested the contractor might be forced into reliance on an implied term and, following *Glenlion,* that would be unlikely to succeed unless some link could be established between the express obligation in respect of requested further information and an implied obligation in respect of other necessary information.

Delay in issue – contractor's failure

Clause 7(4)(b) states that if failure of the engineer to issue drawings, specifications or instructions is caused in whole or part by failure of the contractor after due notice in writing to submit drawings, specifications and other documents he is required to submit, the engineer shall take such failure into account in taking any action under clause 7(4)(a), i.e. allowing extra cost and time.

The important phrase to note in this clause is 'after due notice in writing'. It seems that the contractor will not be penalised for his own late supply unless there has been such notice. Whether a first request by the engineer for the contractor to supply further documents under clause 7(2) is sufficient notice for the purposes of clause 7(4)(b) is debatable.

5.7 Documents to be kept on site

Clause 7(5) requires one copy of drawings and specifications supplied to the contractor and one copy of documents provided by the contractor in respect of permanent works design to be available on site at all reasonable times for inspection and use by the engineer and his staff.

The clause does not expressly say that the obligation to keep such documents on site falls on the contractor. It can probably be implied from the wording although, except on small projects, the engineer is likely to have his own site copies of documents.

5.8 Permanent works designed by the contractor

Clauses 7(6) and 7(7) relate to permanent works designed by the contractor. They detail the obligations on the contractor in respect of design information and the supply of operation and maintenance manuals.

By virtue of clause 8(2) the contractor is not responsible for the design or specification of any part of the permanent works except as expressly provided in the contract. Clauses 7(6) and 7(7) do not come into play unless the contract does so expressly provide. By themselves they say nothing as to whether for any particular contract the contractor has responsibility for design of any part of the permanent works and they certainly do not impose any such responsibility.

Design information

Clause 7(6)(a) requires the contractor to submit to the engineer for acceptance such drawings, specifications, calculations and other information as necessary to satisfy the engineer that the contractor's design generally complies with the requirements of the contract.

There is a change of wording here from clause 7(6)(a) in the Sixth edition which requires the contractor to submit for 'approval' information necessary to satisfy the engineer 'as to the suitability and adequacy of the design'. These words could be taken to imply some fitness for purpose obligation and to be in conflict with the reasonable skill and care obligations expressed in clause 8(2). The amended wording in the Seventh edition avoids the problem.

The Conditions say nothing about when the contractor is to submit his design information for acceptance nor, perhaps more importantly, what is to happen if the contractor's design is not accepted. To remedy this, supplementary clauses are necessary if the Seventh edition is used with any substantial amount of contractor's design.

Operation and maintenance manuals

Clause 7(6)(b) requires the contractor to submit to the engineer for acceptance, operation and maintenance manuals together with as-completed drawings in sufficient detail to enable the employer to operate, maintain, dismantle, re-assemble and adjust the works incorporating the contractor's design.

The clause goes on to state that no certificate under clause 48 covering any part of the permanent works designed by the contractor shall be issued until the manuals and drawings have been submitted to and accepted by the engineer.

For a variety of reasons contractors frequently encounter problems with the timely supply and acceptance of operation and maintenance manuals and problems then arise in obtaining certificates of substantial completion. On the face of it, clause 7(6)(b) debars the engineer from issuing a certificate prior to acceptance of the manuals and drawings but there is a potential conflict with clause 48(3) which requires the engineer to issue a certificate for any substantial part of the works occupied or used by the employer. A certificate so given takes effect from the date of the contractor's request and it is doubtful if the condition precedent of engineer's acceptance under clause 7(6) takes effect in such circumstances.

5.9 *Responsibility for contractor designed works*

Clause 7(7) deals with two different aspects of works designed by the contractor: firstly, the contractor's responsibilities under the contract; secondly, the engineer's responsibility for integration and co-ordination.

Contractor's responsibility

The first part of clause 7(7) states that acceptance by the engineer in accordance with clause 7(6) shall not relieve the contractor of any of his obligations under the contract.

This provision forms part of a set of provisions emphasising that the engineer's acceptances, approvals, consents etc. do not relieve the contractor of any obligations or responsibilities:

- clause 2(1)(d) – no relief from obligations generally or in respect of correctness or accuracy
- clause 7(7)   – no relief in respect of accepted designs of permanent works
- clause 14(9) – no relief in respect of programme and methods of construction

As between the employer and the contractor the intention of these provisions, and similar provisions found in other standard forms, is to prevent the contractor evading responsibility for his own defaults/mistakes by passing responsibility to the engineer and in the process avoiding liability to the employer and/or retaining rights of claim under the contract. As between the employer and the engineer they may be intended to offer some protection to the engineer in the event of acceptance, consent or approval being given to erroneous data and design.

The effectiveness of such provisions cannot be assured. Complications arising from the drafting of other clauses of the contract can lead to unexpected results and in some cases apparent departure from the legal principle that a party should not benefit from its own errors. Thus in *Simplex Concrete Piles Ltd* v. *Borough of St Pancras* (1958) the contractor was held to be entitled to payment for variations necessitated by defective work; and in *Shanks and McEwan (Contractors) Ltd* v. *Strathclyde Regional Council* (1994) the contractor was held to be entitled to payment for repair works necessitated by defective design.

As to protection for the engineer, there is a string of cases from *Holland Hannen & Cubitts* v. *Welsh Health Technical Services Organisation* (1985) to *London Underground* v. *Kenchington Ford* (1998) and including many others, showing the architect/engineer's vulnerability to actions by the employer arising out of defective design and/or defective work by the contractor.

In considering the particulars of clause 7(7) one point to note is that it is only 'Acceptance by the Engineer in accordance with sub-clause (6)' which is covered by the clause. That sub-clause is concerned with general compliance of the contractor's design – the detail relating to 'proper and adequate construction' comes under clause 7(2). Another point is that clause 7(7) is concerned with design information provided after the award of the contract and it does not apply to contractor's designs accepted prior to award and incorporated into the contract. In respect of such designs supplementary clauses or warranties are needed.

Integration and co-ordination

The second part of clause 7(7) states that the engineer shall be responsible for the integration and co-ordination of the contractor's design with the rest of the works.

This is in keeping with the principle established in *Moresk Cleaners Ltd* v. *Thomas Henwood Hicks* (1966) that an architect or engineer employed to design construction works is generally responsible for all the design unless there is an agreement with the employer to the contrary.

Nevertheless, because the burden of integrating and co-ordinating the contractor-designed portion of the works with the engineer-designed portion can be sizeable and the potential for contractor's claims considerable, it was not unusual, under the Sixth edition, to see the clause amended to read 'The Contractor shall be responsible for the integration and co-ordination of the Contractor's design with the rest of the Works'.

Chapter 6
Contractor's obligations/responsibilities for design and construction

6.1 Introduction

ICE Seventh edition, like its predecessors, groups clauses 8 to 35 under the heading 'General Obligations'. The group covers a wide range of matters, mostly relating to the contractor's duties, but not all of which can properly be described as obligations. Clause 12, for example, covers the contractor's entitlements in respect of unforeseen conditions, its place in the group being logical only to the extent that it provides exceptions to obligations in clause 11 (inspection of the site). And with the exception of clause 8 few of the clauses in the group can be said to cover 'general' obligations. The group is in fact no more than a collection of specific obligations not covered elsewhere in the Conditions and, clause 8 excluded, it would be better headed 'miscellaneous obligations'.

Clause 8

Clause 8 is of a general nature although its first part, 8(1), adds little in legal terms to the obligations of the contractor expressed in the offer in the form of tender 'to construct and complete the said Works in conformity with the said Drawings, Conditions of Contract, Specification and Bill of Quantities' and in the undertaking in the form of agreement 'to construct and complete the Works in conformity in all respects with the provisions in the Contract'. The importance of clause 8 lies more in clause 8(2) which deals with design responsibility and clause 8(3) which deals with responsibility for site operations.

Obligations and responsibilities

Clause 8(1) carries the marginal note 'Contractor's general responsibilities'. This is not the best of wording because the clause clearly deals with the contractor's obligations. The distinction between obligations and responsibilities may not always be obvious but in construction contracts the differences can readily be illustrated. Thus, the contractor has an obligation to contract the works but he carries responsibility for damage to the works. The employer has an obligation to give possession of the site but he has responsibility for the defaults of his employees. In the context of construction contracts the words obligations and responsibilities are not always interchangeable.

Although the point is highlighted by the disparity between the marginal note and the text of clause 8(1) it is not of much significance in respect of that clause since the Conditions provide at clause 1(3) that marginal notes are not to be taken into account in interpretation of the contract. Greater significance lies in respect of clause 8(2) which refers in both the marginal note and in the text of the clause to 'design responsibility'. The point which is considered in more detail in section 6.2. below, is whether obligations 'to design' and responsibility 'for design' amount to the same thing or are different.

6.2 Design obligations, responsibilities and liabilities

Design is the accumulation of ideas and details which go into the production of an artefact or the construction of a project.

Three key questions apply to all contracts in which there is some element of design:

- how is the obligation to undertake design allocated between the parties?
- how is the responsibility for effectiveness of the design to be allocated between the parties?
- what standard of liability attaches to the party responsible for design? Is it skill and care or fitness of purpose?

Design obligations

Standard ICE Conditions of Contract have traditionally been drafted on the basis that the employer, through the engineer, undertakes the design of the permanent works and carries the responsibility for that design unless expressly stated otherwise – a proviso intended to have limited and not wholesale effect. This policy, that the engineer designs and the contractor constructs, permeates the entire structure of the Conditions and is fundamental to the roles of the engineer and the parties.

In 1992 the ICE Design and Construct Conditions of Contract were published based on amendments to the ICE Sixth edition. These Conditions have not attracted a great deal of attention or use – quite possibly because they are not wholly convincing as a standard form where the contractor takes the lead in developing the employer's requirements and carrying the responsibilities which flow from that. Too much remains from the Sixth edition.

The Seventh edition follows the traditional policy that the contractor undertakes the design of the temporary works but is entitled to look to the engineer for the design of the permanent works – unless expressly stated otherwise. This is generally well understood and evident from the drafting of the Conditions and supporting documents such as the form of tender and form of agreement. The word 'design' is conspicuously absent from these documents and from the statement of the contract's general obligations in clause 8(1). And clauses 7(6) and 8(2) make it clear that the contractor has no obligation or responsibility for the design of the permanent works except as expressly stated otherwise in the particular contract.

There can be some difficulties applying this policy in practice. Temporary works are sometimes incorporated into the permanent works and the engineer will not always be in the best position to design some of the specialist elements of the permanent works. Unless these difficulties are identified in advance and addressed in the contract documents, there is little doubt that under the standard Conditions the risks associated with the design (of permanent works) are with the employer rather than with the contractor.

Responsibility for design

In the ordinary run of things the party undertaking the obligation to design carries responsibility for that design. That follows from basic legal principles – not least, the duty of care carried by the designer. One of the dangers of ICE Conditions (including the Seventh edition) is that the ordinary rule does not always apply.

For instance, if as is commonplace, the contractor, or his subcontractors or suppliers, undertakes design of part of the permanent works and there is no express statement in the contract that the contractor is to be responsible for that design then, by virtue of clause 8(2), as between the contractor and the employer the responsibility for the design falls on the employer.

Engineers need to be particularly careful when producing contract documents to ensure that design responsibility for specialist works such as piling and electrical and mechanical plant is expressly allocated to the contractor if so intended. It is easy to overlook the effect of clause 8(2) and, by way of example, easy to overlook the point that the length of a pile may be a design matter – albeit one undertaken by an operative on site.

The complications arising from the operation of clause 8(2) were illustrated in the case of *Shanks & McEwan (Contractors) Ltd* v. *Strathclyde Regional Council* (1994) The contract which was under the ICE Fifth edition had a special provision that 'all tunnel and shaft segments ... shall be of approved design and shall be supplied by an approved manufacturer'. There was a further special provision that: 'The contractor shall be responsible for the adequacy of the design insofar as is relevant to his operations'. The engineer approved the tunnel segments as designed by the manufacturer but cracks appeared after installation and the engineer gave instructions for repairs. The contractor claimed payment for the repairs as a variation. In opposing the claim the employer argued that the contract imposed a warranty on the contractor that the tunnel segments would be capable of withstanding the design loads. On the basis that the cause of the cracking was under-design and not mishandling or the like by the contractor; that the tunnel segments when installed formed part of the permanent works; and that the special conditions did not impose responsibility for the design of the permanent works on the contractor; the court held that under clause 8(2) responsibility for the under-design remained with the engineer and his instruction for repairs was a change in quality which qualified as a variation under clause 51.

Standard of liability

Liability for design may arise from:

- the express terms of the contract
- terms implied by common law
- terms implied by statute.

This applies to both design undertaken by the engineer for the employer and design undertaken by the contractor.

Standards of liability may be either on a skill and care basis or a fitness for purpose basis. When the skill and care basis applies, negligence must be proved to establish breach of duty – whether in contract or in tort. Where the fitness for purpose basis applies, all that is required to establish liability is proof of the contention that something is not fit for purpose. In short, fitness for purpose is an absolute test of liability which is far stricter than the skill and care test.

As a general rule contracts for the provision of services (including design) are on a skill and care basis and contracts for the supply of goods are on a fitness for purpose basis. The position of a contractor who designs and constructs was clarified in the case of *IBA v. EMI* (1980) where Lord Scarman said (in relation to a television mast which had collapsed shortly after erection):

'In the absence of a clear, contractual indication to the contrary, I see no reason why one who in the course of his business contracts to design, supply and erect a television aerial mast is not under an obligation to ensure that it is reasonably fit for the purpose for which he knows it is intended to be used. The Court of Appeal held that this was the contractual obligation in this case, and I agree with them. The critical question of fact is whether he for whom the mast was designed relied upon the skill of his supplier (i.e. his or his subcontractor's skill) to design and supply a mast fit for the known purpose for which it was required.'

Many standard forms of construction contract do now have (after *IBA v. EMI)* a clear contractual indication that the contractor's liability for design is not on a fitness for purpose basis. For instance, the standard building form with contractor's design, JCT 81, gives the contractor like liability 'as would an architect or, as the case may be, other appropriate professional designer holding himself out as competent to take on work for such design'.

There is good reason for including such a provision in the contract. Professional indemnity insurance for design work normally covers only skill and care liability and a contractor taking on design with fitness for purpose liability could find himself without cover whether he did the design himself or employed a consultant.

ICE Conditions, Seventh edition included, do not have any statement on contractor's design liability as positive as that found in JCT 81. However, by the statement in clause 8(2) of the Seventh edition that the contractor shall exercise all reasonable skill, care and diligence in designing any part of the permanent works it is reasonably clear that the intention is to exclude any implied term on fitness for purpose.

Generally, therefore, the position under the Seventh edition is that whether it is the engineer or the contractor who undertakes design of the permanent works, the duty is to use reasonable skill and care – a duty assessed by reference to the performance of a competent practitioner. However, there are exceptions as shown by the comments of Lord Justice Neill in the case of *Hawkins* v. *Chrysler (UK) Ltd* and *Burne Associates* (1986) where an employee of Chrysler who had slipped on a wet shower floor was suing for damages. He said:

'I should say a few words however, on the question of the implied warranty. It is clear from the decision of this court in *Greaves & Co (Contractors) Ltd* v. *Baynham Meikle & Partners* (1975) 1 WLR 1095, 4 BLR 56, that in certain circumstances facts may establish the existence of an implied warranty by an independent consulting engineer that a building or structure designed by him will be fit for a stated purpose.

The responsibility undertaken by the consulting engineer in such a case may, it seems, extend to the suitability of materials specified by him for incorporation into his design.

It is also clear that a professional man may be liable on the basis of an implied warranty of fitness for purpose if his contract extends not merely to the design of an article, but also to its supply or manufacture. (See, for example, *Samuels* v. *Davis* (1943) KB 526.) Furthermore it is now established that a contractor, who has agreed to design and erect a building or other structure, may be liable for breach of an implied warranty that the building or structure is fit for a particular purpose, even though the contractor took no part in the design work, which was carried out by a specialist subcontractor (see *IBA* v. *EMI* (1980) 14 BLR 1).

The question which arises in the instant case, however, is whether such a warranty of fitness for purpose is to be implied where there are no special facts; where there is no evidence that the professional man has undertaken any special responsibilities; and where his contract does not involve the manufacture or supply by him of any article or structure.

I recognise that it can be strongly argued that there is no logical basis for drawing a distinction between on the one hand the responsibilities of a contractor who designs and erects a building, and on the other hand the responsibilities of an engineer of the consultant who merely designs it.

Furthermore, there will be cases where the contractor will be liable to the employer for faulty design, and will then wish, in third party proceedings, to recover the sum he had paid to the employer from the independent consultant who in fact designed the building in question.

I have come to the firm conclusion, however, that it is not open to this court, except where there are special facts and special circumstances, to extend the responsibilities of a professional man beyond the duty to exercise all reasonable skill and care in conformity with the usual standards of his profession. There are many authorities which establish that this is the accepted duty of a professional man.'

6.3 Contractor's general obligations

Clause 8(1) of the Seventh edition states the contractor's general obligations – albeit that the marginal note says general responsibilities.

To construct and complete

Clause 8(1)(a) states that the contractor shall, subject to the provisions of the contract, construct and complete the works.

This general obligation to construct and complete can be derived from the form of tender or the form of agreement and it is only by the phrase 'subject to the provisions of the Contract' that the clause makes a useful addition to the terms of the contract.

So far as the obligation to construct is concerned, the phrase probably means nothing more than that the contractor is to construct the works in accordance with the contract – that is, in accordance with the drawings specification, instructions of the engineer and any relevant clauses of the Conditions.

As to the obligation to complete, the phrase is more obviously a proviso to the effect that in certain circumstances the contractor is relieved of his obligation to complete – for example, in the event of the operation of clause 40(2) (suspension lasting more than three months), clause 63 (frustration and war), or clause 64 (default of the employer).

To provide all labour, materials etc.

Clause 8(1)(b) states that the contractor shall, subject to the provisions of the contract, provide all labour, materials, equipment, temporary works etc. required for construction and completion so far as the necessity for providing the same is specified in or reasonably to be inferred from the contract.

Clauses such as this are found in many standard forms but it is doubtful if they add much to the obligation to complete. The New Engineering Contract manages successfully without such a clause. If anything the clause creates uncertainty on whether the contractor must in all instances comply with instructions and/or variations issued by the engineer or whether the contractor is entitled to refuse on the grounds that the instructions and/or variations involve requirements which are neither specified nor reasonably to be inferred from the contract.

The identical clause from the Sixth edition (and similar clause from the Fifth edition) is often quoted in defences to contractor's claims with the argument that the clause requires the contractor to include in his rates and prices for all foreseeable resources. However, it is doubtful if the clause has much relevance to claims. It is concerned with the provision of resources – not with payment for them. The sufficiency of rates and prices is covered in clause 11(3)(b) and by its proviso '(unless otherwise provided in the Contract)' the contractor is able, by various routes, to claim additional payments. For example, even if certain resources are evidently necessary for construction and completion of the works, but the work is not properly billed, the contractor is not disentitled to the benefit of corrections of errors in the bill of quantities under clause 55(2).

6.4 *Design responsibility*

Clause 8(2) states firstly that the contractor shall not be responsible for the design or specification of any part of the permanent works except as expressly provided in the contract, or of any temporary works design supplied by the engineer. It concludes with the statement that the contractor shall exercise all reasonable skill, care and diligence in designing any part of the permanent works for which he is responsible.

The legal difficulties relating to this clause have been considered in sections 6.1 and 6.2 above and in summary the position is:

- unless otherwise stated in the contract the engineer has the obligation to design the whole of the permanent works.
- the contractor only has an obligation to design parts of the permanent works as expressly stated in the contract.
- the contractor only has responsibility for parts of the permanent works he designs as expressly stated in the contract
- any implication that the contractor's liability for his design of any parts of the permanent works is on a fitness for purpose basis is probably excluded by the express reference to skill and care in clause 8(2).

A small point of interpretation but one of some potential importance is whether the proviso in clause 8(2) 'except as expressly provided' applies to the design of temporary works supplied by the engineer. The answer is probably not, given the way in which the relevant sentence in the clause is constructed.

The importance of clause 8(2) in fixing responsibilities on the parties cannot be overstated. Because it is expressed in negative terminology – the contractor shall not be responsible except as expressly stated – it requires a positive statement in the contract to restore the ordinary legal position that the contractor is responsible for designs which he, his subcontractors or his suppliers, undertake. As seen in the *Shanks* v. *McEwan* case, without such a statement the supplier can have the obligation to design but the engineer can have responsibility for the design.

Engineer's responsibility

Where the engineer has responsibility for design of the permanent works, that responsibility will normally be to the employer under terms of engagement. The question of whether the engineer has a duty of care to the contractor in respect of permanent works design is less settled. The possible existence of such a duty in general terms is considered in Chapter 3, Section 3.1 above. As things presently stand in English law the position seems to be that if the engineer is negligent in his design and the contractor suffers economic loss in consequence, the only remedies for the contractor are against the employer. However, the legal trend suggests that if for one reason or another these remedies against the employer cannot be exercised then the engineer might be vulnerable.

Temporary works design

The contractor's obligation to undertake design of all necessary temporary works is not stated in the Conditions with the clarity which might be expected. It can probably be implied from clause 8(3) (responsibility for site operations) and clause 14(6) (methods of construction). It may even follow from the obligations stated in clause 8(1) (contractor's general responsibilities) although the word 'design' is conspicuously absent.

The reference to 'any Temporary Works design supplied by the Engineer' in clause 8(2) is interesting in that it recognises the possibility of some temporary works design being undertaken by the engineer and it puts responsibility for such design on the engineer – unless expressly conferred on the contractor.

Except as expressly provided

The proviso in clause 8(2) that the contractor is not responsible for the design of permanent works '(except as may be expressly provided in the Contract)' raises the question of precisely what is meant by 'expressly provided in the contract'. Does it mean only something written in the documents prior to the formation of the contract or could it include something incorporated into the contract through the variation provisions?

It is thought unlikely that the engineer through a variation can transfer, wholesale or in part, responsibility for design of the permanent works from himself to the contractor. However, it is possible to envisage a situation where, for example, due to unforeseen ground conditions, piled foundations become necessary and the engineer wishes to use the piling subcontractor's design expertise. A variation order in such a case might include a statement to the effect that the contractor is to be expressly responsible for the design. Whether the contractor is obliged to accept such post award responsibility is questionable.

Some guidance on the point may be available from consideration of clause 58(3) which deals with design elements of services provided under provisional sums or prime cost items. On one interpretation of that clause design obligations cannot be imposed after the contract has been made.

Skill and care

Detailed comment on the skill and care aspects of clause 8(1) is given in Section 6.2 above. The only point to add here is the impact of the change from the Sixth to the Seventh edition found in clause 7(6)(a). The Sixth edition required the contractor to satisfy the engineer as to 'the suitability and adequacy' of the design of any permanent works; the Seventh edition requires that the design 'generally complies with the requirements of the Contract'.

Some commentators argued in respect of the Sixth edition that clause 7(6)(a) came close to imposing fitness for purpose obligations. The revised clause in the Seventh edition cannot be given the same effect.

6.5 *Responsibility for site operations*

Clause 8(3) provides that the contractor shall take full responsibility for the adequacy, stability and safety of all site operations and methods of construction.

Taken by itself the clause appears to place full responsibility on the contractor in the clearest of terms. However the clause, which is identical to clause 8(3) of the Sixth edition, cannot be taken by itself. It needs to be considered in conjunction with other contractual and statutory provisions and in the light of legal decisions on similar clauses.

Under the Conditions the employer has express safety responsibilities under clause 19(2) (employer's responsibilities) and has associated responsibilities under clauses 20 and 22 in respect of 'excepted risks' and 'exceptions'. Clause 8(2) leaves responsibility for the design of temporary works supplied by the engineer with the engineer. Clause 12(6) entitles the contractor to additional cost and time arising from unforeseen physical conditions. Clause 13(3) entitles the contractor to similar recovery in respect of certain instructions.

Under statute, responsibility falls on whoever the Act deems it should fall and private contractual arrangements purporting to state otherwise are of no effect. Thus under the Health and Safety at Work Act 1974 the duties under the Act apply to 'any person', and notwithstanding its wording, clause 8(3) of the Conditions does not in any way diminish the responsibility of the employer under the Act.

The impact of clauses 13 and 51 on clause 8(3)

In the case of *Yorkshire Water Authority* v. *Sir Alfred McAlpine & Son (Northern) Ltd* (1985) the court considered the impact of clauses 13 and 51 on clause 8(2) of a contract under ICE Fifth edition Conditions – a clause similar to clause 8(3) of the Seventh edition.

In that case the contractor was required to submit with his tender for works at Grimwith Reservoir a method statement showing that he would work upstream in constructing an outlet tunnel. This method statement became listed as one of the contract documents. In the event the contractor claimed it was impossible to work upstream and worked downstream. He then claimed he was entitled to a variation under clause 51 and to payment accordingly.

The employer, relying strongly on clause 8(2), argued that the adoption of a new method statement and a new method of working was entirely the contractor's responsibility. It was said that, if performance according to the method statement was impossible, it must be the responsibility of the contractor to provide an alternative method statement, and even if he was entitled to a variation order, he must, by virtue of the words 'full responsibility' bear any extra cost which this variation might involve.

This argument was rejected by Mr Justice Skinner who said that where there was a pre-specified method of construction, clause 8 was relevant only to post-contractual methods submitted under clause 14. The judge went on to say:

'In my judgement, the standard conditions recognise a clear distinction between obligations specified in the contract in detail, which both parties can take into account in agreeing a price, and those which are general and which do not have to be specified pre-contractually. In this case the applicants could have left the programme and methods as the sole responsibility of the respondents under clause 14(1) and clause 14(3). The risks inherent in such a programme or method would then have been the respondents' throughout. Instead, they decided they wanted more control over the methods and programme than clause 14 provided. Hence clause 107 of the specification; hence the method statement; hence the incorporation of the method statement into the contract imposing the obligation on the respondents to follow it save in so far as it was legally or physically impossible. It therefore became a specified method of construction by agreement between the parties, who must be taken in my judgement, to have had the provisions in clause 8 in mind as relevant to any programme subsequently submitted under clause 14(1). No such programme was submitted or demanded, presumably because the applicants were content with the control over the programme afforded by the specified method statement. Clause 8(2) is only relevant in the context of the present agreement (and I emphasise those words "in the context of the present agreement") to such part of the method or programme as is submitted, or may be submitted, post-contractually under clause 14. Here there was a specified sequence or method of construction. If the variation which took place was necessary for the completion of the works because of impossibility within clause 13(1), then, in my judgement, the respondents were entitled to a variation order with the consequent entitlement to payment of the value of such variation as is provided in clause 51(2) and clause 52.'

The impact of clause 12 on clause 8(3)

In the case of *Humber Oil Terminals Trustee Ltd* v. *Harbour and General Works (Stevin) Ltd* (1991), again under ICE Fifth edition Conditions, the Court of Appeal had to consider whether the contractor's obligation under clause 8(2) affected his rights of claim under clause 12 for unforeseen conditions. The contract was for the construction of mooring dolphins and the contractor selected and used a jack-up barge equipped with a fixed crane. As the crane was lifting and skewing a concrete soffit member it became unstable and collapsed. The contractor submitted a claim under clause 12 that he had encountered physical conditions which could not have been foreseen. The employer disputed the claim and argued that even if the contractor did encounter such conditions, he had no claim under clause 12 by virtue of the provisions of clause 8(2).

The arbitrator, the judge at first instance and the Court of Appeal all held that the contractor succeeded on both point 1 – that he had met unforeseen physical conditions – and point 2 – that the claim was not excluded by clause 8(2). This is how the Court of Appeal reached its decision on point 2:

'As to point 2, both the arbitrator and the judge disposed of the point very briefly. The [employer's] argument is a simple one. Clause 8(2), they say, imposes an unqualified full responsibility for the adequacy, stability and safety of all site operations and methods of construction. Here they say the operation or method of construction, to wit by Stevin 73, was inadequate unsafe and unstable. Therefore the consequences are wholly a matter for the contractor and had nothing to do with their employers.

There is in clause 8(2) no exception such as is to be found in clause 8(1) and in other clauses in the contract where expressions such as 'subject to the provisions of the contract' appear. [Counsel for the employer] referred us to a number of such provisions. There is thus, says he, no room for the operation of clause 12. Clause 8(2) is unqualified and the fact that it is so unqualified is reinforced by the presence of qualifications in such other conditions.

Against this it is contended that clause 12 is also unqualified, save as to weather conditions and the like, and it makes no commercial sense, so it is said, to construe the contract as giving no effect to clause 12 in a situation such as the present.

The matter is in my view one largely of first impression. For my part, however, I am unable to accept a construction in which such inadequacy, unsafety or instability as occurred was due to unforeseen physical conditions intended to be excluded by clause 8(2) from the operation of clause 12. The construction which is put on it by the [employer] involves that in situations such as this where, as a result of the finding on point 1, the collapse was due to unforeseen physical conditions, clause 12 will have been emasculated to the point of disappearance.'

Lord Justice Nourse, agreeing, said this:

'I agree with my Lord that the second question is really one of impression. I do not feel able to say, as [counsel for the contractor] has suggested, that this is a case where there was no inadequacy or instability of site operations or methods of construction within clause 8(2) of the ICE Conditions. On the whole I think that there was. On the other hand, I cannot construe clause 8(2) as applying to a case where the inadequacy or instability is brought about by the contractor's having encountered physical conditions within clause 12(1). That I think was the instinctive view of the arbitrator, who dealt with this question briefly. Like the judge, I have not found it an easy question, but like him and my Lord I would on balance decide it in favour of the contractors.'

Improving clause 8(3)

As is apparent from the above cases on Clause 8(2) of the Fifth edition, the near identical clause 8(3) of the Seventh edition may not always live up to the expectations given by first reading. What appears to be a clear cut allocation of responsibility is subject to many grounds of attack.

It is known to be a matter of concern to many engineers and employers that clause 12 can be interpreted such that it cuts across the principle that the contractor should be free to select his own methods of working but in doing so must take responsibility for them. It can mean that the contract price as finally determined is dependent on how susceptible the contractor's chosen methods of working are to unforeseen conditions. And that in turn can put pressure on the engineer in deciding whether or not to give consent to particular aspects of the contractor's method under clause 14.

An extreme method of rectifying the situation is for the employer to delete clause 12 from the Conditions. Less extreme, and more common, is to amend clause 8(3) by the addition of such words as 'notwithstanding other provisions in the contract'.

Chapter 7
Form of agreement and performance security

7.1 Introduction

Clause 9 (form of agreement) and clause 10 (performance security) of the Seventh edition are both identical to their counterparts in the Sixth edition. However the form of bond which now backs up clause 10 is an improved and modernised version of the earlier ICE bond.

The reason for including clauses 9 and 10 in the section of Conditions headed 'General Obligations' rather than in the section 'Contract Documents' is possibly that an ICE contract can operate perfectly well without completion of either a form of agreement or a form of bond. Whether or not either is required is normally a matter for the employer to determine. To that extent, clauses 9 and 10 can be seen as optional, but if the employer seeks their use the clauses impose obligations on the contractor to co-operate in their operation.

7.2 Form of Agreement

Clause 9 provides that the contractor, if called upon to do so, shall enter into and execute an agreement, to be prepared at the cost of the employer, in the form annexed to the Conditions.

The form of agreement so annexed to the Seventh edition is substantially the same as the form used with the earlier versions of ICE Conditions and it does no more than formally record that the parties have entered into an agreement that the contractor will construct the works in return for payment by the employer.

Timing of the agreement

It is apparent from the very inclusion of clause 9 in the Conditions that execution of the form of agreement is something which follows, rather than precedes or is concurrent with, the formation of the contract. Nevertheless, where a form of agreement is completed it becomes, by clause 1(1)(e) of the Conditions, a contract document.

The prospect that there may be no execution of a form of agreement is evident from the penultimate paragraph of the form of tender: 'Unless and until a formal Agreement is prepared and executed this Tender together with your written acceptance thereof, shall constitute a binding Contract between us'. And it is similarly evident from the concluding words of clause 1(1)(e) 'Contract means the Conditions of Contract Specification . . . and the Form of Agreement (if completed)'.

It is generally accepted from the wording of the form of tender and clause 1(1)(e), if not from general legal principles, that a binding contract for the construction of the works exists between the parties from the time of the employer's acceptance of the contractor's tender. However, in the *Yorkshire Water* case referred to in Chapter 6, the judge after hearing arguments on the interpretation of similar documents under an ICE Fifth edition contract took a different view. He said:

> '[Counsel for the contractor] argues that the contract was formed when the letter of acceptance was posted. But [counsel for the employer] asks, "A contract to do what?" He argues that it is merely a contract to contract. In my judgement, it is plainly a provisional document and it contemplates no assumption of obligations by either party to perform any of the works until a formal contract has been concluded.'

In the *Yorkshire Wat*er case the point did not amount to much because the form of agreement was executed within four weeks of acceptance of the tender, but for many civil engineering contracts where the form of agreement follows months and sometimes years after the start of the works, the point, if taken and upheld, could be greatly significant. It is doubtful if it would be upheld in the courts but it is not for adjudicators or arbitrators to conclude that cases have been wrongly decided.

Effect of the agreement

A contract formed by ordinary offer and acceptance is known as a 'simple' contract. A contract executed under a deed (e.g. by the ICE form of agreement) is known as a 'speciality' contract.

The differences are principally that a promise is binding under a deed; that consideration is not strictly necessary; that facts stated in a deed cannot be denied; and, most importantly, that limitation periods are extended. The effect of the latter under English law (and the Limitation Act 1980) is that the time for bringing an action for breach of contract becomes 12 years instead of the usual 6 years.

In construction contracts extended limitation periods are usually seen as, and are, of benefit to the employer in respect of actions for defective work. Without a deed such actions have to be commenced within 6 years of the breach occurring, whereas with a deed 12 years is allowed. There is some authority from legal decisions (e.g. *William Tomkinson & Sons Ltd* v. *Church of St Michael in the Hamlet* (1990)) that for construction work the date of breach can be taken as the date of completion – on the principle that prior to that date the contractor has the opportunity and the obligation to correct any defects to bring the works to completion. However, the position is not absolutely certain.

Although extended limitation periods are rarely to the contractor's advantage there is no reason in principle why, if there is a deed, the contractor's claims for damages for the employer's breaches of contract cannot be brought within the extended period. There are, of course, endeavours in most standard forms to close down the final account within a year or so of completion but as the unusual cases of *Whittal Builders Co Ltd* v. *Chester-le-Street District Council* (1984) and (1987) show, such endeavours are not always successful.

The cases are unusual in that in both, the formalities of putting the contracts under seal (the practice at the time for making a speciality contract) were not completed; in both cases agreements had been reached on final accounts; and in both cases it was accepted that the 'late' claims for damages brought by the contractor, Whittal, were statute barred unless the 12 year limitation period applied.

In the first case Whittal succeeded because the judge decided that the contract was a speciality contract under the Limitation Act 1980 (notwithstanding that it had not been actually impressed with the company's seal) and that agreement of the final account did not preclude Whittal from bringing claims for damages not covered by the agreement. In the second case Whittal failed because a different judge decided (in respect of a different contract) that the parties had effectively settled all claims in the final account agreement or, effectively, Whittal was stopped from bringing its damages claims by its 5 year silence before raising its claims. On the speciality contract point, the second judge decided that as the contract was not sealed it could not be considered to be sealed by way of legal fiction but since at the time the parties entered into the contract both assumed it would be under seal, each was estopped by convention from questioning the truth of the assumption.

Obligation to execute the agreement

Clause 9 puts an obligation on the contractor to enter into an agreement if called upon to do so.

Good practice requires that the contractor should be notified of what is intended at the earliest opportunity and ideally at tender stage so that the contractor can make corresponding arrangements with proposed subcontractors.

Failure by the contractor to execute an agreement if called upon to do so would be a breach of contract – with interesting potential for argument as to what damages might flow from such a breach. It is unlikely however, that refusal would be serious enough to be taken as a default entitling the employer to terminate under clause 65.

7.3 *Performance security*

Clause 10 states that if the contract requires the contractor to provide security for the proper performance of the contract he shall obtain and provide for the employer, within 25 days of the award of the contract, security in a sum not exceeding 10% of the tender total. The clause further requires that security shall be provided by a body approved by the employer; that security shall be in the form of a bond annexed to the Conditions; and that the contractor shall pay the cost of the security unless the contract provides otherwise.

Bonds generally

Performance bonds can broadly be classified as on-demand bonds or conditional bonds. On-demand bonds entitle the holding parties to require payment up to the amount of the bond as a matter of right – without proof of any default on the part of the party providing the bond or of loss by the holder. Such bonds are not far removed from uncashed cheques in the hands of the holder and are usually obtained by the providing party from their bankers on payment of a fee and adjustment of account facilities (e.g. reduction of overdraft limits). Conditional bonds entitle the holder to payment only on occurrence of the qualifying conditions stated in the bond (usually default of one kind or another) and/or proof of loss up to the amount claimed (within the limits of the bond). Such bonds are obtainable from specialist insurance companies.

On-demand bonds are generally disliked by contractors and subcontractors in the UK construction industry. They are considered to carry too much risk and to be too financially onerous. For employers however they have the advantages of certainty and simplicity and they are sometimes specified as the only acceptable form of bond.

Given the significant differences between the two types of bond it might be expected that it would be apparent from the wording of each whether it was on-demand or conditional. Unfortunately, until recently that has not always been the case. The language of bonds has been old fashioned and obscure. In the Hong Kong case of *Tins Industrial Co Ltd* v. *Kono Insurance Ltd* (1987) the judge, commenting on a bond identical to the old ICE bond, said:

> 'Now the trouble about this bond is that it is written in thirteenth century (or earlier) language. It is archaic language and therefore difficult to read and to understand.'

Dissatisfaction on the wording of bonds came to a head in the mid-1990s when decisions in a series of high profile cases confounded the construction industry. In *Perar BV* v. *General Surety and Guarantee Ltd* (1994) the Court of Appeal held, in relation to a bond under a building contract, that insolvency was not a default and the holder was not entitled to the benefit of the bond –

notwithstanding that it was precisely for this eventuality that the bond was required. In *Trafalgar House Construction (Regions) Ltd* v. *General Surety and Guarantee Ltd* (1994), in relation to a bond of the old ICE type given under a civil engineering subcontract, the Court of Appeal held that it was payable on demand – notwithstanding that its plain meaning, if not in plain wording, suggested it was conditional. That decision was overturned by the House of Lords in 1995 who held that the bond amounted to a guarantee and that proof of damage and not mere assertion was required before liability under the bond arose.

By then it was evident that new style bonds were required. Sir Michael Latham in his 1994 report *Constructing the Team* considered how performance bonds should be utilised in the UK construction industry and recommended (in Chapter 5) that:

- they should be drafted in comprehensible and modern language
- they should not be on-demand and unconditional, but should have clearly defined circumstances set out in them for being called
- if the circumstances/conditions provided for in the bond are fulfilled, the beneficiary should be able to obtain prompt payment without recourse to litigation
- they should have a clear end date.

In 1999 the Institution of Civil Engineers published its new form of bond – the form now included in the Seventh edition. It carries the title 'ICE Form of Default Bond'.

7.4 ICE Form of Default Bond

The new ICE bond, drafted in collaboration with lawyers S. J. Berwin & Co, is written in plain language and is clearly a conditional bond. Its key functions are:

- it is a condition precedent to payment by the surety, that the employer serves on the surety a copy of the certificate issued by the engineer under clause 65(5) of the Conditions confirming the difference between the costs of completing the works and any amounts due to the contractor
- the surety has the right to refer any dispute on the amount payable under the bond to adjudication
- the bond expires on the earlier of the date stated in a certificate of substantial completion and 'the final expiry date' – a date to be inserted in the schedule to the bond
- the bond can only be assigned by the employer with the prior consent of the surety and the contractor.

The latter point avoids dispute on whether the benefit of the bond is transferable – a point considered in relation to a different form of bond in the case of *De Montfort Insurance Co PLC* v. *Lafferty* (1997) where it was held that a guarantor was not released from its obligations as a result of novation of a building contract. However, not all employers will be content to leave the restriction in place and some will no doubt make amendments to ensure that some allowance for novation is retained.

The point on date of termination of the bond seems designed to bring the life of the bond to an end at substantial completion at the latest. This will be welcomed by the contractors who frequently found that the old style bonds ran until the issue of the defects correction certificate. However it is questionable whether the wording in the bond – 'this Bond shall expire on the earlier of the date stated in a certificate of substantial completion' – fully recognises that under ICE Conditions there can be certificates of substantial completion for parts of the works as well as for the whole.

The importance of compliance by the engineer and the employer with the formalities necessary for invoking the bond cannot be overstated. In *Paddington Churches Housing Association* v. *Technical and General Guarantee Company Ltd* (1999) the employer under a building contract sought to draw on a bond after the contractor had gone into liquidation. The judge found that the bondsman had no liability until the amount had been calculated as prescribed in the contract.

7.5 *Cost of the bond/non-provision of bond*

Clause 10 of the Conditions provides that if the contract requires the contractor to provide security he shall provide such security within 28 days of the award of the contract. It goes on to state that the contractor shall pay the cost of the security unless the contract provides otherwise.

It seems to be implicit from this that the contractor is entitled to know before he submits his tender whether or not a bond is required. Frequently provision for a bond will be made in the bill of quantities. The contractor should be careful, however, in providing a bond as a post contract requirement without first clarifying who is to pay the cost. In *Perini Corporation* v. *Commonwealth of Australia* (1969), after formal acceptance of the tender, the employer required the contractor to provide a bond. The contractor was unable to recover the cost. The judge said:

'The reality of the situation in my opinion is that the plaintiff and the defendant agreed to the provision by the plaintiff of an additional guarantee but that they did not make any agreement at all with respect to the liability of one side or the other side for the cost of doing so. It is in my opinion simply a matter upon which the parties have not expressed any agreement and for that reason the claim of the plaintiff on this point must fail.'

Non-provision of bond

Failure by the contractor to provide a bond if one is required under clause 10 will clearly be a breach of contract. At the least that would give the employer entitlement to damages so as to cover the cost of the employer arranging alternative cover.

In some standard forms there is express provision for the employer to determine the contractor's employment if a bond is not supplied as required. That is not the position in ICE Conditions but arguably the breach may be sufficiently serious to justify common law determination. Support for this is to be found in a South African case, *Swartz & Son (Pty)* v. *Wolmaransstadt Town Council* (1960).

7.6 Dispute resolution upon security

Clause 10(2) expands upon the dispute resolution procedures in the standard ICE form of bond which read as follows:

'(4) If the Surety objects to the contents of or entitlement to issue a certificate under clause 65(5) of the Contract in respect of which the Employer seeks payment from him the Surety shall have the right to refer the matter to adjudication in accordance with the adjudication provisions contained in sub-clauses 66(6) to 66(8) of the Contract as if the Surety were a party to the Contract in place of the Contractor.

(5) Any adjudication under Clause 1(4) shall be commenced by the Surety within 14 days of receipt by the Surety of the documents referred to in Clause 1(2) and the Surety shall have no right to refer the matter to adjudication after that time.

(6) If the content of or entitlement to issue the Certificate under clause 65(5) of the Contract in respect of which payment is sought by the Employer is or has been the subject of an adjudication between the Employer and the Contractor under the Contract (in respect of which both parties have made submissions to the adjudicator) the Surety agrees to be bound by the result of such adjudication and shall have no right to refer the matter to adjudication under Clause 1(4).

(7) In the case of an adjudication under Clause 1(4) payment by the Surety shall be made within 7 days of the decision in such adjudication.'

Employer deemed to be a party

Clause 10(2)(a) states that the employer shall be deemed to be a party to the security for the purpose of doing everything necessary to give effect to its dispute resolution provisions.

The joint effect of clause 1(4) of the form of bond and clause 10(2)(a) of the Conditions is that the surety can commence adjudication proceedings in the name of the contractor against the employer and the employer is bound to participate as though in contract with the surety.

Decisions on discharge of security without prejudice

Clause 10(2)(b) states that any decisions, award or other determination concerning the date of discharge of the security shall be without prejudice to the resolution of any dispute between the employer and contractor under clause 66.

Under clause 5 of the form of bond, the bond expires on the date stated in a certificate of substantial completion of the works – presumably meant to be the certificate of substantial completion for the whole of the works. The intention of clause 10(2)(b) of the Conditions seems to be that neither the employer nor the contractor should be bound in disputes between themselves as to the date of completion by any date which is determined in relation to expiry of the bond.

Chapter 8
Site conditions and sufficiency of tender

8.1 Introduction

Clauses 11 and 12 of the Conditions are closely related to the extent that in broad terms clause 11 requires the contractor to satisfy himself as to the adequacy of his tender and clause 12 provides relief from the obligations imposed by clause 11 in the event that the contractor encounters conditions which could not reasonably have been foreseen.

The drafting of clause 12 has remained fairly constant through the various editions of the ICE Conditions and there is nothing new in the Seventh edition which represents any major policy change. Clause 11 however lacks the same stability. In the Fifth edition (and earlier) the clause was concerned only with the contractor's obligations to inspect the site and to satisfy himself on the adequacy of his tender. In the Sixth edition, in wording which left much to be desired, it effectively imposed an obligation on the employer to make available to the contractor all information on the site. In the Seventh edition the clause more or less reverts to the position in the Fifth edition, although there are details of the new wording in relation to information 'obtainable' by the contractor which may reduce the scope for claims under clause 12.

Responsibility for site conditions generally

In the absence of express terms to the contrary the general rule is that the contractor takes the risk of site conditions. For the contractor to reverse that rule and establish a right to be paid for dealing with adverse or unforeseen site conditions the contractor has to prove entitlement on the basis of an implied term or entitlement on the basis of duty of care or mis-representation.

Under ICE Conditions the presence of express terms, in particular clause 11 (information), clause 12 (unforeseen conditions) and clause 51 (variations), leaves little room for terms to be implied. But even under simpler forms of construction contract where less is stated on the obligations and responsibilities of the parties, the courts have traditionally been reluctant to imply terms cutting across the principle that the contractor is under an obligation to satisfy himself of the circumstances in which the work is to be undertaken. See, for example, *Thorn* v. *Corporation of London* (1876) and *Bottoms* v. *York* (1892).

Duty of care

The question of whether the employer has a duty of care to the contractor to supply accurate and/or sufficient information as a matter of law (as opposed to under a term of contract) is not straightforward. There are potential differences between failures alleged in respect of positive acts (i.e. the supply of inaccurate information) and failures alleged in respect of omissions (i.e. information withheld or not disclosed). There also appear to be, from decisions in various English, Commonwealth and USA cases, differences of approach to the application of the legal principle most commonly relied on, the Hedley Byrne principle – a principle derived from the case of *Hedley Byrne* v. *Heller* (1964) where it was held that an innocent misstatement causing financial loss can give rise to liability in tort.

Thus in *Dillingham Construction* v. *Downs* (1972) an Australian court, whilst accepting that the *Hedley Byrne* principle could exist between the contracting partners, held that in the absence of clear misrepresentation the employer was not liable for failing to disclose records of underground mine workings held in its archives. And in *Morrison-Knusden International* v. *Commonwealth of Australia* (1972) the court, whilst accepting the difficulties of contractors obtaining site information for themselves during the tender period and in holding that disclaimers on information provided by the employer were not effective, declined to say if a duty of care existed. In *Morrison Knusden* v. *State of Alaska* (1974) the court declined to hold that the employer had a duty of disclosure in respect of information about the site made known to the engineer at tender stage by one of the tenderers. However, in *Brown & Huston* v. *York* (1985) an employer was found liable for the engineer's negligence in omitting water levels from the soils information given to tenderers.

Misrepresentation

Contractual representation may be either fraudulent or innocent.
Of fraudulent misrepresentation it was said in *Derry* v. *Peek* (1889):

'fraud is proved when it is shown that a false representation has been made (1) knowingly, (2) without belief in its truth, or (3) recklessly careless whether it is true or false.'

The remedies for fraudulent misrepresentation are termination of the contract and/or damages on an indemnity basis, i.e. damages restoring the position as though the contract had not been made.

Disclaimers as to the accuracy of information given to tenderers provide no defence to an action for fraudulent misrepresentation. Thus in *Pearson* v. *Dublin Corporation* (1907) it was said:

'no one can escape liability for his own fraudulent statements by inserting in a contract a clause that the other party shall not rely upon them'.

In that case the size of an undersea wall which the contractor was to use was incorrectly marked on drawings by the engineer who had taken no steps to verify the accuracy of the dimensions. It was held that a clause in the contract requiring the contractor to satisfy himself on all dimensions could give protection only against honest mistakes but the engineer's conduct amounted to fraudulent misrepresentation.

For innocent misrepresentation, the Misrepresentation Act 1967 introduced a statutory cause of action for misrepresentations made during contractual negotiations. To avoid liability the party which has made the representation has to prove it had reasonable grounds for believing the facts put forward were true. The remedy for innocent misrepresentation is damages.

Contractual clauses excluding or limiting liability for such misrepresentation are of no effect unless deemed reasonable under the criteria of Section 11(1) of the Unfair Contract Terms Act 1977.

The leading construction case on innocent misrepresentation is *Howard Marine and Dredging Co Ltd* v. *Ogden & Sons (Excavations) Ltd* (1977). Ogden hired from Howard two barges to dispose of spoil at sea. Each barge was quoted as being capable of carrying 850 cubic metres. When asked at a meeting what that amounted to in tonnage terms Howard's representative incorrectly said 1600 tonnes per barge. The correct figure was nearer to 1200 tonnes but Ogden did not discover this until well into the contract. Disputes arose, with Ogden refusing to pay Howard's accounts in full and Howard eventually withholding its barges. Howard sued on its unpaid bills and Ogden counter-claimed for damages on the basis of duty of care (under *Hedley Byrne*) or alternatively misrepresentation. The Court of Appeal held that there was no warranty with contractual effect as to the payload of the barges and accordingly the matter fell to be determined under the Misrepresentation Act 1967. On that it was held there was misrepresentation as a matter of fact; that Howard had failed to establish reasonable grounds for putting forward incorrect figures; that Howard could not avoid liability by reliance on the exclusion clause in the charter-party; and that Howard was liable in damages.

Lord Justice Bridge gave the following warning about the effect of the Misrepresentation Act 1967:

'If the representee proves a misrepresentation which, if fraudulent, would have sounded in damages, the onus passes immediately to the representor to prove that he had reasonable ground to believe the facts represented. In other words the liability of the representor does not depend upon his being under a duty of care the extent of which may vary according to the circumstances in which the representation is made. In the course of negotiations leading to a contract the statute imposes an absolute obligation not to state facts which the representor cannot prove he had reasonable ground to believe.'

That warning and the general impact of the Misrepresentation Act 1967 should be noted by employers who seek to transfer the risk of unforeseen conditions to the contractor under ICE Conditions by deleting clause 12 from the contract – a practice of growing popularity. What tends to happen is that the contractor, deprived of his contractual entitlement, looks for another remedy – and frequently finds it in misrepresentation.

Sufficiency of tender generally

A general principle of English law is that the contractor is under an obligation to satisfy himself of the circumstances of the work to be undertaken. In Halsbury's *Laws of England* it says:

> 'It is the duty of a contractor, in the exercise of common prudence, before making his tender, to inform himself of all particulars concerning the work, and particularly as to the practicability of executing every part of the work contained in the plans, drawings and specification according to the specific terms and conditions. Ignorance on his part when making his tender will not excuse him from performing his contract. Even if it be the usage of contractors to rely on the specification without examining it for themselves, or employing a skilled person to do so on their behalf, such a usage cannot excuse him. The fact that the contractor has agreed to erect a building or construct works as described in the plans and specification, implies that he understands such plans and specification, and he cannot escape liability by setting up that he exercised ordinary care and skill and yet failed to understand them.'

Early examples of the law in practice are:

- *Williams* v. *Fitzmaurice* (1858) where floorboards were omitted from the specification of a house and it was held that the contractor was responsible for supplying and fixing floorboards within his price
- *Sharpe* v. *San Paulo Railway* (1873) where it was held that an extra 2 million cubic yards of excavation gave no entitlement to extra payment.

Application of the law to lump sum contracts is comparatively straightforward compared with its application to contracts under ICE Conditions where rates and prices are the basis of the contract and re-measurement is undertaken to determine the final contract value. The question for such contracts is frequently what should the contractor have allowed for in his rates and prices. The answer is sometimes to be found in the details of the method of measurement. If there is no such method the answer may have to be derived from a wider view of the contractual provisions.

Unforeseen conditions

There is comparatively little legal authority on what circumstances constitute unforeseen conditions. In part this is because disputes on such matters have traditionally been put to arbitration and determined by findings of fact on which there is no appeal.

In the case of *Ceredigion County Council* v. *Thyssen Construction Ltd* (1999) where application was made for leave to appeal against an arbitrator's award, the judge, in refusing leave, said this:

> 'The application of clause 12 is heavily dependent on the facts of each particular case and it would be difficult to lay down any general principles for an arbitrator. Those who drafted Clause 12 and those parties who chose to include it in their contracts conferred on an engineering arbitrator a very wide discretion with which the Court had in general no wish to interfere.'

One of the few cases under ICE Conditions where comment is made on unforeseen conditions is *C.J. Pearce Ltd* v. *Hereford Corporation* (1968) where the precise location of an old sewer which the contractor had to cross was unknown. The sewer collapsed into the contractor's excavation. It was held that even had the contractor claimed under clause 12, which he did not, his claim would have failed since the condition could have been foreseen and a right to claim would in effect remove any incentive for the contractor to take care.

It is difficult to disagree with the logic of this decision in so far as something which is known about cannot by ordinary standards be said to be unforeseen. But it does illustrate the difficulties of establishing firm rules for all situations. Had it been stated that the old sewer was on a particular line then it could have been unforeseen had it been encountered on another line. Thus, the vagueness of information increases the risk on the contractor and decreases the value of the unforeseen conditions clause. But this may not be what is intended by clause 12 in commercial terms. The purpose of transferring risk from the contractor to the employer is not simply for the contractor's benefit. The employer should get lower tender prices the more of the risk he carries.

This difficulty of maintaining a balance between analytical analysis of wording and commercial intention is readily apparent in the test for payment for unforeseen conditions stated in clause 12 of the ICE Conditions – 'could not reasonably have been foreseen by an experienced contractor'. This is no doubt intended to be an objective test but it is not clear how it encompasses the actual knowledge of a particular contractor. The point is often made that the more a particular contractor knows of the difficulties of a site the less likely it is that his tender will be successful.

A practical method by which a tenderer can overcome the problems of special knowledge is to write to the engineer during the tender process with a query designed to reveal his knowledge and to put all tenderers on an equal footing by the engineer's reply.

When unforeseen conditions claims do arise it is not unknown for engineers/employers to reject claims on the basis that unsuccessful qualified bids from other tenderers show that the particular conditions were not only foreseeable but were actually foreseen by other experienced contractors. The danger for the employer in this is that it can open the door to claims from the contractor on a breach basis (where the contract provides for the employer to make known to tenderers information on the site) or alternatively on a duty of care basis.

A further question which sometimes arises in respect of unforeseen conditions is to what extent the term applies to conditions which pre-exist the formation of the contract and/or to conditions which arise from events after formation. The question was considered in *Holland Dredging (UK) Ltd* v. *Dredging and Construction Co Ltd* (1987) where the Court of Appeal overturned a surprise decision by the judge at first instance that claims under clause 12 applied only to supervening events (i.e. those occurring after formation of the contract) – a decision which, had it been allowed to stand, would have left little point in the clause.

Meaning of physical conditions

Unforeseen conditions are generally taken to be, in the context of construction contracts, unexpected ground conditions on the site. It is something of an oddity, if not a matter of serious concern, that clause 12 of the ICE Conditions in the Seventh and earlier editions refers in its wording neither to the ground nor the site. The clause refers instead to physical conditions (and artificial obstructions) encountered during the carrying out of the works.

The consequences of this wording were revealed in the case of *Humber Oil Trustees Ltd* v. *Harbour & General Works (Stevin) Ltd* (1991). In that case the question arose whether the contractor had encountered physical conditions within the scope of clause 12. As a 300 tonne crane on a jack-up barge was placing precast soffit units on piles the barge became unstable and collapsed causing extensive damage to the works, plant and equipment. The barge was a total loss and had to be replaced. There was much delay and extra cost.

The contract was under the ICE Fifth edition Conditions and the contractor claimed under clause 12 that collapse of the barge and its consequences was due to encountering physical conditions which could not have been foreseen by an experienced contractor. The dispute went to arbitration.

The arbitrator gave an award in favour of the contractor finding that although the soil conditions were foreseeable, clause 12 contains no limitation on the meaning of 'physical condition'; that a combination of strength and stress, although transient, can fall within the terms; and that in this case an unforeseeable condition had occurred.

The employer appealed maintaining that the question should be not whether the collapse could have been foreseen, which it clearly could not, but whether physical conditions could reasonably have been foreseen. The judge upheld the arbitrator's award but gave leave to appeal. The arguments advanced for the employer before the Court of Appeal would certainly have found favour with many engineers – namely that a physical condition is something material, such as rock or running sand, and that applied stress is not a physical condition nor is it something which can be encountered. The Court of Appeal, however, rejected the arguments and dismissed the appeal.

Lord Justice Parker dealt with the arguments as follows:

'Mr Dyson [Counsel for the employer] submits that the physical condition of the soil, which was found by the arbitrator to be foreseeable, really concludes the matter and that applied stress is not and cannot be any part of the physical condition.

Attractive as his argument appears to be at first sight, I cannot accept it. The arbitrator was in my view saying that the general soil conditions were foreseeable and well able to stand the applied loads and stresses. There was, however, here a peculiar characteristic which could not have been reasonably foreseen, namely a liability to shear at a much lower loading than had already been withstood.

The matter may perhaps be put in this way. General soil conditions were known and were foreseeable and foreseen. Such soil conditions would not have resulted in a shear failure. There was thus an unforeseeable condition.

Suppose that the Contractor, just before the event, had been informed by the engineer that some further information had just arrived showing that Stevin 73 would collapse because of a special feature of the soil under the leg, which nobody had hitherto known about. In my view the Contractor would then have encountered a physical condition which was not reasonably foreseeable. He would then have made proposals under clause 12, which the engineer might have approved, or indeed even directed the Contractor to take.

As to his submission that applied stress cannot be a part of a physical condition, this in my view is not so. The soil conditions which prevail at any moment when one is considering operations such as foundations or any other operation which puts weight on the soil is in effect the load-bearing capacity of the soil conditions. A particular condition of soil may, for example, be well known safely to sustain without shear 1,000 tonnes. If in fact there is settlement at a load of 300 tonnes what does it show? In my view, surely, that there was an unknown, foreseeable fault which was plainly a physical condition.'

Lord Justice Nourse agreeing said:

'The arbitrator found that there must have been a very unusual combination of soil strength and applied stresses around the base of leg 2 of the barge just before the failure occurred. He found as a fact that that state of affairs could not reasonably have been foreseen by an experienced contractor. That finding cannot be re-opened in this court. Accordingly the first question which we have to decide is whether this very unusual combination of soil strength and applied stresses was, as both the arbitrator and Judge Fox-Andrews have held, a physical condition encountered by the Contractors within clause 12(1) of the ICE Conditions.

The principal submissions of Mr Dyson, for the Employers, are to this effect. He says that a physical condition is something with a material, intransient existence, such as rock or running sand. An applied stress is not a physical condition nor, moreover, is it something which can be encountered. Accordingly, the only physical condition which here fell within clause 12(1) was the soil itself, whose nature could, as the arbitrator found, reasonably have been foreseen by an experienced contractor.

I reject these submissions for the following reasons. First, I agree with Mr Blackburn, for the contractors, that there is nothing to restrict the application of clause 12(1) to intransient, as distinct from transient, physical conditions. Indeed the express reference to weather conditions, albeit by way of exclusion, suggests the contrary. Secondly, while I would agree that an applied stress is not of itself a physical condition, we are not concerned with such a thing in isolation, but with a combination of soil and an applied stress. Thirdly and most significantly, as Butler-Sloss LJ pointed out during the course of the argument, it is impossible to speak of a contractor encountering any form of ground, be it rock, running sand, soil or whatever, without recognising that stress of one degree or another will have to be applied, at any rate notionally, to the ground, which will in turn behave, at any rate notionally, in one way or another; no doubt passively in the case of rock, actively in the case of running sand and perhaps unpredictably in the case of soil. In other words, for the purpose of clause 12(1), you cannot dissociate the nature of the ground from an actual or notional application of some degree of stress. Without such an application you cannot predict how the ground will behave. In the present case I would say that the condition encountered by the contractors was soil which behaved in an unforeseeable manner under the stress which was applied to it, and that that was a physical condition within clause 12(1).'

The decision in *Humber Oil* has attracted a good deal of attention – from lawyers concerned as to its logic; from employers wishing to limit its consequences; and from contractors hoping to exploit its opportunities. But the cause of the problem lies not in the legal decision but in the wording of clause 12. Some employers now amend clause 12 substituting 'ground conditions on site' for 'physical conditions'.

8.2 *Provision and interpretation of information*

Clause 11(1) of the Conditions provides firstly, that as between the employer and the contractor, and without prejudice to clause 11(2) (contractor's obligations to inspect the site), information on:

(a) the nature of the ground and subsoil and hydrological conditions, and
(b) pipes and cables in, on or over the ground

obtained by, or on behalf of the employer from investigations undertaken relevant to the works shall only be taken into account to the extent that such information was made available to the contractor before submission of his tender.

The second part of clause 11(1) states that the contractor shall be responsible for the interpretation of all such information for the purpose of constructing the works and for any design which is the contractor's responsibility under the contract.

The first part of clause 11(1) is significantly changed from the corresponding clause in the Sixth edition where the employer was deemed to have made available all information obtained from investigations relevant to the works. That clause caused some concern because it appeared to impose obligations on the employer which could inadvertently be breached by non-disclosure.

'As between the Employer and Contractor'

The purpose of the opening words of clause 11(1) is not immediately obvious. The words hint at some relationship other than the employer/contractor relationship where information on the site obtained by the employer may be of interest. There may be such a relationship, for example, with the end user of the site, but its existence is of no relevance to the terms of the main contract.

'without prejudice to sub-clause (2)'

The intention of these words is presumably to detach the contractor's obligations to make his own enquiries on the nature of the site from the supply of information by the employer.

'nature of the ground' etc.

The references in clauses 11(1)(a) and 11(1)(b) to nature of the ground, subsoil, hydrological conditions, pipes and cables match those in the August 1993 corrigenda to the Sixth edition.

Obtained from investigations 'relevant to the Works'

The employer may be in possession of information on the site from a variety of sources other than investigations undertaken relevant to the works. It is not clear how, if at all, such information fits into clause 11(1). Essentially the clause is about what 'is to be taken into account' and to the extent that the clause restricts that to information obtained from investigations relevant to the works, it leaves open what importance can be attached to information otherwise obtained.

'shall only be taken into account'

These are enigmatic words. Taken into what account? Taken in by whom? One explanation is that they mean taken into account by the contractor during the tender process. But, by itself, it serves little purpose to state in the Conditions that the contractor shall only take into account in his tender information supplied before submission of his tender. That goes without saying.

An attractive and more plausible explanation is that the words refer to assessment of the contractor's entitlements in respect of variations, claims and the like. Thus the engineer, in considering a clause 12 claim for unforeseen conditions, shall only take into account information made available to the contractor before submission of his tender.

Again this may seem to be obvious but it could have application to clause 12 claims when there is a question of whether 'unforeseen' means unforeseen at the time of tender or unforeseen at commencement of the actual work on site. That is to say, a clause 12 claim should not be disallowed on the basis that at the time of commencement of the relevant work the conditions were foreseeable in the light of information provided by the employer after submission of the tender.

'made available to the Contractor'

'Made available' suggests more than mere inclusion in the tender documents. It could cover information available for inspection at the employer's or engineer's offices.

'before submission of his tender'

The cut-off date for the relevance of information supplied by the employer is stated in clause 11(1) not by reference to the date of formation of the contract but by reference to submission of the tender. This suggests that the information, and the clause, are to be seen as applying to financial/valuation matters rather than to the wider obligations and responsibilities. But it is questionable if this fits in with the scope of the second part of clause 11(1).

Contractor responsible for interpretation of information

The second part of clause 11(1) is unchanged from the Sixth edition. The provisions of this part seemed ill-fitting in the Sixth edition and they look even more so in the Seventh edition. They require the contractor to be responsible for interpretation of information made available before submission of the tender:

- for the purposes of constructing the works, and
- for any design which is the contractor's responsibility.

'such information'

One difficulty is in understanding what is meant by 'such information' in the phrase in clause 11(1) – 'The Contractor shall be responsible for the interpretation of all such information'.

The information referred to in the first part of the clause is information:

- on the nature of the ground etc.
- obtained by or on behalf of the employer
- from investigations relevant to the works.

In the first part of the clause, information is only to be taken into account to the extent that it was made available to the contractor before submission of his tender. This limitation, which is explainable when the clause is put to financial purposes cannot sensibly apply to the second part of the clause which concerns matters of design and construction. It can be presumed therefore that the words 'such information' in the second part do not draw in the 'taken into account' limitations of the first part.

By the same logic it is difficult to see why the contractor's responsibility for interpreting information for construction or design of the works should be restricted or purposefully linked to information obtained by the employer from investigations undertaken relevant to the works. It may be that no restriction is intended – in which case 'all such information' would be better replaced by 'all information'.

8.3 Inspection of the site

Clause 11(2) of the Seventh edition is the same as clause 11(2) of the Sixth edition with the minor change that the words 'and hydrological conditions' are added at the end of clause 11(2)(a) of the Seventh edition.

The clause provides that the contractor shall be deemed to have:

- inspected and examined the site, its surroundings and available information
- satisfied himself so far as is practicable and reasonable before submitting his tender as to:
  (a) the form and nature of the ground, sub-soil and hydrological conditions

(b)   the extent and nature of the work and necessary materials
(c)   the means of communication with and access to the site and any required accommodation
- obtained for himself all necessary information as to risks, contingencies and all other circumstances influencing or affecting his tender.

The purpose of clause 11(2) is to detail the actions that the contractor is deemed to have taken in gathering knowledge of the site in order to satisfy himself as to the sufficiency of his tender.

'the Contractor shall be deemed'

It will not normally be open to the contractor to seek relief from his obligations under clause 11(2) on the basis that he did not, or was not able to, inspect and examine the site. However, when entry to the site is strictly controlled the employer or the engineer need to arrange access for tenderers if the deemed provision in clause 11(2) is to remain effective.

'inspected and examined'

The words 'inspected and examined' give little indication in themselves of the extent of the contractor's obligation but it is apparent from later parts of the clause which refer to subsoil and hydrological conditions that the contractor may be deemed to have done more than make a visual inspection.

'surroundings'

The contractor's obligations in respect of 'surroundings' are dealt with in the Conditions (i.e. as between the contractor and the employer) principally in clause 22 (damage to persons and property); clause 29 (interference with traffic and adjoining properties); and clause 30 (avoidance of damage to highways).

The inclusion of 'surroundings' in clause 11(2) links up with those clauses and additionally covers such matters as the availability of off-site working space, the potential for off-site activities and the possibility of flooding.

'information available'

One of the more contentious aspects of the contractor's obligation to investigate the site is how far the contractor is required to go in researching information. It sometimes happens that when disputes on site conditions occur, information is obtained from archives, libraries and other sources and is put forward as being information which was available to the contractor and which should have been considered.

The test for what is required by way of research is, it is suggested, to be found in the purpose of the clause – which is the collection of information for tendering.

'so far as is practicable and reasonable'

This proviso recognises the difficulties and limitations faced by tenderers in investigating the site. The circumstances of each case will determine what is practicable and reasonable.

Although the proviso applies to ground conditions, nature of the work, availability of materials, access to the site and accommodation, the only express relief for the contractor in the event of under-estimation of costs and difficulties is under clause 12 – which deals only with unforeseen physical conditions and artificial obstructions. In some circumstances the contractor might gain relief for other matters through clause 13 (instructions) or clause 51 (variations) but otherwise the contractor is bound by his rates and prices notwithstanding that he may have done everything practicable and reasonable.

8.4 Basis and sufficiency of tender

Clause 11(3) states:

- in 11(3)(a) – that the contractor shall be deemed to have based his tender on his own inspection and examination and on all information whether obtainable by him or made available by the employer
- In 11(3)(b) – that the contractor shall be deemed to have satisfied himself before submitting his tender as to the correctness and sufficiency of the rates and prices inserted in the bill of quantities and which shall (unless otherwise provided in the contract) cover all his obligations under the contract.

Clause 11(3)(a)

Clause 11(3)(a) differs from the corresponding clause in the Sixth edition where the contractor was deemed to have based his tender on the information made available by the employer and on his own inspection and examination. The change in wording to 'and on all information whether obtainable by him or made available by the Employer' may make it much harder for the contractor to claim for the unforeseen. 'All information obtainable' by the contractor has potentially wider scope than the information which the contractor is obliged to collect under clause 11(2).

Clause 11(3)(b)

The obligation for the contractor to satisfy himself on the correctness and sufficiency of his rates and prices is more of an exhortation than a burden. In the event that the contractor fails in his duty he carries the consequences. It is difficult to see how the possibility of damages for the employer in the event of breach of contract could arise.

A more trying aspect of the clause is whether the requirement for the rates and prices to cover all obligations under the contract is intended to be taken compositely or singularly. There is a case for saying that all rates taken individually shall be sufficient for the work involved, since ICE Conditions are not far removed from a schedule of rates contract and the rates have relevance to the valuation of variations. However, it is only compositely that the rates and prices can be said to cover all the contractor's obligations under the contract.

'unless otherwise provided in the Contract'

The proviso in clause 11(3)(b) is an essential piece of wording in giving the Conditions sensible commercial effect. Without the proviso the requirement for the rates and prices to cover all the contractor's obligations under the contract would negate many other clauses of the contract – in particular, clause 55(2) which requires the engineer to correct errors and omissions in the bill of quantities.

8.5 *Adverse physical conditions and artificial obstructions*

The principal purpose of clause 12 of the Conditions is to give the contractor relief from the obligations imposed by clause 11 in respect of sufficiency of tender in the event that certain unforeseeable conditions are encountered.

Clause 12(1)

Clause 12(1) provides that if during the carrying out of the works the contractor encounters physical conditions (other than weather conditions or conditions due to weather conditions) or artificial obstructions which in his opinion could not reasonably have been foreseen by an experienced contractor he shall give written notice to the engineer as early as practicable.

This part of clause 12 is concerned not so much with the contractor's entitlements but with his obligation to give notice to the engineer when unforeseen conditions are encountered. This obligation exists irrespective of whether the contractor intends to make a claim for delay or extra cost and presumably it is to put the engineer on notice of potential difficulties. However, to the extent that that is the purpose, as distinct from setting the parameters for the claim provisions which follow in later parts of clause 12, it is not helpful to the engineer that there are exclusions in respect of weather related conditions and that notice is only required in respect of the unforeseen.

Failure by the contractor to give notice has consequences in respect of claims but it is also a breach of contract which could theoretically give rise to liability for damages. In practice, however, the contractor's loss would be the employer's gain and it is difficult to envisage circumstances where an employer's loss could be proved.

Comment has been made in section 8.1 above on the scope of the term 'physical conditions' and the absence from clause 12(1) of any reference to the ground or to the site. Artificial obstructions are usually taken to be buried, man-made objects, and clause 12 can, in appropriate circumstances, serve as an effective claim clause for statutory undertaker's apparatus which is found to be out of position. However, much depends on the accuracy of location indicated by any information supplied or obtainable.

'If during the carrying out of the Works the Contractor encounters...'

The opening phrase of clause 12(1), particularly the part 'during the carrying out of the Works', seems innocuous if not superfluous unless it is to be taken as indicating that clause 12 does not operate after completion. However questions can arise as to meaning and application of the phrase when the contractor discovers unforeseen conditions during investigations made after the award of the contract in connection with the design of temporary works and/or the design of any permanent works for which the contractor is responsible. The key question is – are these 'conditions encountered during the carrying out of the works'? The secondary questions are – does 'carrying out' include design? Do 'the Works' include design? Does 'encounter' include discovery in advance of construction? The decision in the case of *Norwest Holst* v. *Renfrewshire* (1996) mentioned in Chapter 2 and discussed in some detail in Chapter 9 illustrates the difficulty of analysing such questions. It is suggested, although not with much confidence, that the term 'carrying out' is wide enough to include design, and similarly that the term 'encounters' can include for pre-discovery.

Weather related conditions

The exclusion for weather conditions has the effect of putting the risks from adverse weather on the contractor. Most standard forms take the same approach but the New Engineering Contract has a secondary option clause which, if included in the contract, transfers the risks of extreme 10 year weather conditions to the employer.

The exclusion for conditions due to weather conditions is probably intended to apply to the immediate effects of weather such as flooding after a rainstorm. It is doubted if it should be applied to longstanding conditions which may have their origin in weather conditions.

'could not reasonably have been foreseen'

There are differences of opinion on whether the foreseeability test is wholly objective or whether it allows for the special knowledge of a particular contractor.

In order to put all tenderers on the same footing it is suggested that the test should be wholly objective and that the phrase 'in his opinion' entitles the tenderer/contractor to take the test as expressed literally. Consider for example the position of a contractor who foresees from his special knowledge that the engineer's design will not work in the particular ground conditions – for example, that piles will be needed for a bridge abutment. Clearly he should not include in his tender for piles because he is entitled to be paid for them as a variation. All he can do is price the tender on the same basis as other tenderers.

Not infrequently the action taken by the engineer under clause 12(4) (action by the engineer) will be a fair indication of what could, or could not, have been reasonably foreseen. If conditions were not foreseen by the engineer it may seem reasonable to assume that the contractor should not be adjudged to have better foresight. And that argument is often advanced by contractors although to the extent that it carries any weight it is usually in connection with the effect of the unforeseen on the design of the permanent works or the specification of particular methods of construction.

'by an experienced contractor'

The foreseeability test is that applied to an experienced contractor. Not infrequently that seems to be overlooked when parties call expert evidence in formal dispute proceedings as to what could or could not have been foreseen. Eminent geologists and academics may well be the best people to explain and interpret ground investigations but their knowledge and practice is likely to be on a different plane to that expected of a contractor, experienced or otherwise.

The point was made by Mr Justice Webster in the case of *Wimpey Construction Ltd* v. *Poole* (1984) who had this to say about a consultant of world renown who gave expert evidence:

'He is without doubt an outstandingly brilliant exponent of the complexities of soil mechanics and his work in that field has received international acclaim and recognition. He applies, however, both to himself and to others, the highest possible standards; and it can fairly be said that he is generous with his criticism. Few people connected with the case escaped it. For these reasons, and because his experience has given him little contact with the ordinary day to day problems of designing structures in soil, I am able to place little if any reliance upon his evidence as to the standards to be expected of an ordinarily competent designer.'

Clause 12(2) – intention to claim

Clause 12(2) provides that if the contractor intends to make any claim for additional payment or extension of time arising from any condition or obstruction as stipulated in clause 12(1), he shall at the same time as giving notice, or as soon thereafter as reasonable, inform the engineer in writing specifying the condition or obstruction to which the claim relates.

The requirement for notice to be given 'as soon as may be reasonable' differs from the requirement in clause 53(1) (additional payments) for notice to be given within 28 days. Perhaps it should be assumed that reasonable notice under clause 12 will always be less than 28 days.

Clause 12(3) – measures being taken

Clause 12(3) requires that when giving notification or information in accordance with sub-clauses (1) or (2), or as soon thereafter as practicable, the contractor shall give details of any anticipated effects of the condition or obstruction and of the measures he has taken, is taking or is proposing to take together with their estimated costs and the extent of the anticipated delay or disruption.

The clause does not impose a firm obligation on the contractor to decide what action should be taken. Although such an obligation probably rests on the contractor for methods of construction it is not for the contractor to decide what should be done where design changes are involved. Consequently the contractor should be wary of taking premature responsibility for clause 12 difficulties.

Clause 12(4) – action by the engineer

Clause 12(4) provides that following receipt of any notification under sub-clauses (1) or (2) or receipt of details in accordance with sub-clause (3) the engineer may if he thinks fit:

(a)   require the contractor to investigate and report upon the practicality, cost and timing of alternative measures which may be available
(b)   give written consent to measures notified under sub-clause (3) with or without modification
(c)   give written instructions as to how the physical conditions or artificial obstructions are to be dealt with
(d)   order a suspension under clause 40 or a variation under clause 51.

The engineer's action at this stage is not in itself an admission of employer's liability except in respect of variations, but there is practical difficulty for the engineer in holding this position. Whichever option is chosen it seems to imply acceptance of liability. The engineer therefore may be reluctant to act until he has formed a view on liability.

The option of giving instructions is additional to that of ordering a variation and it may be that one is intended to apply to temporary works and the other to permanent works. But the engineer should be careful in giving instructions for temporary works, whether liability under the clause is accepted or not.

The ordering of a suspension under clause 40 does not necessarily entitle the contractor to recover his costs – the suspension may be necessary for the proper execution or for the safety of the works. This is therefore one option which may provide the engineer with a measure of breathing space whilst he appraises the situation.

Clause 12(5) – conditions reasonably foreseeable

Clause 12(5) states that if the engineer decides that the physical conditions or artificial obstructions could in whole or in part have been reasonably foreseen by an experienced contractor, he shall inform the contractor in writing as soon as he has reached that decision but the value of any variation previously ordered under sub-clause (4)(d) shall be ascertained in accordance with clause 52 and included in the contract price.

'if the Engineer shall decide'

A curious feature of the delegation powers conferred by clause 2(4)(c) is that the engineer cannot delegate under clause 12(6) (delay and extra cost) but can delegate under clause 12(5). Thus the engineer's representative can be given the power to decide if conditions are foreseeable but only the engineer can decide if they are unforeseeable. Putting aside the scope for confusion in this it is not obvious why rejection of a clause 12 claim should be of lower standing than acceptance.

No time limit is set in clause 12(5) on informing the contractor that notified conditions are considered to have been foreseeable and that in effect the contractor's claim is rejected. The requirement is that the contractor shall be informed as soon as the decision is made.

Delay in notifying the contractor could lead to a variety of secondary claims and given the engineer's project management style duties to the employer it is not appropriate that decisions on clause 12 claims should be left for lengthy consideration.

The express provision that any variation ordered 'previously' shall be included in the contract price seems to imply that if the engineer has taken any other action under clause 12(4) before deciding that the contractor has no right of claim, the contractor cannot recover any 'previous' costs related to such action. However, it is not thought to mean this since there will clearly be cases where the contractor has contractual rights to payment. It probably means only that there can be no going back on the valuation of a variation whatever the engineer's findings.

Clause 12(6) – delay and extra cost

Clause 12(6) provides that where an extension of time or additional payment is claimed, the engineer shall, if in his opinion such conditions or obstructions could not reasonably have been foreseen:

(a) determine any delay which the contractor has suffered, and
(b) determine the amount of any costs which may reasonably have been incurred by the contractor (together with a reasonable percentage additional therefore in respect of profit).

The clause further provides that the engineer shall notify both the contractor and the employer of his determinations and that any delay so determined shall be considered for an appropriate extension of time.

Clause 12(6) is the only part of clause 12 which differs in the Seventh edition from the Sixth edition. The first change is the inclusion in the Seventh edition at clause 12(6)(a) of the requirement for the engineer to determine delay. In the Sixth edition the requirement was only to determine any entitlement to extension of time.

This is a worthwhile change which brings clause 12 into line with clause 44 where the first step for the engineer is to consider delay – and, only after that, to consider whether an extension of time is due. Moreover it recognises that delay can be costly to the contractor even if there is no entitlement to extension of time because of non-criticality or concurrency.

Arguably the clause might be further improved if it said 'delay and disruption' rather than simply 'delay'.

The contractor's entitlement to a reasonable percentage for profit on costs reasonably incurred was first introduced in the Sixth edition; previously the entitlement had been to profit only on additional work and plant. The change is retained in the Seventh edition notwithstanding that elsewhere in the Conditions, for example in clauses 13(3) and 31(2), profit is only allowable on additional permanent or temporary work.

The second change in clause 12(6) from the Sixth to the Seventh edition is the inclusion at the end of clause 12(6) of the Seventh edition of the paragraph:

'Any delay so determined shall forthwith be considered under Clause 44(3) for an appropriate extension of time and the Contractor shall subject to Clause 53 be paid in accordance with Clause 60 the amount so determined.'

The part of this relating to payment was first introduced into the Sixth edition in the August 1993 corrigenda thereby closing a surprising gap in clause 12 in that previously, although the clause implied the contractor was to be paid certain costs, it did not actually say so. The new clause covers both time and money.

Chapter 9
Instructions and impossibility

9.1 Introduction

This chapter is concerned principally with clause 13 of the Conditions and the manner in which the clause deals with the engineer's powers to give instructions and the effects of legal or physical impossibility on the contractor's obligations.

The Seventh edition version of the clause is the same as that in the Sixth edition with the exception that whereas the Sixth edition refers in clause 13(1) to instructions from the engineer or the engineer's representative, the Seventh edition refers to instructions from the engineer or 'his duly appointed delegate'.

Content of clause 13

Clause 13(1) has provisions relating to:

- legal or physical impossibility
- the contractor's obligation to construct and complete the works
  - in strict accordance with the contract
  - to the satisfaction of the engineer
- the contractor's obligation to comply with instructions.

Clause 13(2) requires the contractor's materials, equipment and labour, and the mode, manner and speed of the works to be acceptable to the engineer.

Clause 13(3) details the contractor's entitlements to recover delay and extra cost for certain instructions and provides that if any instruction requires a variation it shall be deemed to have been given pursuant to clause 51.

General comment on clause 13

Clause 13 has expanded through various editions of ICE Conditions to its present state. It apparently started as a simple provision on legal or physical impossibility and clause 13 of the Fourth edition was, in its entirety, much the same as is now clause 13(1) of the Seventh edition.

The Fifth edition took certain provisions from clause 46 of the Fourth edition and repositioned them as clause 13(2) – where they now remain, largely unchanged in the Seventh edition.

Clause 13(3) was new to the Fifth edition and its inclusion and the uncertainty of its wording was much criticised by leading construction lawyers of the day. Nevertheless it survived and is repeated, albeit with some modification, in the Sixth and Seventh editions.

Clause 13 as it now stands is of wider application than to matters relating to legal or physical impossibility. It provides rules for instructions generally and as such is one of the most important clauses in the Conditions.

9.2 *Instructions generally*

Two important aspects of any construction contract are the contract administrator's (engineer's) powers to give instructions and the contractor's obligations to comply with instructions. Following these are questions of the status of instructions and in particular in what circumstances they can be taken as variations; questions of the contractor's entitlement to be paid for compliance with instructions; and questions of instructions given orally (when the contract requires they should be in writing) and instructions not given in circumstances where they are required.

Legal examples

By way of example the following are a few of the better known cases, from the many on instructions, which illustrate that disputes on instructions are widespread within the construction industry and have been so for a long time:

- *Brodie* v. *Corporation of Cardiff* (1918) where the House of Lords ruled that an arbitrator was entitled to award payment to a contractor for extras notwithstanding that the contract provided for payment only on the engineer's written instruction and the engineer had refused to give an instruction.
- *Simplex Concrete Piles* v. *Borough of St Pancras* (1958) where it was held that an architect's letter, accepting the contractor's proposals for changes to foundations to overcome faulty work by a piling subcontractor, amounted to an instruction involving variations for which the contractor was entitled to be paid.
- *Neodox Ltd* v. *Borough of Swinton and Pendlebury* (1958) where it was held:
  - where the contract gave the engineer power to determine the method by which the works were to be executed and where there were alter-native methods possible under the terms of the contract, the contractor had no claim if the engineer chose one of the envisaged methods of working even though that choice was unreasonable.
  - where a contract set out no specific method of carrying out particular

operations necessary to complete the works and merely provided that the works should be carried out under the engineer's directions and to his satisfaction, a direction by the engineer intimating the manner in which the operations should be carried out in order to satisfy him could not be a 'variation of or addition to the works'.

- *Crosby & Sons Ltd* v. *Portland Urban District Council* (1967) where it was held that an engineer's requirement that the contractor should use 'Staveley' pipes when the specification provided 'Stanton's or Staveley's pipes' was a variation.
- *C J Pearce Ltd* v. *Hereford Corporation* (1968) where it was held that advice given by an engineer to a contractor who was in difficulties in crossing an old sewer did not amount to instructions

See also the *Yorkshire Water* and the *Holland Dredging* cases discussed in Chapter 20 where it was held in both cases that where instructions were required, the contractor was entitled to be paid as though a variation had been ordered.

It needs, of course, to be recognised that the decisions in the above cases and in others are frequently heavily dependent on the facts, but nevertheless it is possible to detect general rules to the effect that engineers should be careful not to give advice which could be construed as an instruction and that contractors are not necessarily deprived of payment for work which amounts to a valid variation even in the absence of an instruction or a formal variation order.

Instructions under ICE Conditions

Under ICE Conditions, Seventh edition included, the engineer is empowered under various clauses to give instructions or to state requirements on particular matters. The list is extensive. But apart from clause 51 (variations) the engineer has no expressly stated general power to give instructions – although such a power can perhaps be implied from the wording of clause 2(1), duties and authority of the engineer; clauses 13(1) and 13(2), works to be to the satisfaction of the engineer and to be acceptable to the engineer; and clause 36, quality of materials and workmanship.

The general obligation of the contractor to comply with the engineer's instructions and requirements is most clearly stated in clause 13(1) although it may be implied from the statement of the contractor's general obligations in clause 8(1).

A common cause of friction in the operation of ICE Conditions is the extent to which the contractor is entitled to instructions when he considers the specification is unachievable, or the engineer's design deficient or unworkable, or specified methods impossible. That brings into play the relationship between the various clauses of the Conditions and in particular that between clause 13 (instructions) and clause 51 (variations). Thus if the works are impossible to perform as specified the contractor is entitled to a variation whether or not the engineer orders a variation. In *Yorkshire Water Authority* v. *Sir Alfred McAlpine & Son (Northern) Ltd* (1985), a case under ICE Fifth edition, it was held that the incorporation of a method statement into the contract imposed an obligation on the contractor to follow it, save insofar as it was legally or physically impossible. The judge said:

> 'Here there was a specified sequence or method of construction. If the variation which took place was necessary for the completion of the works because of impossibility within clause 13(1), then, in my judgement, the respondents were entitled to a variation order with the consequent entitlement to payment of the value of such variation as is provided in clause 51(2) and clause 52.'

9.3 Impossibility generally

At common law the contractor who undertakes to construct and complete the works warrants expressly, or impliedly, that he can do so. If he fails he is liable in damages. Thus it was said in *Taylor* v. *Caldwell* (1863):

> 'Where there is a positive contract to do a thing, not in itself unlawful, the contractor must perform it or pay damages for not doing it, although in consequence of unforeseen accidents, the performance of his contract has become unexpectedly burdensome, or even impossible.'

Similarly at common law, and in the absence of express contractual provisions to the contrary, the employer does not warrant that the works can be built. The classic case is *Thorn* v. *Corporation of London* (1876) where the contractor was to take down an old bridge and build a new one.

The design prepared by the employer's engineer involved the use of caissons which turned out to be useless. The contractor completed the works with a different method and sued for his losses on the grounds that the employer warranted that the bridge could be built to the engineer's design. The House of Lords held that no such warranty could be implied.

The *Thorn* case was followed by *Tharsis Sulphur & Copper Company* v. *McElroy & Sons* (1878) where a contractor who had agreed to supply and erect girders to a specified thickness subsequently found that the girders could not be manufactured to that thickness because of a tendency to twist. The contractor manufactured the girders to a greater thickness and claimed the extra cost. The House of Lords ruled that the claim failed.

Contractual provisions

Most construction contracts contain provisions which give some relief to the contractor when it is legally or physically impossible to construct and/or complete the works. In ICE Conditions the relief is found in clause

13(1) in the qualifying phrase 'save insofar as it is legally or physically impossible'.

Provisions such as those found in clause 13 need to be distinguished from provisions covering frustration (a subject discussed later in this book in Chapter 29). Frustration is when supervening events (events occurring after formation of the contract) which are unforeseen and beyond the control of the parties, make performance impossible.

In the case of *Holland Dredging (UK) Ltd* v. *Dredging & Construction Co Ltd* (1987) Lord Justice Purchas made the point in connection with a dispute under a contract similar to ICE Conditions that:

> '...When taken in conjunction with clause 64 (frustration), clause 13(1) must relate to something short of a frustrating event, where supervening events are concerned.'

Meaning of physical impossibility

There are surprisingly few cases which provide guidance on the meaning of physical impossibility. In the *Yorkshire Water* case, which is authority for the proposition that if the contractor's method statement is incorporated as a contract document and subsequently found impossible to perform, the contractor is entitled to a variation, the matter of whether or not there was impossibility was left for the arbitrator to decide on the facts.

The most commonly cited authority for what is meant by 'physical impossibility' is the judgement in *Turriff Ltd* v. *Welsh National Water Authority* (1980). In that case problems arose with tolerances for precast concrete sewer segments and eventually the contractor abandoned the works on the grounds of impossibility. The employer argued that impossibility meant absolute impossibility without any qualifications and since there was no absolute impossibility the contractor was in breach. To sustain this argument the employer contended that the contractor would be obliged to depart from the specification if that was necessary to overcome the impossibility.

The judge declined to accept that impossibility meant absolute impossibility and he held that the works were impossible in an ordinary commercial sense. On the proposition that the contractor was obliged to depart from the specification the judge made the point that if absolute impossibility was the test for impossibility then an equally strict construction of the contract would apply. So what was physically impossible would relate to the precise limits of the specification and that could not be relieved by a deviation on the part of the contractor.

The following extracts from the judgment of Judge William Stabb QC show how the arguments of the case were examined. The judge said:

> 'But the real issue is as to the construction of the words "physically impossible" in the light of those authorities which establish the common law position. [Counsel for the employer] urges me to construe it as having

an absolute meaning, so as to give it certainty. "Physically" is coupled with "legally" and he contends that as "legally impossible" must be absolute in the sense that there can be no degrees of legal impossibility, so "physical impossibility" should be read in the same way. Furthermore, he contends that if the contractor finds it absolutely impossible to comply with his obligation to execute, complete and maintain the works in strict accordance with the contract, as stated in clause 13 of the contract ("contract" meaning the general conditions, specification, drawings, bill of quantities, tender and contract agreement) then the contractor, in pursuance of his overall obligation to execute and complete the works, is excused from strict adherence to the provisions of the contract, i.e. he is not only entitled but is required to break the contract in order to render possible what was otherwise impossible, and in that event he will be excused such a breach or, by virtue of clause 13, such a departure from the specification will not constitute a breach.

As to the word "impossibility", I was referred to various dictionary definitions, which in the end show, as one would expect, that something which is physically impossible is something which cannot be done according to the laws of nature. In one sense, this could mean absolutely impossible in that, from the material or physical point of view, the end result could not be achieved whatever the steps taken to try and get round or over the difficulty which was said to render the performance physically impossible. On the other hand, it could be construed as "impossible to perform in compliance with the specification and drawing" which is part of the contract. If that be the correct interpretation, [counsel for the employer] contends that a contractor, when faced with a physical impossibility resulting from the employer's design or specification, is under an obligation to render possible that which the employer had by contract made impossible, and, it would seem, that the contractor must redesign or respecify some other mode of working, not in compliance with the contract, in order to achieve the end result.

[Counsel for the employer] contends that the absolute meaning avoids uncertainty, but it seems to me that if such a construction be right, the contractor, when faced with what, from all practical points of view, appears to be an absolute impossibility, would still be faced with the uncertainty of knowing whether there was not some means of getting round the obstacle. It must be very rare to find some instance in which human ingenuity, coupled with wholesale deviation from the specified contract works, cannot achieve completion.

One can readily understand the Sale of Goods Act type of case, where the subject matter of the contract turns out to have become non-existent. In such instances, no contract has come into existence. But here, there is an express provision for excusal in the event of physical impossibility, and what [Counsel for the employer] contends is that given its absolute meaning, manufacturing and laying were not beyond the bounds of possibility and jointing was rendered possible by relaxation of the speci- fication. Alternatively, if manufacturing and jointing was impossible to achieve in accordance with the contract, then clause 13 would excuse a breach, or non-compliance with the contract on the part of the contractor if he had to resort to such means in order to overcome the impossibility. I am bound to say that I find such a rigid construction difficult to accept, although the linking of legal with physical impossibility makes it difficult to avoid. I think that the answer must be that if absolute impossibility is to be the test, then equally a strict construction of the contract must also be applied, in the sense that what is physically impossible in accordance with the precise limits of the specification and drawings, which are expressly made part of the contract, should not be permitted to be considered possible, because it can be rendered such by a deviation from the contract on the part of the contractor.'

And, later in the judgment:

'Turriff's contractual obligation was to manufacture, lay and joint the units in accordance with the drawings and specification. I have already indicated that it was, in that strict context, absolutely as well as practically impossible successfully to joint them. It was not, plainly, absolutely impossible to manufacture the units to the required dimensions and tolerance, but in the ordinary competitive commercial sense, which the parties plainly intended, I am satisfied that it was quite impossible for [the manufacturer] to achieve the degree of dimensional accuracy required.'

In short, the *Turriff* case is authority that physical impossibility within the meaning of clause 13 has to be interpreted from a practical commercial viewpoint, that it does not mean impossibility in the literal sense, and it can include commercial impossibility.

Impossibility and pre-existing events

The point is made above that frustration can only apply to supervening events and that it does not apply to pre-existing events. The question of whether 'impossibility' in the context of clause 13(1) can apply to both supervening events and events existing at the time of the contract is another matter. For practical reasons there would seem to be little doubt that it should apply to both but it may be worth noting that the judge at first instance in the *Holland Dredging* case said this:

'The words in parenthesis in sub-clause (2) of clause 11 tie in with the general proposition that, where a contractor bears risk of a site, he must, if he so wishes, include in his price for that risk . . .
   In these circumstances no claim can lie under condition 12. As regards condition 13 the same consideration applies so far as "physically impossible" is concerned as in the case of condition 11. Only a supervening

event resulting in physical impossibility would enable a contractor successfully to advance a claim.'

The Court of Appeal overruled the judge in respect of clause 12 but did not obviously do so in respect of clause 13. This might support an argument that impossibility arising from design or the like is not covered by clause 13 but the decision in the *Yorkshire Water* case seems to be against this proposition.

Impossibility and design responsibility

ICE Conditions such as the Seventh edition are drafted on the basis that generally the engineer is responsible for design. That being so clause 13 (instructions) and clause 51 (variations) fit together in enabling the contractor to recover costs and delay arising from impossibility in the design.

In the ICE Design and Construct Conditions clause 13 is omitted with the intention that the contractor should not benefit from defaults in his own design.

Users of traditional ICE Conditions, Fifth, Sixth and Seventh editions, need to give thought when conferring responsibility for design on the contractor how clause 13 and other clauses of the Conditions are intended to operate in such circumstances. There is a basic legal rule that contracts must be considered as a whole and it may be arguable that clause 13 with contractor's design responsibility is to be construed differently from clause 13 with engineer's design responsibility. But if the same interpretation applies in both cases any transfer of design responsibility to the contractor may not be as complete as it seems.

Some guidance on this is given by the case of *Norwest Holst Construction Ltd* v. *Renfrewshire Council* (1996). The contract was under ICE Conditions, Fifth edition, and by a special clause the contractor, Norwest, was responsible for design of piled foundations of a railway under-bridge. Norwest in conjunction with its piling subcontractor produced a design which met the specified requirements but which was impossible to construct because of certain restraints. The Scottish court which dealt with the matter was asked to decide a number of questions which can be paraphrased as follows:

(1) did the contractor provide a warranty that he could implement his obligations under the contract?
(2) did the proper interpretation of the contract incorporate design of the works within the meaning of 'the Works' as defined in the contract for the purposes of clauses 8(2) and 13(1)?
(3) was the engineer contractually responsible for resolving the issue of impossibility?

Put briefly the answers given by the court to these questions were:

1. No
2. No
3. Yes.

The case has not been widely reported and for that reason the following rather lengthy extracts from the judgment are given:

(1) 'The starting point of the debate before us was the proposition that, if contractors undertake to design and build certain works, in this case the bearing piles, to specified criteria, the contractors are to be taken to have satisfied themselves in advance that this can be done for the price and if they find that they cannot in fact carry out what they have undertaken to do, then they will be in breach of contract. Reference was made to *Tharsis Sulphur and Copper Co* v. *McElroy* (1878) 5 R (HL) 171; *Thorn* v. *London Corporation* (1876) 1 App Cas 120; *Wilson* v. *Wallace* (1859) 21 D 507 and *Gillespie* v. *Howden* (1885) 12 R 800. As a general proposition, that was not disputed by counsel for Norwest. The issue which divided the parties was, rather, whether the opening words of clause 13(1) altered the position.

Tharsis Sulphur and the other cases simply spell out and interpret the contractor's obligations under a contract to construct works. These obligations can be varied by the parties and it is well known that in clause 13(1) of the ICE Conditions the contractor's obligation has been qualified. In particular the contractor does not simply undertake to construct, complete and maintain the works in strict accordance with the contract to the satisfaction of the engineer, but undertakes to do so "save in so far as it is legally or physically impossible". Norwest pray this qualification in aid. They say that the bearing piles could not be built within the original confines of the cofferdam. Therefore they were not required to construct and complete the works, including the design and construction of the bearing piles, if it was impossible to do so. Renfrewshire argue that on its wording the qualification in clause 13(1) applies only to Norwest's obligation to construct and does not impinge upon their obligation to design the bearing piles. So far as that is concerned, their obligation is unqualified and, since they failed to perform it, they were in breach of contract.'

(2) 'I accept, of course, that in certain contexts the "works" must include the contractor's work in designing the bearing piles.... So, for instance, I have no doubt that when, in terms of the Form of Tender, Norwest offered to "carry out and complete the whole of the said Works", the works in question included the design of the piling. But one always requires to interpret any phrase not only in the contexts of the contract as a whole but also in its particular context...

Even from a purely linguistic point of view it is difficult to accept that design can be included in something which the contractor is to "construct". There is a more important point, however.

The basic obligation of the contractor is set out in clause 8(1) which says that "The Contractor shall subject to the provisions of the Contract construct complete and maintain the Works...". In framing that basic obligation the draftsman uses the same phrase as is used in clause 13(1) ("the Contractor shall construct complete and maintain the Works"), though there the obligation is qualified in a different way. It seems to me that the phrase must have the same scope in each of these clauses. But, as used in the parties' contract in this case, the phrase in clause 8(1) cannot be intended to cover Norwest's obligation to design the bearing piles since that obligation is set out in Particular Condition 141.'

(3)   'Where, as here, the contractor has taken on responsibility for the design of part of the permanent works, then that is reflected in clause 8(2) since the exception then comes into play. But that is quite different from saying that the contractor's design responsibilities are one of the obligations envisaged in clause 8(1). They are not. Since Norwest's design responsibilities are not within the scope of their obligations to "construct complete and maintain the Works" in clause 8(1), they cannot in my view be within their obligation to "construct complete and maintain the Works" in clause 13(1).'

(4)   'It is, of course, true that that sub-clause [8(2)] envisages that a contractor may be responsible for "the design ... of the Permanent Works", but that very phrase shows that the design is something different and apart from the "Permanent Works" as that phrase is used in that clause.'

(5)   'In my view, however, it cannot be right to say, even in the case of an aspect of the permanent works which a contractor designs, that clause 13(1) does not apply to them. After all, in this case Particular Condition 141 is concerned only with the contractor's responsibilities for designing the piles and for incorporating them into the permanent works. It does not contain his obligation to construct the piles. His obligation to construct them must therefore be part of his obligations to construct and complete the contract works as a whole and these obligations are to be found in clauses 8(1) and 13(1). The piles are only one part – though an important part – of the contract works and plainly the engineer must be able to give instructions and directions which cover the construction, completion and maintenance of the contract works as a whole, including the piles. That is ensured by the provision in clause 13(1). It follows that, even where the contractor designs part of the works, clause 13(1) applies to that aspect of the works. The result is that Norwest were obliged to construct the piles but not in so far as it was physically impossible to do so.'

(6)   'On the approach which I have taken, Norwest's design responsibilities for the piles are not covered by the terms of clause 13(1), but their obligations to construct, complete and maintain the works, including the piles, are covered by clause 13(1) and the qualification which it contains. In some cases that difference might give rise to a practical issue, but on the basis of what we were told about the facts of this particular case I do not believe that it does. Both sides were agreed that Norwest, or their

subcontractors, had been able to come up with a design for the piles which met the criteria in Particular Clause 137. The problem was that, even though they could be designed, the bearing piles could not be constructed as part of the permanent works so long as the design of the cofferdam remained as in the original drawings and specification. And by virtue of clause 13(1) the contractor had given no warranty that he could construct the piles or any other part of the works.'

(7) 'Even though the fact that the piling could not be constructed emerged at the time when the piling was being designed, in my view that is irrelevant. The reality is that Norwest were telling the engineer that part of the works which had been specified in the contract could not be constructed and the reason for that lay in the specification of the abutments which were not part of Norwest's design responsibility. In that situation, because of the terms of clause 13(1), Norwest would come under no contractual obligation to construct the piling and they can have had no responsibility for devising a solution to the difficulties, although in fact, along with their sub-contractors, they put forward a number of possible ways round the problem. The reality was therefore that, if [the employer] wanted the contract works to be completed, they or the engineer acting on their behalf required to devise a solution and to give an appropriate direction or instruction. Moreover in terms of clause 51 the engineer has the responsibility to order "any variation to any part of the Works that may in his opinion be necessary for the completion of the works".'

9.4 *Work to be to the satisfaction of the engineer*

Clause 13(1) provides:

- save insofar that it is legally or physically impossible
- the contractor shall construct and complete the works in strict accordance with the contract
- to the satisfaction of the engineer
- and shall comply with and adhere to the engineer's instructions
- on any matter connected therewith
- whether mentioned in the contract or not
- the contractor shall take instructions only from the engineer or subject to clause 2(4) his appointed delegate.

The opening words 'Save insofar that it is legally or physically impossible' are clearly intended to qualify the contractor's obligation to construct and complete. What is less clear is whether the contractor is excused further performance or simply entitled to a variation to make the works capable of performance. Given the references which follow in clause 13(1) to compliance with instructions, the linkage in clause 13(3) to clause 51(variations), and the absence of any provisions in clause 13 dealing with the effects of premature cessation of performance, it seems unlikely that the contractor is entitled to take the clause as granting rights of termination comparable with those in clause 63 (frustration) or clause 64 (employer's default). It might in extreme circumstances cover abandonment of the works, and perhaps this is reflected in clause 65(1)(e) by the phrase 'has abandoned the contract without due cause'. But more commonly the effect of qualification of the contractor's obligation will be to oblige the engineer to vary the works or for the employer to accept something less or different than specified.

Engineer's instructions

The phrase 'instructions on any matter connected therewith' is not without ambiguity. The question is whether 'connected therewith' relates to matters which are legally or physically impossible or to matters generally on the construction and completion of the works. If the proper interpretation of clause 13(1) is the former, the scope of the clause and of clause 13(3) is quite narrow. However, it is doubted if this is the proper interpretation. The structure of clause 13(1) leads more obviously to the conclusion that matters 'connected therewith' are matters of general application.

The phrase in clause 13(1) 'whether mentioned in the Contract or not' has to be reconciled with clause 2(1)(c) which states that the engineer has no authority to amend the contract, and with clause 51 which empowers the engineer to order variations which are either necessary or desirable. This suggests that the purpose of the phrase is to empower the engineer to give instructions as he thinks fit on matters relating to construction and completion of the works, for instance in respect of the contractor's methods or temporary works. Although this may seem to breach the division of responsibility between the contractor and the engineer/employer given elsewhere in the contract it is not, when considered in connection with clause 13(3) which entitles the contractor to recovery of the unforeseen cost and delay of complying with the instruction, an inequitable provision.

The final words of clause 13(1) 'duly appointed delegate' correct a drafting error in the Sixth edition where the corresponding phrase was 'the Engineer's Representative'. That phrase could have been taken to confer delegated powers on the engineer's representative independently of clause 2(4).

9.5 *Mode and manner of construction*

Clause 13(2) supplements the requirement in clause 13(1) that the works are to be in strict accordance with the contract to the satisfaction of the engineer by requiring:

- the materials, equipment and labour, provided by the contractor, and
- the mode, manner and speed of construction
- are to be acceptable to the engineer.

The purpose of this clause is not immediately obvious. The quality of materials and workmanship is dealt with in clause 36 and the removal of unsatisfactory work and materials in clause 39. The removal of unsatisfactory subcontractors and labour is dealt with in clauses 4 and 16. The contractor's programme and his methods are dealt with in clause 14. His rate of progress is dealt with in clause 46. The contractor's general obligation to construct and complete the works in accordance with the contract is stated in clause 8 and repeated in clause 13(1).

It may be that clause 13(2) is a superfluous remnant from an earlier edition of the Conditions (see under 'general comment' in the introduction to this chapter) or it may be that it lives on with a particular purpose. If it is the latter, the most likely explanation is that the clause is intended to emphasise that the test as to whether the contractor is fulfilling his obligations in accordance with the contract is the engineer's view of what is acceptable.

It is doubted if clause 13(2) should be taken as entitling the engineer to give instructions independently of clause 13(1) or at all. It is only instructions given under clause 13(1) which attract the benefits of clause 13(3).

9.6 Delay and extra cost

The provisions and procedures of clause 13(3) can be summarised as follows:

- if the engineer issues instructions
- pursuant to clause 5 or clause 13(1)
- which involve the contractor in delay or disruption
- causing him to incur cost beyond that foreseeable by an experienced contractor at the time of tender
- the engineer is to determine any extension of time to which the contractor is entitled
- the contractor shall be paid the amount of such cost as is reasonable
- except to the extent that such delay or extra cost results from the contractor's default
- profit shall be added to cost in respect of additional permanent or temporary work
- if the instruction requires a variation it shall be deemed to have been given pursuant to clause 51.

Application to clause 5

Clause 5 expressly states that instructions given by the engineer to explain and adjust ambiguities or discrepancies shall be regarded as instructions issued in accordance with clause 13.

Application to clause 13(1)

If the engineer issues instructions pursuant to clause 13(1) they are likely to be either resolving matters of impossibility or related to matters which are not expressly covered in other clauses of the Conditions. It would be inappropriate, for instance, for the engineer to order variations through clause 13(1) rather than through clause 51 given the different mechanisms for valuation. Even more so it would be inappropriate for the engineer to give instructions on the removal of unsatisfactory work under clause 13(1) rather than under clause 39 given the link between clause 13(1) and clause 13(3).

Delay and disruption

The clause refers to delay and disruption to the contractor's 'arrangements or methods of construction'. These are very general words – 'Arrangements' are not necessarily restricted to site based activities, and 'methods of construction' although mentioned in clause 14 are not obviously covered by any defined term.

'cost beyond that reasonably to have been foreseen'

Note that it is the amount of cost which could not reasonably have been foreseen – not the giving of the instruction. This is not the easiest of tests to apply. For example, if the contractor foresaw at tender stage ambiguities or impossibility in the specification/contract documents he could likewise have foreseen the issue of an instruction under clause 5 or clause 13(1). But what cost should he have foreseen in respect of that instruction and how should he have allowed for it in his tender given that the works are subject to remeasurement in accordance with the standard method of measurement?

Presumably cost which is recovered through the valuation of variations or by valuation of measured works is not cost recoverable under clause 13(3). That leaves not the cost of the work itself but the cost of the delay and disruption. And this perhaps is where clause 13(3) has potentially wide scope in giving the contractor entitlement to recover delay and disruption costs arising from instructions.

Reasonable cost

The contractor's entitlement is stated as 'the amount of such cost as may be reasonable'. This differs from the entitlement in clause 12(6) to 'costs which may reasonably have been incurred'. It suggests a more general approach to assessment of the allowable cost.

Contractor's default

When clause 13(3) first appeared in the Fifth edition it did not include the exclusion 'except to the extent that such delay and extra cost result from the Contractor's default'. This apparently left the employer liable to reimburse the contractor for claims even where the contractor was at fault. The clause was corrected in the Sixth edition and remains so in the Seventh edition.

Nevertheless there is still some uncertainty as to how the clause should operate in that the timing of any instruction may be critical to the amount of delay, disruption and extra cost and although the contractor may not be in default in regard to the need for the instruction, he may be in default in respect of its timing – see for example the requirement for 'adequate notice' in clause 7(3).

Variation deemed to have been given

The intention of the final sentence of clause 13(3) seems to be that to the extent an instruction pursuant to clause 5 or clause 13(1) involves a variation, that variation although it may be not stated as such, is to be treated as a variation and presumably valued accordingly. Whether the delay and disruption effects can be valued separately under clause 13(3) is non too clear – that may have been the case under the Sixth edition but the new procedure in clause 52 for the valuation of variations by quotation makes it less likely in the Seventh edition.

Chapter 10
Programmes and methods of construction

10.1 Introduction

Clause 14 sets out the contractor's obligations to submit for the engineer's acceptance/consent, programmes for the works and details of methods of construction, and the engineer's obligations in dealing with the same. The structure of the clause remains as in the Sixth edition:

- 14(1) – programme to be furnished
- 14(2) – action by the engineer
- 14(3) – provision of further information
- 14(4) – revision of programme
- 14(5) – design criteria
- 14(6) – methods of construction
- 14(7) – engineer's consent
- 14(8) – delay and extra cost
- 14(9) – responsibility unaffected by acceptance or consent.

Changes

There are no major drafting changes to clause 14 between the Sixth and Seventh editions and with the exception of the following minor changes the wording is identical throughout the nine sub-clauses:

- 14(4) – the further period for submission of a revised programme is changed from 'as the Engineer shall allow' to 'as the Engineer may allow'
- 14(8) – the contractor's entitlement to payment is changed from 'such sum in respect of the cost incurred as the Engineer considers fair in all the circumstances' to 'the amount of such cost as may be reasonable except to the extent that such delay and extra cost result from the Contractor's default'
- 14(9) – 'Acceptance by the Engineer of the Contractor's programme' is changed to 'Acceptance (or deemed acceptance) by the Engineer of the Contractor's programme'.

Policy/status

The absence of significant change between the Sixth and Seventh editions suggests that the draftsmen took the view that the major changes made between the Fourth, Fifth and Sixth editions have resulted in a set of provisions which can stand inspection and are satisfactory for some years to come. It would have been no surprise if there had been more change, not because the Sixth/Seventh editions are particularly inadequate or defective, but because the trend over recent years in most construction contracts has been to expand clauses on programmes and methods in pursuit of better project management.

There is, however, a danger in such expansion of promoting the status of programmes and method statement towards the status of contract documents. The policy in traditional ICE Conditions of Contract has always been that programmes and method statements are documents which follow the making of the contract and although they set out the intentions of the contractor they do not bind either the contractor or the employer as do contract documents. This policy is reflected in the drafting not only of clause 14 but also of other clauses of the Conditions.

Other standard forms adopt different policies. The New Engineering Contract effectively gives the contractor's programme contractual status by using the accepted programme as the basis for evaluating additional time and money. The IChemE model forms adopt a similar approach by requiring the purchaser (employer) to perform his obligations so as to enable the contractor to work to his programme.

It can be said that the policy of ICE Conditions leads to uncertainty as to the status of clause 14 programmes and method statements and it is certainly the case that in disputes on claims and delay, arguments abound on what is to be taken from the clause 14 documents. But it can also be said that the contractor should be free to perform his obligations to construct and complete the works within the time allowed as he sees fit, subject to compliance with any specified requirements, and that burdening the contractor and the employer with additional obligations arising from contractually binding programmes and method statements is to the benefit of neither party.

10.2 *Programmes and method statements generally*

A contractor's performance is measured by the employer and by the engineer, not just in terms of the quality of the work constructed but also by the manner in which the works are executed. Contractors who perform to programme and on time attract repeat business; contractors who fail in these respects frequently disrupt the employer's plans, become embroiled in claims and are likely to be regarded as unreliable.

The manner in which a contract deals with the contractor's obligations to carry out and complete the works and his liability for defaults is therefore of keen interest to all parties. The contract, by its provisions, sets the scene for the judgement of the contractor's performance.

Progress

It is debatable whether, having been set a date for commencement and a date for completion, the contractor's rate of progress in between the two should be of any concern to the employer or the engineer. The point was argued in the case of *Greater London Council* v. *Cleveland Bridge & Engineering Co* (1986) where it was held that as a general principle, in the absence of specific provisions on progress, it is for the contractor to plan and perform his work as desired during the contract period. The contractor's primary obligation is to complete the work within the contract period. If he does that he cannot be said to have failed to exercise due diligence and expedition.

More recently in the case of *Pigott Foundations Ltd* v. *Shepherd Construction Ltd* (1994), it was held that a piling subcontractor was entitled to plan and perform his work as he pleased provided that he finished within the time allowed in the subcontract.

Most contracts do, however, contain some obligations on progress, usually stated as obligations to proceed with due diligence and/or expedition. And some state that failure in respect of these obligations is a default which entitles the employer to terminate the contract.

As to obligations to perform to programme such obligations will not be implied and it is a matter of construction of the contract whether in a particular case there are any such obligations. Traditional ICE Conditions seek to avoid the imposition of such obligations but ad-hoc amendments, special conditions and/or the incorporation of tender programmes and method statements, can significantly alter the position.

Tender programmes and method statements

Few construction contracts make any mention of tender programmes or method statements in their standard text. That, however, does not stop many employers (and/or their engineers) requiring tenderers to include details of their methods and their intended programme with their tender. The reasons for this include:

- obtaining information to be used in the appraisal of tenders
- obtaining information to assist the employer and/or engineer in fulfilment of their own obligations
- obtaining information to assist the employer and/or engineer in co-ordinating the contract works with other activities
- obtaining information in the hope of controlling the contractor's operations and his scope for claims.

The value of this information to the employer and/or engineer depends upon whether it is formally incorporated into the contract so as to bind the appointed contractor. But the danger of incorporating tender programmes and method statements into contracts is that they bind both parties not just the contractor.

Thus in the case of *Yorkshire Water Authority* v. *Sir Alfred McAlpine & Sons (Northern) Ltd* (1985) it was said by Mr Justice Skinner (of an ICE Fifth contract into which a tender method statement had been incorporated):

'In my judgement, the standard conditions recognise a clear distinction between obligations specified in the contract in detail, which both parties can take into account in agreeing a price, and those which are general and which do not have to be specified pre-contractually. In this case the [employer] could have left the programme and methods as the sole responsibility of the [contractor] under clause 14(1) and clause 14(3). The risks inherent in such a programme or method would then have been the [contractor's] throughout. Instead, they decided they wanted more control over the methods and programme than clause 14 provided. Hence clause 107 of the specification; hence the method statement; hence the incorporation of the method statement into the contract imposing the obligation on the [contractor] to follow it save in so far as it was legally or physically impossible. It therefore became a specified method of construction by agreement between the parties.'

The decision in the *Yorkshire Water* case, although criticised in Hudson's Building and Engineering Contracts (Eleventh edition, page 897) as being inconsistent with the contractor's normal obligations, was noted with approval in the Court of Appeal ruling in *Holland Dredging (UK) Ltd* v. *Dredging & Construction Co Ltd* (1987) and was followed in *Havant Borough Council* v. *South Coast Shipping Co Ltd* (1996). Employers and engineers should, therefore, take the greatest care in incorporating tender documents emanating from the contractor, other than the form of tender and the priced bill of quantities, into the contract as formal contract documents.

A practice used to avoid doubt as to whether tender programmes and method statements are to become contract documents is to state in the instructions to tenderers that such documents are for tender appraisal purposes only – but even then care must be taken to ensure that the documents are not accidentally incorporated.

Approval of programmes

It is sometimes thought that if the contractor's programme is approved by the engineer the approval itself in some way gives the programme contractual status and makes it binding on the parties. Contractors are more inclined to advance this proposition than employers and frequently do so to support claims for delay and disruption.

However, most contracts expressly make the point that approvals or consents by the engineer do not relieve the contractor of any of his obligations under the contract – for example, clause 14(9) of ICE Seventh edition.

And it is not obvious why there should be an implied term that approval binds the employer. Perhaps the most that can be said is that although approval of a programme will not ordinarily create additional contractual obligations, it may be a factor in determining what is a reasonable time for the performance of obligations.

Shortened programmes

Programmes showing completion in a shorter time than that allowed in the contract frequently cause concern to engineers who suspect such programmes are more likely to be used as a platform for claims than as a guide to intended progress.

The matter was considered in the case of *Glenlion Construction Ltd* v. *The Guinness Trust* (1987) where the judge was asked to rule whether there was an implied term in a JCT 63 contract that, where the programme showed early completion, then the employer, through his architect, was obliged to perform so as to enable the contractor to carry out the work in accordance with the programme and to complete early. The judge held that such a term could not be implied and he made the point that one party cannot unilaterally change the obligations of the other party.

However, it must be emphasised that this decision related only to an implied term and the case does not apply to contracts where there are express terms relating the employer's obligations to programmes such as are found in the New Engineering Contract, MF/1, and the IChemE model forms.

In short, the *Glenlion* case rules out claims on shortened programmes based on implied terms but leaves open the question of the impact of shortened programmes on contractual provisions.

Programmes as the basis of assessment for delays and costs

Because programmes only create contractual obligations when they are either incorporated as contract documents or are expressly linked with obligations in the contract they do not ordinarily form the legal basis of claims for delay and extra cost.

This can work to the benefit of contractors as well as to employers. In the case of *Walter Lawrence & Sons Ltd* v. *Commercial Union Properties (UK) Ltd* (1984) a dispute arose as to whether the contractor was entitled to an extension of time for inclement weather. The architect in correspondence had said: '...It is our view that we can only take into account weather conditions prevailing when the works were programmed to be put in hand, not when the works were actually carried out...'

The contractor refuted this and claimed that his progress relative to programme was not relevant to his entitlement to an extension. It was held by the judge that the effect of the exceptionally inclement weather was to be assessed at the time when the works are actually carried out and not when they were programmed to be carried out.

However, programmes are useful as evidence in establishing proof of delay and disruption, but even here they are not always the best evidence. That is found by comparing progress during the delay/disruption period with actual progress during a non-affected period; a method judicially approved in the case of *Whittall Builders Ltd* v. *Chester-le-Street District Council* (1985).

Contractor's failure to produce or revise a programme

If a contract requires the contractor to produce and submit for approval a programme within a specified time and requires the contractor to revise his programme when instructed to do so it will, on the face of it, be a breach of contract if the contractor fails to do either.

The problem then is this – does the employer have any legal remedy for the contractor's breach or does the engineer have any powers under the contract to oblige the contractor to comply?

As a general rule there are few legal remedies or contractual powers. The employer's legal remedies for breach depend on proof of loss and this would be difficult to establish for such a breach. Specific contractual remedies and powers addressing the problem are conspicuously absent from most contracts. In extreme cases it might be possible to terminate the contractor's employment under the contract for default – for example, under clause 65 of ICE Seventh edition on the grounds of breach of obligation under the contract – but the circumstances would have to be appropriate to the gravity of such an action.

An exception to the general rule is found in the New Engineering Contract. Under the terms of that contract, the contractor is entitled to receive only 25% of the amounts due on interim payments until an approved programme is in place.

10.3 Clause 14 programmes

The parts of clause 14 relevant to programmes are:

- 14(1)(a) – submission of programme
- 14(1)(c) – further submission following rejected programme
- 14(2)(a) – acceptance of programme
- 14(2)(b) – rejection of programme
- 14(2(c) – request for further information
- 14(3)   – provision of further information
- 14(4)   – revision of programme
- 14(9)   – contractor's responsibility not affected by acceptance of programme.

The general scheme is that the contractor shall submit a programme within 21 days of the award of the contract and if this is rejected by the engineer within 21 days he shall resubmit within 21 days thereafter and so on until there is an accepted programme.

Submission of programme

Clause 14(1)(a) requires that within 21 days after the award of the contract the contractor shall submit to the engineer for his acceptance a programme showing the order in which he proposes to carry out the works having regard to the provisions of clause 42(1).

Clause 14(1)(c) states that should the engineer reject any programme under clause 14(2)(b) the contractor shall within 21 days of such rejection submit a revised programme.

Clause 14(3) requires that the contractor shall within 21 days (or such other period as the engineer may allow) of a request from the engineer for further information, provide such further information, failing which the relevant programme shall be deemed to be rejected.

The reference in clause 14(1)(a) to the provisions of clause 42(1) is to ensure the programme has regard to any contractual prescriptions on possession of the site or the order in which the works are to be constructed.

Format of programme

There is nothing in the Conditions, clause 14 or elsewhere, prescribing the form or detail of the contractor's programme. Taken literally clause 14(1)(a) requires nothing more than illustration of an order of procedure. Other clauses of the Conditions indicate that at the very least the order should be linked to a timetable but if even more is required, such as a network with critical path analysis or resource schedules, such requirements are normally stated in the specification or other project specific documents.

Acceptance/rejection of programme

Clause 14(2) requires that the engineer shall within 21 days of receipt of the programme:

(a)   accept the programme, or
(b)   reject the programme, or
(c)   request further information
    –   to clarify the programme, or
    –   to substantiate the programme, or
    –   to satisfy the engineer on its reasonableness having regard to the contractor's obligations.

The clause concludes by stating that if none of the above actions is taken within 21 days the engineer shall be deemed to have accepted the programme.

The final part of clause 14(3) endeavours to maintain the submission/acceptance sequence by stating that within 21 days of receipt of further information the engineer shall accept or reject the programme.

Revision of programme

Clause 14(4) provides that if it appears to the engineer at any time that the actual progress does not conform with the accepted programme the engineer shall be entitled to require the contractor to produce a revised programme showing modifications to the programme as necessary to ensure completion within the stated or extended time for completion.

The contractor is required to submit his revised programme within 21 days of such further period as the engineer may allow.

Clause 14(4) needs to be considered in conjunction with clause 46 (rate of progress) and to be exercised by the engineer with caution. The key part of the clause is that the contractor can be required to revise his programme in the light of actual progress to show how he intends to complete on time. It is questionable if the clause is intended to operate where the contractor's actual progress is ahead of programme. But in any event the engineer would be unwise to call for a revised programme under clause 14(4) without first having fully considered the contractor's entitlement to extension of time under clause 44. To call for a revised programme showing completion within a lesser time than the contractor might at that time be entitled to could be seen as constructive acceleration if not ordered acceleration.

Responsibility not affected by acceptance

Clause 14(9) states that acceptance of the contractor's programme (or deemed acceptance) shall not relieve the contractor of any of his duties or responsibilities under the contract.

This is intended to dispose of the proposition that an accepted programme binds the parties. However, it does not go so far as stating that acceptance of a programme has no relevance to the legitimacy of the contractor's claim. Acceptance may be relevant to any test of reasonableness which has to be applied.

10.4 Clause 14 method statements

The point is made earlier in this chapter that the parties should be careful to distinguish between method statements intended to be contractually binding and method statements to be submitted under clause 14.

The parts of clause 14 relevant to method statements are:

- 14(1)(b) – submission of general description
- 14(5)   – design criteria
- 14(6)   – methods of construction
- 14(7)   – engineer's consent to methods
- 14(8)   – delay and extra cost
- 14(9)   – contractor's responsibility not affected by consent.

General description

Clause 14(1)(b) requires that the contractor at the same time as submitting his programme (i.e. within 21 days of award) shall also provide:

- in writing for the information of the engineer
- a general description of proposed arrangements and methods of construction.

Note that what is required by clause 14(1)(b) is a general description. Any requirement for detail follows in clause 14(6). Note also that clause 14(1)(b) omits any reference to the information being provided for the engineer's consent – but that follows in clause 14(7).

The detailed provisions of clauses 14(2) and 14(3) have no application to the contractor's methods and accordingly there is no deemed consent if the engineer fails to respond to the contractor's proposals. But again see clause 14(7) for the obligation on the engineer to make some response.

Design criteria

Clause 14(5) requires the engineer to provide to the contractor such design criteria relevant to the permanent works (or any temporary works designed by the engineer) as necessary to enable the contractor to submit detailed methods of construction.

The provision in clause 14(8)(b) entitling the contractor to claim if he incurs unavoidable cost because of limitations imposed by design criteria supplied under clause 14(5) should be noted carefully by engineers. Usually such design information will be available at the time of tender and that is when it should be given to the contractor to minimise claims.

Methods of construction

Clause 14(6) states that:

- if requested by the engineer
- the contractor shall submit at such times as the engineer may reasonably require

- in such detail as the engineer may reasonably require
- information on methods of construction (including temporary works and equipment)
- which the contractor proposes to use
- and resulting calculations of stresses, strains and deflections in the permanent works
- to enable the engineer to decide if the works can be constructed:
  - in accordance with the contract, and
  - without detriment to the permanent works when completed.

The primary purpose of this clause, if not the only purpose, is that the engineer should be able to assess whether the contractor's proposed methods are compatible with the integrity of the permanent works. Hence the requirement in clause 14(5) for the engineer to provide relevant information on the permanent works design.

The words which conclude clause 14(6) 'Permanent Works when completed' are a little odd since concerns as to damage/detriment to the permanent works might rightfully apply at earlier stages.

Engineer's consent

Clause 14(7) requires that the engineer shall inform the contractor, in writing, within 21 days after receipt of the information submitted in accordance with clauses 14(1)(b) and 14(6) either:

(a)   that the contractor's methods have his consent, or
(b)   in what respects in his opinion they fail to meet the requirements of the contract or will be detrimental to the permanent works.

The clause concludes by stating that if clause 14(7)(b) applies (methods considered detrimental), the contractor shall make such changes as necessary to obtain the engineer's consent, and that the contractor shall not change methods already given consent without first obtaining further consent – which shall not be unreasonably withheld.

It is easier to see the relationship between clause 14(7) and clause 14(6) than that between clauses 14(7) and 14(1). Under clause 14(1) all that the contractor is obliged to provide is a 'general description'.

It is not clear on what grounds, other than safety or non-compliance with specified requirements, that the engineer can withhold consent to methods which have no influence on the permanent works. To do so could amount to a variation under clause 51.

Note that under clause 14(7) the engineer is obliged to respond within 21 days to the contractor's proposals. Unlike the provisions on programmes there is no deemed consent and clause 14(8) expressly states the contractor's entitlements when there is delayed consent. Engineers should be careful therefore in requesting information under clause 14(6) to ensure that they have the ability and capacity to deal with it under clause 14(7).

Delay and extra cost

Clause 14(8) provides that:

- if the contractor suffers unavoidable delay or extra cost because of:
  (a) unreasonable delay by the engineer in giving consent to proposed methods of construction, or
  (b) limitations imposed by design criteria supplied by the engineer under clause 14(5) or by requirements of the engineer under clause 14(7)
- which could not reasonably have been foreseen by an experienced contractor at the time of tender
- then the engineer shall take such delay into account in determining any extension of time due
- and the contractor shall be paid the amount of such costs as may be reasonable
- except to the extent the delay and extra cost arises from the contractor's default
- profit shall be added in respect of additional permanent or temporary work.

This clause applies only to delay in giving consent to methods and not to any delay in accepting the contractor's programme – that delay is covered by deemed acceptance.

Responsibility unaffected by consent

Clause 14(9) includes the engineer's consent to the contractor's methods with the engineer's acceptance of the contractor's programme in stating that neither relieve the contractor of his duties or responsibilities under the contract.

Chapter 11
Supervision, setting-out and safety

11.1 Introduction

Clauses 15 to 19 of the Conditions deal with the contractor's obligations and responsibilities in respect of site supervision, setting-out of the works, safety and security of the works.

There are no changes of any significance between the Seventh edition clauses and the corresponding Sixth edition clauses.

11.2 Contractor's superintendence

Clause 15(1) requires firstly that the contractor shall provide all necessary superintendence during the construction and completion of the works and for as long thereafter as the engineer reasonably considers necessary. It then requires that such superintendence shall be given by sufficient persons having adequate knowledge of the operations to be carried out for the satisfactory and safe construction of the works. The knowledge so required is expressly to include knowledge of:

- methods and techniques required
- hazards likely to be encountered
- methods of preventing accidents.

The overriding obligation from clause 15(1) is for the contractor to provide all necessary superintendence for the satisfactory and safe construction and completion of the works. This encompasses technical expertise and regards for safety. It cannot be assumed that 'superintendence' in this context is restricted in meaning to the provision of adequately trained site personnel. It could well include head office support.

To some extent the clause duplicates obligations stated in other clauses of the Conditions or obligations which could reasonably be inferred from them, e.g. clause 13(2) on the mode and manner of construction and clause 19 on safety and security. But by specifically stating obligations on superintendence the clause provides a focal point for the engineer if there is dissatisfaction with the contractor's performance and action is contemplated under clause 40(1) – suspension of works, or clause 65(1) – termination.

In the event of failings of individuals the engineer has at his disposal less draconian measures than suspension or termination of the works – he can order the removal of individuals from site under clause 16. For corporate failings, however, the engineer does not have lesser express powers – a situation which sometimes leads to exasperation and long-running conflict. It is not unusual to see the inadequacies/failings of the contractor in respect of superintendence put forward as grounds of opposition to delay and disruption claims but the weakness of this approach is that it is an 'after the event' complaint the merits of which are difficult to assess in liability and financial terms.

Quite apart from any contractual obligations on superintendence, the contractor can have a duty to warn – see, for example, the cases of *Plant Construction plc* v. *Clive Adams Associates and JMH Construction Services Ltd* (1999) and (2000) where it was held that the contractor's implied duty of skill and care extended to a duty to warn. It is doubtful if the contractor could escape this duty on the basis that he had undertaken works beyond his abilities.

'so long thereafter as ... necessary'

The obligation under clause 15(1) for the contractor to provide super-intendence for as long after construction and completion of the works as the engineer reasonably considers necessary is presumably intended to apply to superintendence during the defects correction period. It is difficult to see on what basis it can be extended thereafter.

The contractor's entitlement to payment for the costs of extended super-vision (if any) will depend on whether the requirement arises from default or from instructions/variations of reimbursable effect. No entitlement flows automatically from clause 15(1).

Contractor's agent

Clause 15(2) provides that:

- the contractor or a competent and authorised agent or representative is to be constantly on the works
- the agent/representative is to be approved of in writing by the engineer
- approval may at any time be withdrawn
- the contractor/agent/representative shall give his whole time to super-intendence of the works
- the agent/representative shall be in full charge of the works
- the agent/representative shall receive on behalf of the contractor directions and instructions from the engineer or, where appropriate, from the engineer's representative
- the contractor or the agent/representative shall be responsible for the safety of all site operations.

This clause is usually taken to mean that the contractor is obliged to nominate an employee or similar to be in charge on site full time. By tradition such a person was called the 'site agent'. With changing times that title is going out of fashion and, in the process, the oddities of wording of clause 15(2) are all too obvious.

Taken literally this clause makes no distinction between the contractor and the agent. Either is required to be constantly on the works. Either is responsible for safety. It is most unlikely, however, that the intention of the clause is to pass any of the contractor's corporate obligations to an employee, or in law, to individuals, agents/representatives or whatever else they might be called.

It is also doubtful if, taken literally, the clause does impose an obligation on the contractor to have a full time site agent.

As to responsibility for the safety of site operations, that is a matter of legal obligation which cannot be settled by statements in contracts purporting to put all responsibility on one firm or individual.

Withdrawal of approval

The statement in clause 15(2) that the engineer's approval to the contractor's agent/representative may be withdrawn at any time, although expressed without reservation, is almost certainly subject to an implied test of reasonableness or subject to the test for the removal of the contractor's employees found in clause 16.

11.3 Removal of contractor's employees

The first part of clause 16 states that the contractor shall only employ, or cause to be employed, in the construction, completion and superintendence of the works, persons who are careful, skilled and experienced in their trades and callings.

Taken literally this appears to prohibit the employment of trainees and apprentices on site but it is unlikely that this is intended.

Removal of employees

The second part of clause 16 states that the engineer shall be at liberty to object to and require the contractor to remove from the works any person who in the opinion of the engineer:

- misconducts himself
- is incompetent or negligent in the performance of his duties
- fails to conform with particular provisions regarding safety
- persists in conduct prejudicial to health or safety.

Such persons shall not be re-employed on the works without the provision of the engineer.

The grounds for removal are all related to individual rather than corporate failings and in that sense the second part of clause 16 is not directly related to any breach by the contractor of the first part of the clause. If there is any corporate breach the engineer's/employer's options other than persuasion are limited to extreme remedies such as suspension or termination or, in the case of subcontractors, exercise of the engineer's powers under clause 4(5) of the Conditions.

Removal orders against individuals need to be given with care and only on firm evidence of one or more of the failings listed in clause 16. Engineers should not assume that because it is the contractor who is required to effect the actual removal from site they are protected from suit by aggrieved individuals.

11.4 *Setting-out*

Clause 17 on setting-out is in three parts:

- 17(1) states the contractor's responsibilities
- 17(2) deals with errors
- 17(3) relates to checking and the maintenance of setting-out markers.

Contractor's responsibilities

Clause 17(1) states that the contractor shall be responsible for:

- the true and proper setting-out of the works
- the correctness of the position, levels, dimensions and alignment of all parts of the works
- the provision of all necessary instruments, appliances and labour for setting-out.

Although this clause puts the burden of setting-out on the contractor it does not require the contractor to fix the required locations of the various parts of the works. These are design matters for the engineer unless the contract expressly provides otherwise. This can put to test the meaning of the phrase 'true and proper setting-out of the works'. The contractor's setting-out may be true and proper on the basis of information provided but not true and proper of what is actually required or is necessary for architectural, technical or land ownership considerations.

Errors in setting-out

Clause 17(2) provides that:

- if at any time during the progress of the works
- any error shall appear or arise in the position, levels, dimensions or alignment of any part of the works
- the contractor on being required so to do by the engineer
- shall at his own cost rectify the error to the satisfaction of the engineer
- unless such error is based on incorrect data supplied in writing by the engineer or the engineer's representative
- in which case the cost of rectifying the same shall be borne by the employer.

This clause is probably intended to apply to errors which are departures from base data rather than to errors which relate to design. However, on its wording it seems to cover both.

When setting-out errors of either kind arise there is frequently dispute as to the extent (if any) of the contractor's obligation to check the setting-out information provided by the engineer for its accuracy and, in particular, to cross-check drawings for consistency. It is not uncommon for dimensions to be correctly stated on one drawing but incorrectly stated on another. The Conditions do not expressly put an obligation on the contractor to check the engineer's drawings for accuracy/consistency and it is questionable how far the contractor is under an implied duty to do so. In any particular case it may come down to whether the error was so apparent that under the contractor's implied duty of skill and care the error should have been recognised and avoided.

'incorrect data supplied in writing'

Contractors should be careful about accepting oral answers to queries on setting-out. The right to payment for costs of rectifying errors only applies to incorrect data supplied in writing.

'by the Engineer or the Engineer's Representative'

The specific inclusion in clause 17(2) of reference to the engineer's representative may be an attempt to exclude from the scope of the clause setting-out information provided by assistants. Clearly that will not work where the assistant has delegated powers and it is questionable whether it works even in the absence of delegation having regard to the provisions of clause 2(5)(b) that where an assistant gives instructions for the purpose of assisting the engineer's representative in his duties, such instructions are deemed to have been given by the engineer's representative.

Cost of rectification

The recoverable cost of rectification under clause 17(2) will normally be cost as defined by clause 2(5) – a definition which allows for overheads but not for profit. However, where the rectification work is putting right what amounts to a design error, the contractor should be able to get the work valued either as a variation or as an instruction under clause 13(3) which provides for profit on additional work.

Errors discovered after completion

Clause 17(2) applies to errors discovered 'at any time during the progress of the Works'. It is arguable that this phrase does not include the defects correction period and if that is the case errors discovered during that period are to be dealt with under clause 49.

Errors discovered after the issue of the defects correction certificate could in some circumstances entitle the employer to damages for breach of contract but it would of course have to be proved that the contractor was in breach and that the employer had suffered loss.

In the case of *Ruxley Electronics and Construction Ltd* v. *Forsyth* (1995), where it was discovered that a swimming pool detailed to be built to a depth of 7 feet 6 inches had only been built to 6 feet 9 inches, the House of Lords held:

- that the proper application of the general principle that where a party sustains loss by virtue of breach of contract he is so far as money can do it to be placed in the same situation in respect of damages as if the contract had been performed, was not the monetary equivalent of specific performance but required the court to ascertain the loss that the plaintiff had in fact suffered by reason of the breach.
- the cost of reinstatement was not the only possible measure of damage for defective performance under a building contract and is not the appropriate measure of damage where the expenditure would be out of all proportion to the benefit to be obtained, even if the alternative measure of value, diminution in value, would lead to only nominal damages because there was no diminution in value.
- while the court was not concerned with what a plaintiff might do with damages if awarded, the plaintiff's intentions were relevant to the question of reasonableness which arose at the stage of considering whether damages should be awarded. The judge's findings of fact as to Mr Forsyth's intentions to the effect that he had no intention of rebuilding the pool were relevant because they showed that he had lost nothing except the difference in value if any.

The principles of this case would apply where it was belatedly found that the contractor had set out the works marginally in the wrong location.

Checking and maintenance of setting-out

Clause 17(3) provides firstly that the checking of any setting-out line or level by the engineer or the engineer's representative shall not in any way relieve the contractor of his responsibility for the correctness of the setting-out; and secondly that the contractor shall carefully protect and preserve all bench-marks, sight rails, pegs and other things used in setting out the works.

The application of the first part of clause 17(3) can be contentious. Sometimes setting-out is done jointly by the contractor's and engineer's staff; sometimes for practical reasons the contractor places more reliance on the engineer's checking than the Conditions contemplate.

It is unlikely that the engineer owes a duty of care to the contractor in checking setting-out but the engineer will normally have a duty of care to the employer such that liability could fall back on the engineer if the costs of rectifying errors were beyond the contractor.

The requirement for the contractor to protect and preserve his setting-out is presumably intended to ensure that the engineer is not impeded in checking the contractor's work. To the extent that the contractor increases his own costs by lack of protection that is for his own account. But where the lack of protection increases the engineer's/employer's costs that is some-times seen as a claim/counteraction by the employer as damages for breach of contract.

11.5 Boreholes and exploratory excavations

Clause 18 provides that, if at any time during the construction of the works, the engineer shall require the contractor to make boreholes or to carry out exploratory excavation, such requirement shall be ordered in writing and shall be deemed to be a variation under clause 51 unless a provisional sum or prime cost item in respect of the work has been included in the bill of quantities.

The intention of the clause is fairly clear – that the employer will pay for boreholes etc. taken during the progress of the works to enable the engineer to finalise his design. The problem is that the contractor has general obligations to take boreholes etc. to avoid damage to buried apparatus and there is frequently no clear distinction between what the contractor does for his own purposes and what is required for the engineer's purposes. The contractor may be able to argue with some force that the information from his own boreholes is being used by the engineer for design purposes free of charge. In such circumstances it is suggested that even if the engineer is unwilling 'to require' the contractor to make boreholes the contractor may be able to include the cost in any variation which results from his discoveries.

11.6 Safety and security

Clause 19(1) states in very general terms the contractor's responsibilities in respect of safety and security. Clause 19(2) similarly states the employer's responsibilities.

The draftsmen make the point in the Guidance Notes to the Seventh edition that liability for safety is imposed by the Health and Safety at Work Act 1974 and other statutes and regulations and that it would be inappropriate and potentially misleading for liability to be detailed in the Conditions. This is sensible policy because under statute, liability falls on whoever the Act says it shall fall on and private contractual arrangements purporting to say otherwise are of no effect.

Contractor's responsibilities

In summary, clause 19(1) requires that the contractor throughout the progress of the works shall:

- have full regard for the safety of all persons entitled to be on the site
- keep the site and works in an orderly state appropriate to the avoidance of danger
- provide and maintain all lights, guards, fencing, warning signs, watching as necessary or as required by the engineer, engineer's representative or other competent authority.

Employer's responsibilities

In summary, clause 19(2) requires that if the employer carries out work on the site with his own workpeople, he shall:

- have full regard for the safety of all persons entitled to be on the site
- keep the site in an orderly state appropriate to the avoidance of danger.

Clause 19(2) concludes by requiring the employer to impose similar duties on other contractors he employs on the site.

In the Sixth edition there was a specific reference in clause 19(2) to the employer carrying out work or employing other contractors 'under clause 31'. That reference, which could mistakenly have been thought to limit the employer's responsibility, is wisely dropped from the Seventh edition.

Safety and duties of care

In *Clayton* v. *Woodman* (1962) a bricklayer was injured when a wall in which he was cutting a chase collapsed on him. The bricklayer sued, amongst others, the architect who had rejected his advice that it would have been better to pull the wall down. The Court of Appeal overturned the trial judge's finding that the architect was liable. Lord Justice Pearson said:

'Now it is quite plain, in my view, both as a general proposition and under the particular contract in this case, that the builder, as employer of the workman, has the responsibility at common law to provide a safe system of work, and he also has imposed on him under the Building Regulations the responsibility of seeing that those regulations are complied with, so that everything is as safe for the workman as it reasonably can be. That is the responsibility of the builder, and it is important that that responsibility should not be overlaid or confused by any doubt as to where the builder's province begins or some other person's province ends in that respect. The architect, on the other hand, is engaged as the agent of the owner of the building for whom the building is being erected, and his function is to make sure that in the end, when the work has been completed, the owner will have a building properly constructed in accordance with the contract and plans and specification and drawings and any supplementary instructions which the architect may have given. The architect does not undertake (as I understand the position) to advise the builder as to what safety precautions should be taken or, in particular, as to how he should carry out his building operations. It is the function and the right of the builder to carry out his own building operations as he thinks fit, and of course, in doing so to comply with his obligations to the workman.'

In *Clay* v. *Crump* (1963) a wall left standing after a demolition contractor had cleared a site collapsed on a builder's site hut killing two workmen and injuring another. The architect had accepted advice from the demolition contractor that the wall was safe to leave standing without examining it himself. The architect, the demolition contractor and the building contractor were all found liable by the Court of Appeal, the liability of the architect arising from his breach of duty to ensure the site was safe for the main contractor to enter after the demolition work.

In *Plant Construction* v. *Clive Adam Associates* (1999) the motor company, Ford, employed Plant to design and build certain rigs at its research centre at Dunton, Essex. Plant employed Clive Adam to provide consultancy services and a second defendant, JMH Construction Services, as a subcontractor. During construction an employee of Ford gave instructions on propping a roof to JMH. The propping was inadequate and the roof collapsed. Plant settled Ford's claims against it and sued Clive Adams and JMH for its losses. The Court of Appeal held, amongst other things:

- JMH was contractually obliged to carry out the temporary support works instructed by Ford
- JMH was under an implied obligation to carry out its duties with the skill and care of a competent contractor and given that JMH recognised that the support works were dangerous its duty carried with it an obligation to warn on the risks.

Chapter 12
Care of the works and insurances

12.1 Introduction

Clauses 20–25 of the Conditions outline the obligations and responsibilities of the parties for care of the works, accidents, damage to third party property and requirements on insurance cover. In short they deal with the allocation of risks and the insurance of risks.

There are no changes of significance from the Sixth to the Seventh edition and the clauses retain the general position that the contractor is responsible for care of the works and damage or injury to third parties and must carry appropriate insurance cover.

Complexities of application/interpretation

Although in themselves, the provisions of clauses 20–25 seem reasonably straightforward, and certainly no more complex than the provisions in other standard forms, there are good reasons why they should be treated with caution by civil engineers, project managers and the like unless they have specialist knowledge and training. Insurance has its own terminology; insurance law its own peculiarities; and the outcome of cases on damage and application of insurance provisions a constant capacity to surprise.

Engineer's duty of care

Given the difficulties it is understandable that engineers to contracts are sometimes inclined to leave insurance matters to others they regard as better equipped to look after the employer's interests. This is sound policy to the extent that an engineer should always ensure that the employer receives the best professional advice. But in so far that an engineer may have by the terms of his appointment a general duty to advise the employer on all contractual matters the engineer should not assume that the employer has of his own accord recognised and understood the obligations and implications of the contract. To do so is to invite a charge of negligence

In the case of *William Tomkinson & Sons Ltd* v. *Church of St Michael in the Hamlet* (1990), under a JCT Minor Works contract, it was held that an architect who failed to advise the employer of certain risks and the need to insure against them was in breach of his duty of care.

And in *Pozzolanic Lytag Ltd* v. *Brian Hobson Associates* (1998) it was held that a civil engineer appointed as project manager had a duty to ensure that all insurances required of the contractor were in place. As the judge said:

'If a project manager does not have the expertise to advise his client as to the adequacy of the insurance arrangements proposed by the contractor, he has a choice. He may obtain expert advice from an insurance broker or lawyer. Questions may arise as to who has to pay for this. Alternatively, he may inform the client that expert advice is required, and seek to persuade the client to obtain it. What he cannot do is simply act as a "post box" and send the evidence of the proposed arrangements to the client without comment.'

12.2 Risks and insurances generally

Risks

Standard forms vary in the extent to which they identify and deal with particular risks. The employer's responsibility for his own property and its contents is one area of significant variance. But generally matters to be considered include:

- damage to the works prior to completion
- damage to the works after completion
- faulty materials and workmanship
- thefts and vandalism
- design defects
- damage to the employer's property
- damage to third party property
- consequential losses from damage
- injuries to the contractor's employees
- injuries to the employer's employees
- injuries to third parties.

Some of these can be grouped together for drafting purposes.

Allocation of risks

The policy which underlies most standard forms is that risks should be allocated to the party best able to control them. Thus take-over of the works by the employer is usually seen as a watershed in respect of the works. Up to that time the contractor has care of the works and is generally responsible for damage, whereas afterwards the employer becomes responsible – subject to the proviso that the contractor is responsible for any damage he causes whilst remedying defects.

Responsibility for damage or injury to third parties usually follows the cause but damage to the employer's property is the employer's risk in some contracts.

The contractor is almost invariably responsible for the quality of work and carries the risks of faulty workmanship and materials. Responsibility for defective design generally falls on the party which undertook the design but that is not always the case.

Excepted risks

Excepted risks, or the employer's risks as they are called in some contracts, are those risks which are expressly excluded from the contractor's responsibility. Typically they include:

- acts or omissions of the project manager, employer or his servants
- use or occupation of the works
- damage which is the inevitable or unavoidable consequence of the construction of the works
- war, riots and similar non-insurable events.

Broadly the excepted risks fall into three categories:

- fault or negligence of the employer
- matters under the control of the employer
- matters not the fault of either party.

The logic of the first two categories is obvious enough. The argument for the third category, where it applies, is that the employer is the party better able to carry the risk.

Limitations on liability

Some contracts place limitations on the liability of the contractor to the employer for his acts and defaults. Such limitations, however, apply only between the contractor and the employer and they do not protect the contractor against third party claims.

Insurances generally

Certain insurances are required by law, for example, motor insurances and employer's liability. In addition construction contracts invariably impose insurance requirements on one or both parties to ensure that funds are available to meet damage claims and to facilitate completion of the works.

Some forms specify only the insurances which the contractor must carry. Other forms place obligations to insure on both parties.

Common insurance provisions

The common insurance provisions of construction contracts are:

- the contractor is responsible for care of the works until completion
- the contractor must insure the works to their full replacement cost
- the contractor must indemnify the employer against claims for injury to persons or damage to property
- the contractor must insure against that liability.

Other insurance provisions

According to the amount of detail in the insurance clauses of particular contracts, other provisions may cover:

- approval of insurers
- production of documentary evidence
- minimum levels of cover
- maximum levels of excess
- the employer's rights if the contractor fails to insure
- professional indemnity
- joint insurances.

Insurance terminology

Insurance clauses in contracts often use phrases which are not particularly clear in themselves but which have particular meanings to insurers. For example:

- Subrogation
  This is the legal right of an insurer who has paid out on a policy to bring actions in the name of the insured against third parties responsible for the loss.
- Waiver of subrogation
  An agreement by one party's insurers to give up its rights against another party.
- Joint names
  Insurance in joint names provides both parties with rights of claim under the policy and it prevents the insurer exercising his rights of subrogation one against the other.
- Cross liability
  The effect of a cross liability provision in a policy is that either party can act individually in respect of a claim, not withstanding the policy being in joint names. Without such a provision, liability between joint names is by definition not between third parties and is not covered.

- All risks

  An all risks policy does not actually cover all risks since invariably there
  will be exceptions. However, the effect of an all risks policy is to place on
  the insurer the burden of proving that the loss was caused by a risk
  specifically excluded from cover. In contrast, under a policy for a specified
  risk it is the insured who must prove that his loss was caused by the
  specified risk.

Professional indemnity for consultants

Employers or contractors who engage consultants as designers almost
invariably require that they have professional indemnity insurance. The
limits of such insurances vary enormously from modest thousands of
pounds to many millions of pounds. The amount is the prime point of
interest to the client but there are other matters which need to be noted.

Firstly, professional indemnity insurance is usually on a claims made
basis so that the cover is only effective in respect of the year in which the
claim is made. Thus, once a policy lapses there is no cover for past work.
Consequently a contractual requirement for such insurance needs to be
drafted to ensure that cover is maintained for the legal limitation period
rather than merely the construction period as for other insurances.

Secondly, there is the problem that the legal responsibility of a professional designer is limited at common law to the exercise of reasonable
skill and care, and his professional indemnity cover is usually similarly
limited. A claim on a fitness for purpose basis will have no access to such
insurance unless the policy specifically provides for it.

Design cover for contractors

Contractors who undertake design and build contracts usually ensure that
they have insurance cover either through the policies of their design consultants or through special policies for in-house designers.

There is a danger, however, that contractors undertaking work under
conditions of contract such as the ICE Seventh edition, where design is
principally a matter for the employer/engineer, may take on design
responsibilities without arranging any corresponding cover. As a point of
note there is no requirement in the Seventh edition for the contractor to carry
or provide professional indemnity insurance for any design for which he is
expressly responsible under the contract.

Contractor's in-house design insurance

Contractors who undertake in-house design can insure against the negligence of their own designers. The cover is usually defined as being in respect

of a negligent act, error or omission of the contractor in performance of his professional activities.

The need for such insurance arises because a contractor's all risk policy usually excludes design entirely or limits the indemnity to damage caused by negligent design to third party property or construction works other than those designed.

An ordinary professional indemnity policy does not cover the contractor against the problem of discovery of a design fault before completion. At that stage there is no claim against the contractor as there would be against an independent designer. To overcome this, contractors usually seek a policy extension giving first party cover. In effect this amounts to giving the construction department of the contractor's organisation a notional claim against the design department.

The complexities of in-house design claims were revealed in the case of *Wimpey Ltd* v. *Poole* (1984) which concerned a contract under the ICE Fourth edition modified for contractor's design. During the construction of a new quay wall at Vospers Southampton shipyard, movement of the quay wall occurred and extensive remedial works were necessary. The cause was found to be errors by Wimpey's in-house designers in their assumptions on soil mechanics. Wimpey sought to recover the cost of the remedial works from their insurers and set out to prove their own negligence even to the point of advancing the argument that a company of their standing should be judged by a more stringent and exacting task than an ordinary practitioner.

Wimpey failed to establish negligence although they did win the argument that the words 'negligent act, error or omission' should be construed to include any error or omission without negligence.

12.3 Care of the works

Clause 20 of the Conditions is in three parts:

- 20(1) which fixes the time limits during which the contractor carries the risk of care of the works
- 20(2) which states which risks are excepted from the contractor's responsibility
- 20(3) which states how the costs of rectification of loss or damage are to be borne.

Time limits

Clause 20(1)(a) provides, subject to certain exceptions, that the contractor shall take full responsibility for care of the works, materials, plant and equipment from the works commencement date until the date of issue of the certificate of substantial completion for the whole of the works. Thereafter, responsibility passes to the employer.

Note that it is the date of issue of the certificate not the date of comple-tion stated on the certificate which is the relevant date. This avoids the possibility of retrospective liability falling on the employer, but is still not as satisfactory (from the employer's point of view) as the position under the ICE Fifth edition where the transfer was 14 days after the issue of the certificate – thereby giving the employer time to arrange his own insur-ances.

Clause 20(1)(b) states that if the engineer issues a certificate of substantial completion for any section or part of the permanent works the contractor shall cease to be responsible for the care of that section or part from the date of issue of that certificate of substantial completion.

The engineer should be cautious therefore before granting a certificate under clause 48(4) for any part of the works which may be completed but which has not been put into use. The employer should also consider before putting any part of the works into premature use that such use entitles the contractor to a certificate of substantial completion under clause 48(3) if the part is 'substantial'.

But whether or not a certificate is issued, clause 20(2)(a) includes damage due to use or occupation by the employer as an excepted risk. This, however, is not quite the same as a formal transfer of responsibility for a section or part because the test is 'due to' use or occupation and not simply use or occu-pation itself.

It should be noted that clause 20(2)(a) applies only to 'any part of the Permanent Works'. It would seem that if the employer uses the contractor's temporary works as well he might – e.g. a temporary road bridge – the responsibility for care of the works remains fully with the contractor.

Clause 20(1)(c) states that the contractor shall take full responsibility for the care of any work, materials, plant and equipment relating to works undertaken during the defects correction period until such work has been completed.

It would seem from its wording that this clause is intended to operate notwithstanding the employer's continuing use of the relevant part of the works. It is not difficult to envisage circumstances of possible conflict between clauses 20(1)(b) and 20(1)(c).

Excepted risks

Clause 20(2) states, in somewhat curious wording, that the excepted risks are loss and damage to the extent it is due to:

(a)   use or occupation by the employer, his agents, servants or other con-tractors (not being employed by the contractor) of any part of the per-manent works

(b)   any fault, defect or omission in the design of the works (other than a design provided by the contractor pursuant to his obligations under the contract)

(c)   riot, war, invasion, act of foreign enemies or hostilities (whether war be declared or not)
(d)   civil war, rebellion, revolution, insurrection or military or usurped power
(e)   ionising radiations or contamination by radioactivity from any nuclear fuel or from any nuclear waste from the combustion of nuclear fuel, radioactive, toxic, explosive or other hazardous properties of any explosive nuclear assembly or nuclear component thereof and
(f)   pressure waves caused by aircraft or other aerial devices travelling at sonic or supersonic speeds.

Events (a) and (b) above imply some involvement of the employer and to the extent that the employer is the cause of any damage he must take legal responsibility. The remaining items fall into a different category. They imply no fault on the part of the employer. They are generally standard exclusions from insurance policies and purely as a matter of balance of risk in the contract these risks are placed on the employer.

Note that under clause 20(2)(a) relating to use or occupation the excepted risk only applies to loss or damage to the 'Permanent Works'. This suggests that if the employer or other contractors use the temporary works they either do so with the contractor's approval and responsibility remains with the contractor or the contractor has to look for his remedy for any loss or damage elsewhere in the contract – perhaps under clause 31 (facilities for other contractors).

The problem with the wording of clause 20(2) is that instead of describing the above events (a) to (f) as the excepted risks, the clause describes the risks as loss or damages to the extent it is due to one of the events. It would seem therefore that where there are competing causes of damage to the works, liability cannot simply be apportioned between the contractor's risks and excepted risks since by definition in clause 20(2) the excepted risks are already the subject of apportionment.

In clause 20(2)(b) there is a change of wording from the Sixth to the Seventh edition. In the Sixth edition 'any fault defect error or omission in the design of the Works' was qualified by the phrase 'for which the Contractor is not responsible under the Contract'. In the Seventh edition the qualifying phrase is 'other than a design provided by the Contractor pursuant to his obligations under the Contract'.

It is arguable that the Sixth edition had the better wording because it does not follow from clause 8(2) that every design provided by the contractor is his responsibility.

Rectification of loss or damage

Clause 20(3) sets out in three sub-clauses the consequences which follow the allocation of risks as stipulated in clauses 20(1) and 20(2).

Under clause 20(3) the contractor has to rectify at his own cost any loss or damage for which he is responsible so that the permanent works conform in every respect with the provisions of the contract and the engineer's instructions. The contractor is also liable for any loss or damage to the works caused by him in carrying out outstanding works or repairs under clause 49 or searches under clause 50.

It is difficult to see how the contractor's obligation to rectify loss or damage at his own cost to the engineer's instructions can be anything other than in conformity with the provisions of the contract. The wording does not obviously suggest that the engineer can accept work repaired to standards below those required by the contract and if higher standards are instructed by the engineer that would seem to be a variation entitling the contractor to payment.

Note that the final part of clause 20(3)(a) deals with the contractor's liability for loss or damage, not the obligation to repair. To the extent that the clause is in this part referring to damage during the defects correction period this is duplication of the provision in clause 20(1)(c). To the extent that it does not so refer it seems to create a distinction between the contractor's obligation to rectify loss and damage and his liability for such loss or damage which is not consistent with the first part of clause 20(3)(a).

Damage from excepted risks

Under clause 20(3)(b) the contractor is obliged to rectify loss or damage arising from the excepted risks but he does so at the employer's expense and only to the extent required by the engineer.

The clause does not go into the detail of how the engineer's requirements are to be regarded, for example as instructions under clause 13(3) or variations under clause 51, and the payment mechanism may depend on the circumstances.

It follows from clause 20(3)(b) that if the engineer does not require the contractor to rectify loss or damage from excepted risks then the works on completion may not be to the standards required by the contract. The engineer needs to consider the potential impact of this on the performance of the undamaged works and the possible diminution of the contractor's responsibilities of the same.

Apportionment of responsibility

Clause 20(3)(c) provides that where loss or damage is due to a combination of contractor's risk and excepted risk, the engineer shall apportion the costs of rectification.

The provision is in full accord with common-sense but it is not free from legal difficulty or the complexities of insurance rules. Generally where a loss has two causes, one defined as a risk and the other as an exception, an insurer is not liable to pay. The apportionment provision may be intended to overcome this. It is also worth noting however that by definition in clause 20(2) an excepted risk already assumes apportionment of cause.

Clause 20(3)(c) does not expressly require the contractor to rectify damage resulting from combined causes – an omission which raises questions as to whether an obligation to rectify can be implied following clause 20(3)(a) or whether there is need for an engineer's instruction following clause 20(3)(b).

12.4 Insurance of the works

Clause 21 deals with the contractor's obligation to insure the works and the details of the cover which is required.

Obligation to insure

By clause 21 the contractor is to insure the works, materials, plant and equipment against loss or damage in the joint names of the contractor and the employer. The purpose of this is to ensure that the contractor has funds to complete the works in the event of damage and alternatively to protect the employer's investment in the works.

The effect of putting insurance in joint names is to permit the employer to claim directly on the policy but it also has the advantage of preventing the insurer from exercising rights of subrogation between the parties.

The cover is to be for full replacement cost plus 10% to cover additional costs/incidentals such as professional fees, demolition, and the like.

Extent of cover

Clause 21(2) regulates the extent and duration of cover. The insurances shall:

(a) cover all loss or damage from whatsoever cause arising other than the excepted risks
(b) run from the works commencement date to the date of issue of the relevant certificate of substantial completion
(c) cover loss or damage arising during the defects correction period from a cause arising prior to the issue of any certificate of substantial completion
(d) cover loss or damage caused by the contractor in carrying out any work during the defects correction period.

Defective work

Clause 21(2)(c) confirms that the contractor is not obliged to insure against defective work. This is usual if for no better reason than the fact that such insurance is not readily available.

The clause does continue to say 'unless the Bill of Quantities provides a special item for this insurance'. But, of course, if this applies the employer is in effect funding the contractor against what is normally regarded as very much a contractor's risk.

Unrecovered losses

Clause 21(2)(d) is a provision which emphasises the point in clause 21(1) that insurance does not limit the obligations of the parties. It states that amounts not insured or recovered from insurers, whether by excesses or otherwise, shall be borne by the parties in accordance with their respective responsibilities.

This clause needs to be read in conjunction with clause 25 (evidence and terms of insurance) which, amongst other things, allows for controls on excesses.

12.5 *Damage to persons and property*

Clause 22 relates to third party claims. It requires the contractor to indemnify the employer against losses and claims which arise out of or in consequence of the construction of the works or the remedying of defects and which are losses and claims in respect of:

- death or injury to any person, or
- loss or damage to any property other than the works.

The indemnity is to cover all claims, proceedings etc. and all charges and expenses whatsoever. It extends, therefore, to legal costs as well as to sums claimed.

However, the case of *Richardson* v. *Buckinghamshire County Council* (1971) under the ICE Fourth edition is worth noting. A motor cyclist sued the County Council for injuries alleged to have been sustained when he fell off his machine at roadworks. The County Council successfully defended the claim but had to meet their own costs because the motor cyclist was legally aided. They tried to recover their costs from the contractor under clause 22 but it was held by the Court of Appeal that they had not arisen 'out of or in consequence of the execution of the works'.

Exceptions

As with clause 20 there are exceptions to the contractor's indemnity obligations. Clause 22(2) lists the following items as the responsibility of the employer:

(a)   crop damage
(b)   use or occupation of land
(c)   right of the employer to construct on land
(d)   damage which is the unavoidable result of the construction of the works in accordance with the contract
(e)   claims arising from negligence or breach by the employer, his servants, agents or other contractors.

The final point of exception does not expressly mention the engineer and his staff but they may be deemed to be included in the employer's servants or agents.

Unavoidable damage

The exception in clause 22(2)(d) for damage which is the unavoidable result of the construction of the works is frequently invoked and usually contentious. By way of example, a contractor laying pipes in a carriageway may claim that due to the poor condition of the road, damage beyond the net width of the trench was unavoidable; a contractor de-watering, excavating or piling adjacent to buildings may claim that damage was unavoidable; and a tunnelling contractor may claim that ground level subsidence causing damage was unavoidable.

If the contractor can establish that the damage falls within the scope of clause 22(2)(d) he acquires some valuable benefits: he avoids a claim on his own insurance and he is well placed to recover from the employer the costs of delay, disruption and repairs. Clause 22(2)(d) is not intended as a claim clause on par with clause 12 but if there is argument over which party should pay the costs arising from damage the clause takes on much the same function.

In disputes on the application of clause 22(2)(d) one frequent point of argument is the meaning of 'unavoidable'. Is it absolute or subject to practical or commercial tests? Another such point of argument is whether the unavoidability is to be considered by reference to actual working methods (approved, perhaps, and in full conformity with the contract) or by reference to alternative working methods which might have avoided the damage.

The answers to these questions can sometimes be found by examination of the circumstances in the light of the full wording of clause 22(2)(d):

'damage which is the unavoidable result of the construction of the Works in accordance with the Contract.'

Thus, where damage occurs from a specified method of working, properly carried out, it may well, in the context of the clause, be unavoidable. Where damage occurs from working methods of the contractor's own choice, again properly carried out, the contractor will have to show, by expert evidence or the like, that at the very least these were sound and well considered methods and the normal and competent choice for the circumstances.

Note that the exception in clause 22(2)(d) for unavoidable damage relates only to damage. It does not cover death or injury to persons.

Loss or damage due to design

Clause 22(2)(d) does not expressly cover damage which is the unavoidable result of the design of the works – whether that design be the responsibility of the contractor or not. This raises interesting questions as to whether there is any scope within clause 22(2)(d) for considering damage due to design or whether such damage falls within the scope of some other clause of the Conditions.

It is arguable that construction and design are separate tasks/obligations/responsibilities and that if clause 22(2)(d) was intended to provide an exception in respect of unavoidable damage due to design it would say so. But against that it can be said that construction is simply the implementation of design and that clause 22(2)(d) is operative for design as well as construction. The difficulty with that is there is no distinction between engineer's/employer's design and contractor's design in the clause and all liability would fall on the employer.

The answer to the conundrum may be found in clause 22(2)(e) which provides an exception in respect of damage to acts, neglects etc. of the employer, his agents etc. If design for which the employer is responsible comes under this clause there is no need for any separate reference to design for which the contractor is responsible. The exception for design, however, would not be on the same footing as the exception for construction. Damage due to design would not be subject to the 'unavoidable' test but it might be subject to a 'negligence' test.

Indemnity by employer

By clause 22(3) the employer is required to indemnify the contractor against all claims etc. made in respect of the exceptions in clause 22(2).

Shared responsibility

Clause 22(4) on shared responsibility contains two separate provisions, one for the contractor, the other for the employer.

In clause 22(4)(a) the contractor's liability to indemnify the employer is reduced in proportion to the extent that the employer or his agents may have contributed to the damage or injury.

In clause 22(4)(b) the employer's liability to indemnify the contractor in respect of the exceptions in clause 22(2)(e) is reduced in proportion to the extent that the contractor or his agents etc. may have contributed to the damage or injury.

There is no reduction of employer's liability in respect of the other excepted risks in clauses 22(2)(a), (b), (c) and (d). This may be on the basis that there can be no element of contractor's fault for these exceptions but, as discussed above, there is considerable scope for dispute on the exception under 22(2)(d) for 'unavoidable' damage and an all or nothing approach does not always fit the circumstances.

12.6 Third party insurance

Clause 23 provides for insurance against third party risks in much the same way that clause 21 provides for insurance of the works.

Clause 23(1) requires the contractor to insure in joint names against liabilities for death or injury to any person (other than a person in the workforce) and for loss or damage to property (other than the works), arising out of the performance of the contractor.

The contractor is not required to insure in respect of exceptions in clauses 22(2)(a) to (d) but the insurance must cover the exception in clause 22(2)(e). This is because 22(2)(e) relates to the matters specifically addressed in the clause 23(1) insurance.

Cross liability

Clause 23(2) requires the insurance policy to have a cross liability clause. This permits either party to claim as separate insured.

Amount of insurance

Clause 23(3) states, as is usual in civil engineering contracts, that third party insurance is to be for at least the amount stated in the appendix to the form of tender.

12.7 Accident or injury to work people

Clause 24 seeks to relieve the employer from liability for damages or compensation payable to operatives of the contractor and his subcontractors in the event of accident or injury except to the extent such accident or injury results from or is contributed to by any act or default of the employer, his agents or his servants. The contractor is required to indemnify the employer against all such claims for damages or compensation and all other costs of proceedings etc.

In the event of a claim being made against the employer it is unlikely that clause 24 is of any assistance to the employer by way of defence. The general rule is that a contract between A and B has no relevance in an action between

A and C. However, under the indemnity provisions the employer (A) would look to the contractor (B) for recompense of its liability to (C).

There is no express requirement in clause 24 for insurance cover by either the contractor or the employer but employers (in the broad sense) are required by statute to carry insurance against injury to their employees in the course of their employment.

12.8 Evidence and terms of insurance

Clause 25 deals with five issues:

- evidence of insurance
- terms of insurance
- excesses on policies
- failure to insure
- compliance with policy conditions.

Evidence and terms

Clause 25(1) requires the contractor to provide satisfactory evidence prior to the commencement date that the insurances are in force. If required the policies must be produced for inspection.

The clause further provides that the terms of the insurance shall be to the approval of the employer and that the contractor shall, on request, produce receipts of payment of premiums. There is no requirement for insurance to be with an insurer approved by the employer.

Excesses

Clause 25(2) provides that any excesses on policies shall be as stated in the appendix to the form of tender.

The provision is intended to guard against the contractor taking on insurance with high excesses and thereby leaving the employer exposed to the amount of the excess in the event of the contractor's financial failure. Some employers state in the tender documents the maximum excesses they are prepared to accept.

Failure to insure

Clause 25(3) allows the employer to take out insurance when the contractor fails to produce satisfactory evidence that his insurances are in force. The premiums paid by the employer can be deducted from monies due to the contractor or recovered as a debt.

Contractors would be most unwise to allow this state of affairs to develop since the employer would insure in his own name, leaving the contractor exposed to claims from the insurer under his rights of subrogation.

Compliance with policy conditions

Clause 25(4) is a provision requiring both the employer and the contractor to comply with the conditions laid down in the insurance policies and indemnify each other against any claims arising from failure to comply.

The effect of this is that if one party renders the insurance policies void by his actions or omissions, he then stands in the place of the insurer in providing cover.

Chapter 13
Statutes, street works, facilities and fossils

13.1 Introduction

Clauses 26 to 35 of the Conditions cover miscellaneous obligations and responsibilities which, for the most part, relate to the involvement of the parties in the design and construction of the works with outside authorities and third parties.

The clauses are much the same as in the Sixth edition except for the addition of clause 26(4) which provides the contractor with express entitlements when delay or extra cost results from action necessary to bring the works into conformity with statutes etc., and for the inclusion in clause 27(1)(a) of additional obligations on the contractor in respect of street works.

Clause 27 in the Sixth edition was itself significantly changed two years after publication in 1991 by the August 1993 corrigenda and it is the corrigenda clause which now forms the basis of clause 27 in the Seventh edition. However, there appears to be a drafting error in part of the new clause and another part which is redundant.

Delays caused by outside authorities

Within the batch of clauses 26 to 35, various clauses state the contractor's entitlement to recovery of cost and time for delays caused by the engineer and/or employer in dealing with outside authorities.

The clauses do not directly cover the consequences of situations where the contractor is obliged to have dealings with such authorities and then encounters administrative or bureaucratic hold-ups over which neither he nor the employer has any control. The contractor may be able to argue for extension of time under 'other special circumstances' in clause 44(1)(f) but recovery of additional costs is more difficult. It is doubted if it can be argued that the employer warrants timely performance by outside authorities or that failings on the part of such authorities amount to prevention or breach of contract by the employer.

13.2 Notices and fees

Clause 26(1) states that the contractor shall give all notices and pay all fees required to be given or paid by:

- any Act of Parliament
- any regulation or bye-law of any local or statutory authority
- the regulations of public bodies and companies whose property or rights are affected by the works.

The contractor's obligation is subject to the proviso in the opening words of the clause – 'Except where otherwise provided in the Contract'. Instances of this are found in clause 26(3)(c) – contractor not responsible for obtaining any planning provision, and in clause 27(2)(a) – the employer to obtain any street works licences.

It may also be the case that a distinction is to be drawn generally in the application of clause 26(1) between the giving of notices and the obtaining of consents and licences. The Conditions do not cover every eventuality by way of express provision and, depending on the nature of the works, there can be a range of permits/licences required before works can be commenced or progressed. And to the extent that any such permits or licences involve design rather than construction matters it is difficult to see on what contractual or practical basis the responsibility for their acquisition should fall on the contractor – unless the contractor is expressly made responsible for design.

Repayment of fees

Clause 26(2) provides that the employer shall 'repay or allow' the contractor:

- such sums as the engineer shall certify 'to have been properly payable and paid' in respect of fees and
- rates and taxes paid by the contractor in respect of the site or in respect of:
  - temporary structures situated elsewhere but used exclusively for the purposes of the works, or
  - structures used temporarily and exclusively for the purposes of the works.

Although on its wording clause 26(2) seems to be clear enough, application of the clause is frequently a matter of dispute. Not every engineer and employer willingly accepts that rates on the contractor's site compound are repayable by the employer – even though for most contracts this is one of the more obvious examples of proper application of the clause. When it comes to less obvious situations, for instance, rates on permanent buildings used by the contractor, rates on casting yards and the like dedicated to production for the works, or rates on land used for borrow-pit/quarrying for the works, it is often suggested that the prices in the bill of quantities should have made due allowance for all costs.

The test in clause 26(2) for repayment of rates and taxes is essentially 'used exclusively for the purposes of the Works'. The question is whether this is to be taken as meaning exclusively whilst the works are under construction or exclusively over a longer term. The potential variability of circumstances suggests there may be no single correct answer but, if anything, exclusive use during the progress of the works may be what is generally intended.

Conforming with statutes

Clause 26(3) requires the contractor to ascertain and conform in all respects with:

- Acts of Parliament
- regulation and bye-laws of local or statutory authorities
- rules and regulations of public bodies.

The contractor is required to indemnify the employer against all penalties and liability of every kind for breach of any Act, regulation or bye-law. There are, however, two important exemptions to those duties:

- the contractor is not required to indemnify the employer against any breach which is the unavoidable result of complying with the contract or instructions of the engineer – clause 26(3)(a).
- the contractor is not responsible for obtaining planning permission in respect of the permanent works or any temporary works designed by the engineer – clause 26(3)(c).

'unavoidable result'

The exception in clause 26(3)(a) for breach which is 'the unavoidable result of complying with the Contract or instructions of the Engineer' does not correspond exactly with the exception in clause 22(2)(d) in respect of third party liability 'damage which is the unavoidable result of the construction of the Works in accordance with the Contract'. In so far as the clause 22(2) exclusion is expressly related to 'construction' it has less scope for application than the clause 26(3) exclusions, but that apart the differences between the two may not amount to much.

Clause 26(3)(a) can be seen as complementing clause 13(1) where the contractor's obligation to construct and complete the works is qualified by 'save insofar that it is legally or physically impossible'. In short the contractor is not obliged to proceed if what he is required to do by the contract is illegal but if he does unwittingly proceed then he is not required to indemnify the employer of the consequences of breach.

The exception in clause 26(3)(a) to the indemnity given to the employer does not, of course, relieve the contractor of his legal responsibility for any breach of statute or regulation. All that it does is relieve the contractor from his indemnity to the employer.

Notwithstanding the firmly stated requirement in clause 13(1) that the contractor shall comply with and strictly adhere to the engineer's instructions, the contractor has obligations at statute and common law which take priority.

The case of *Eames* v. *North Hertfordshire District Council* (1980) is instructive. In that case a contractor erected a portal frame on made-up ground. Both the architect and the Council's building inspector allowed the work to proceed. Nonetheless the contractor was held liable for breach of a statutory duty (under the Building Regulations) and liable in negligence. Judge Fay QC referred in his judgment to the comment of Lord Wilberforce in *Anns* v. *London Borough of Merton* (1978):

> 'Since it is the duty of the builder, owner or not, to comply with the bye-laws, I am of the opinion that an action could be brought against him in effect for breach of statutory duty by any person for whose benefit or protection the bye-law was made.'

Engineer to issue instructions

Clause 26(3)(b) requires the engineer to issue instructions, and/or order a variation if necessary, if at any time it is found that the contract or instructions given are not in conformity with statutes, regulations, bye-laws etc.

The new clause 26(4) deals with contractor's entitlements in the event that the engineer issues instructions under clause 26(3)(b).

Planning permission

Clause 26(3)(c) provides important clarification on which party is responsible for obtaining planning permissions. The clause states that the contractor is not responsible for obtaining permissions:

- necessary in respect of the permanent works in their final position, or
- necessary for temporary works designed by the engineer in their designated position on site.

The wording of clause 26(3)(c) is slightly different in the Seventh edition from the wording in the Sixth edition but this is not thought to be a matter of policy change. However, to the extent that the change is a matter of style it is questionable if there has been much, or any, improvement. Firstly, there is the enigmatic phrase 'Permanent Works in their final position' and then there is the ill-constructed phrase 'Temporary Works designed by the Engineer in their designated position on Site'. The first may have something to do with storage of components or the like for incorporation into the works but otherwise the implication that the permanent works may not at some stage be in their final position takes some understanding. The second phrase, notwithstanding its wording, is unlikely to have anything to do with where the engineer does any temporary works design; it is more likely to be by way of caution that temporary works designed for one location cannot be relied on in another location.

The concluding statement in clause 26(3)(c) that the employer warrants that all permissions have been or will be obtained in due time gives the contractor good grounds for claiming damages for breach if there is any delay. The Canadian case of *Ellis-Don Ltd* v. *Parking Authority of Toronto* (1978), much quoted as authority on winter working and the recovery of overheads, concerned the failure of the Authority to obtain building permits in time. But note, however, the contractor's specific entitlements in the new clause 26(4).

Delay or extra cost

Clause 26(4) is one of the few completely new clauses in the Seventh edition. It fills the gap in previous editions where there was no specific remedy available to the contractor for any breach by the employer or engineer of the obligations imposed by clause 26(3).

The clause provides:

- if the contractor incurs delay or extra cost arising from:
    - instructions given under clause 26(3)(b), or
    - failure of the employer to comply with clause 26(3)(c)
- then, the engineer shall take such delay into account in determining any extension of time due under clause 44
- and, subject to clause 53, the contractor shall be paid 'such extra cost as may be reasonable except to the extent that such delay or extra cost results from the Contractor's default'.

Where there is a claim founded on instructions under clause 26(3)(b), the closing part of clause 26(4) referring to contractor's default may be critical to the contractor's entitlements. The severity of the consequences of discovering that contractual requirements are illegal may well depend, in terms of delay, disruption and extra cost, on the timing of the discovery relative to the stage of the works and the criticality of any affected work to the programme. That may in turn lead to consideration of the extent to which the contractor has fulfilled his obligations under clause 26(3) to ascertain all applicable laws, regulations, bye-laws etc., and whether there is a case for saying, in particular circumstances, that there is contractor's default.

13.3 *New Roads and Street Works Act 1991*

At the time the Sixth edition of ICE Conditions was first published in 1991 the control of works in highways was governed by the Public Utilities Street

Works Act 1950. Clause 27 of the Sixth edition was drafted to conform with that Act. On passage into law of the replacement Act, the New Roads and Street Works Act 1991, a revised version of clause 27 was produced and issued in the 1993 corrigenda to the Sixth edition.

Clause 27 as now found in the Seventh edition is a modified version of the corrigenda clause. In summary clause 27(1) provides:

- for the purpose of obtaining any licence under the Act for the permanent works, the undertaker shall be the employer who will be the licensee – clause 27(1)(b)
- if the licence contains a prohibition against assignment the contractor shall give the employer all notices required to be given by the undertaker and shall indemnify the employer against any failure to do so.

Comparing the wording of the new clause 27(1) with that in the corrigenda clause it appears that a necessary part of clause 27(1) is missing from the Seventh edition. In the corrigenda, clause 27(1)(b) states (as does the new clause) that the employer is the undertaker for the permanent works and then in clause 27(1)(c) it states, 'For all other purposes the undertaker under the licence shall be the Contractor'.

To make complete sense of the second part of the new part of clause 27(1)(b) relating to assignment, the wording from clause 27(1)(c) of the corrigenda (as stated above) could usefully be added to the end of the first part.

Licences

Clause 27(2)(a) provides that the employer shall obtain any street works licence or any other consents or permissions required for the carrying out of the permanent works and shall supply the contractor with copies including any conditions or limitations.

Clause 27(2)(b) provides that any condition or limitation in any licence obtained after the award of the contract shall be deemed to be an instruction under clause 13. The effect of this is that the contractor becomes potentially entitled to extension of time for any imposed delay and recovery of additional cost.

Notices

Clause 27(3) states that the contractor shall be responsible for notices required before the commencement of any work and that a copy of every notice is to be given to the employer.

For practical and operational purposes it is sensible that the burden of giving notices falls on the contractor; it is less practical that copies shall be given to the employer rather than to the engineer.

Delays attributable to variations

Clause 27(4) states that if any instruction pursuant to clause 27(2)(b) causes delay because of compliance with the notice requirements of clause 27(3) the engineer shall 'in addition to valuing the variations' take any delay into account in determining any extension of time.

It is not clear what purpose clause 27(4) serves since the reference in clause 27(2)(b) to clause 13 seems to give adequate expression of the contractor's entitlements. The reference in clause 27(4) to 'variation' comes from the old clause 27(2)(b) in the Sixth edition and with that clause no longer in the Seventh edition it is arguable that the whole of clause 27(4) is redundant.

13.4 *Patent rights and royalties*

Clause 28 deals briefly with patent rights and royalties for materials. The connection between the two elements of the clause is fairly tenuous except that both relate to the rights of third parties for things incorporated into the works – whether by design or by construction.

Patent rights

Clause 28(1) requires the contractor to indemnify the employer against all claims and proceedings for infringement of any:

• patent rights
• design
• trademarks
• names
• other protected rights.

The indemnity is to be given in respect of:

• contractor's equipment used in connection with the works
• materials, plant and equipment for incorporation into the works.

The clause is subject to the proviso that the contractor is not obliged to indemnify the employer where the infringement results from compliance with the design or specification (unless provided by the contractor) and that in the event of infringement from compliance with a design or specification (unless provided by the contractor) the employer shall indemnify the contractor against all claims and proceedings.

Royalties

Clause 28(2) provides that the contractor shall 'Except where otherwise stated' pay all tonnage and other royalties, rent, payments and compensation, for getting stone, sand, gravel, clay or other materials for the works.

As a general rule under the Conditions the burden of paying suppliers of materials falls on the contractor, so to give some purpose to clause 28(2) it is necessary to consider what is special about the types of materials mentioned in the clause. The answer is that they are all materials which might be obtainable from excavations, borrow pits or quarries on or adjacent to the site. One purpose of the clause may be therefore to put the contractor on notice that in respect of such materials no entitlement for free use should be assumed. That would seem to apply to materials obtained on or off the site and presumably the contractor is to pay the employer for use of site won materials unless the contract states otherwise.

The proviso can also be seen as a possible link back to clause 26(2) (repayment of fees, rates and taxes by the employer) in respect of certain charges such as rates which fall on the employer by virtue of clause 26(2).

13.5 *Interference, noise and pollution*

Clause 29 deals with various aspects of nuisance to third parties – as distinct from damage which is dealt with in clause 22.

Clause 29(1) covers interference with traffic and adjoining properties whilst clauses 29(2), 29(3) and 29(4) cover noise, disturbance and pollution.

Interference

Clause 29(1) requires that all operations for the construction of the works shall be carried out, so far as compliance with the contract permits, so as not to interfere 'unnecessarily or improperly' with:

- the convenience of the public
- access to roads, footpaths, properties
- use of roads, footpaths, properties.

The contractor is required to indemnify the employer against all claims and proceedings 'whatsoever arising' out of, or in relation to, such interference. Note, however, that there is a discrepancy between the obligation on the contractor to avoid interference and the indemnity he must give. The obligation is qualified by the phrases:

- 'so far as compliance with the requirements of the Contract permits' and
- 'not to interfere unnecessarily or improperly'.

The indemnity is not so qualified.

This suggests that even when the contractor is not at fault contractually, in that he is merely complying with the requirements of the contract, he is required to indemnify the employer against claims. There is no provision for a counter-indemnity by the employer in such circumstances as there is elsewhere in the Conditions (including clause 29(4) for noise, disturbance and pollution) to protect the contractor when he is complying with contractual requirements. It is not clear whether this is an unintended omission or whether, as a matter of policy, the contractor is to accept (as between himself and the employer) full liability for all claims relating to interference with traffic and adjoining properties. If the latter is the case the proviso as stated in the first part of the clause on the contractor's obligations in these matters seems to be of little effect.

Noise, disturbance and pollution

Clause 29(2) requires that all work shall be carried out without unreasonable noise or disturbance or other pollution.

This is not expressed in the style of wording used elsewhere in the Conditions – 'the Contractor shall carry out . . . without . . .' which suggests that the requirement may not be aimed solely at the contractor. In practical terms, however, the clause is likely to be of application only when the contractor is in default. The engineer could then, if matters were serious enough, take action under clause 40 (suspension) or in the extreme under clause 65 (termination).

Permissible noise levels will often be stated in the specification or elsewhere in the contract or imposed directly by local authorities. Disturbance can take many forms – vibrations due to piling or blasting are common examples. Pollution could include problems with dust and mud as well as problems with toxic wastes, smells and the like.

Clause 29(3) requires the contractor to indemnify the employer against claims and proceedings except to the extent that noise, disturbance or pollution are the 'unavoidable consequence of carrying out the Works'.

Clause 29(4) requires the employer to indemnify the contractor against claims which are so unavoidable.

There is wide scope for argument when claims arise as to what is 'unavoidable'. And since insurers may be involved in meeting claims, either directly or through indemnities, they will want to be involved in any decisions reached. Indeed, many policies preclude any admission of liability by the parties themselves. The safest course for the parties is to keep their insurers fully informed and to act only on their advice.

13.6 *Damage to highways*

Clause 30 sets out, as between the parties, their respective liabilities under the Highways Act 1980 or, in Scotland, the Roads (Scotland) Act 1984. The Acts allow the highway authority to recover its excess maintenance expenses arising from excessive weight or extraordinary traffic.

Broadly the scheme is that the contractor takes all risks from movement of his equipment and temporary works and the employer takes the risk for movements of materials and manufactured articles.

Clause 30(1) requires the contractor to use every reasonable means to prevent extraordinary traffic and to use suitable routes and vehicles to ensure that any inevitable extraordinary traffic is limited as far as possible and no unnecessary damage is caused to roads and bridges.

Clause 30(2) provides that the contractor must pay the costs of strengthening or improving the roads or bridges to facilitate movement of his equipment or temporary works and must indemnify the employer against all claims.

Clause 30(3) requires the contractor to notify the employer as soon as he becomes aware of any damage caused by the transport of materials or manufactured articles. The employer is to negotiate and settle any claims and to indemnify the contractor. If the claim is due to any failure by the contractor to use reasonable means and limit damage as required by clause 30(1), then the engineer is required to certify the amount due from the contractor to the employer.

As in clause 22, this brings the engineer into the field of the loss adjuster. Almost invariably insurance companies are involved in claims for damage and they are not disposed to take a back seat while employers negotiate and engineers apportion their liabilities.

13.7 Facilities for other contractors

Clause 31 serves a number of purposes. It effectively confirms that the contractor is not entitled to exclusive possession of the site; it obliges the contractor to provide what are termed 'reasonable facilities' to other contractors employed on or near the site by the employer; and it entitles the contractor to additional time and money in the event that the provision of 'reasonable facilities' goes beyond what could reasonably have been foreseen by the contractor at the time of the tender.

Provision of facilities

Clause 31(1) states that the contractor shall afford all reasonable facilities in accordance with the requirements of the engineer or the engineer's representative to:

- other contractors employed by the employer and their workmen
- workmen of the employer
- workmen of authorities or statutory bodies employed in the execution of work not in the contract
- workmen of authorities or statutory bodies employed on any contract entered into by the employer in connection with or ancillary to the works.

Delay and extra cost

Clause 31(2) provides that if compliance with affording reasonable facilities involves the contractor in delay or cost beyond that reasonably foreseeable at the date of award of the contract, the contractor is entitled to:

- an extension of time under clause 44
- payment of the amount of such cost as may be reasonable
- profit on any additional work.

Reasonable facilities

The phrase 'reasonable facilities' in clause 31 is open to a variety of interpretations. It may mean no more than allowing access and space in which to work. It may also include the use of scaffolding, mess rooms, toilets and the like and possibly the supply of power and water.

The uncertainty of the phrase begs the question: what if the contractor complains that the requirements of the engineer are to afford 'unreasonable facilities'? The answer would seem to be that the contractor must comply with any instructions and seek reimbursement through clauses 31(2) or 13(3) as appropriate or clause 51 if a variation can be established.

Statutory undertakers

The references in clause 31(1) to authorities and statutory bodies are probably wide enough to avoid disputes of the kind in *Henry Boot Ltd* v. *Central Lancashire Development Corporation* (1980).

In that case there was an extension of time dispute on a building contract as to whether statutory undertakers, who were laying mains to a building site, were engaged in executing work not forming part of the contract or were carrying out the work in pursuance of their statutory obligations. The decision was that they were involved in the former because they were carrying out the work not because statute obliged them to do so but because they had contracted with the employer to do so.

Not surprisingly since clause 31 specifically refers to statutory undertakers, clause 31(2) is often taken as the starting point for claims for delays caused by statutory undertakers. More often than not the dispute is on what could reasonably have been foreseen at tender. That will depend on what information the contractor was given at the time of tender on the estimated duration of the statutory undertaker's works and the extent of the facilities to be provided. If the engineer gave duration times which were inadequate and nothing was specified on facilities, the contractor starts with a good case.

However, clause 31(2) is not the only or even the best provision for the contractor's recovery for statutory undertakers' delays. Under clause 42(2) for example the employer is required to give possession of the site as necessary to enable the contractor to proceed with the works with due despatch, and many delays caused by statutory undertakers can be seen as depriving the contractor of his rights of possession under that clause. Any failure of possession which involves the contractor in delay or extra cost entitles the contractor to an extension of time and recovery of cost without consideration of 'that reasonably to be foreseen' as in clause 31.

13.8 Fossils

Clause 32 deals with fossils, coins, articles of value, antiquities or other things of interest found on the site. It has remained unchanged through successive editions of ICE Conditions.

The clause has two functions:

- to deal with ownership as between the contractor and the employer
- to deal with removal and disposal.

As to ownership, all things found on the site are deemed to be the absolute property of the employer.

As to discovery and disposal, the contractor is required to take precautions to prevent removal or damage and is to acquaint the engineer immediately upon discovery. The contractor is to carry out at the expense of the employer the engineer's orders for examination and disposal.

The Conditions do not give a defined meaning for 'expense' but in the context of this clause it probably means whatever basis of reimbursement is applicable under the Conditions in the particular circumstances. That might be under clause 12 (unforeseen conditions), clause 13 (instructions), clause 42 (lack of possession), or clause 51 (variations). It might even be day works under clause 52(5).

Clause 32 does not provide expressly for an extension of time in the event of delay but entitlement could be established under 'special circumstances' in clause 44 or through whichever of the above mentioned clauses (12, 13, 42 or 51) might be applicable.

13.9 Clearance of site

Clause 33 requires the contractor to clear away and remove from site 'on completion' all contractor's equipment, surplus materials, rubbish and temporary works and to leave the site in a clean and workmanlike condition to the satisfaction of the engineer.

'Completion' is not a defined term but for the purposes of this clause it is presumably meant to be after the issue of the certificate of substantial completion for the whole of the works or any significant part. It is unlikely to mean at the end of the defects correction period.

13.10 Returns of labour and equipment

Clause 35 of the Conditions details an administrative requirement for returns of labour and equipment. Clause 34 is not used.
Clause 35 states that the contractor shall:

- if required by the engineer
- deliver to the engineer or to the engineer's representative
- a return in such form and at such intervals as the engineer may prescribe
- showing in detail:
  - the number of classes of labour employed on the site
  - information on contractor's equipment as required.

The clause concludes by stating that the contractor shall require sub-contractors to observe the provisions of the clause.

'if required by the Engineer'

Clause 35 leaves it to the engineer to decide if plant and labour returns (as they are usually called) are required. Many contracts contain additional provisions to the effect that weekly, or sometimes daily, returns are mandatory. There is good reason for this. The information on such returns is vital in the analysis of delay and disruption claims when comparisons are being made between the contractor's actual progress and his planned progress – particularly so when the contractor has been required to submit, as is often the case, a fully resourced clause 14 programme.
So potentially important is information on plant and labour returns that it could in some circumstances amount to negligence for the engineer not to require that it be provided, and be provided on a regular basis.

Subcontractors

The reference in the clause to subcontractors can be taken to mean that if the contractor is required to submit plant and labour returns, the contractor shall operate a corresponding system with his subcontractors and then submit consolidated returns.

Chapter 14
Materials and workmanship

14.1 Introduction

In a change which is not seen as of any significance, the heading for clauses 36 to 40 of the Conditions becomes materials and workmanship rather than workmanship and materials as in previous editions. Disappointingly the change does not extend to the relocation of clause 40 (suspension of work) to another section of the Conditions. In its present location the clause gives the superficial impression that it is related to stoppages for materials and workmanship problems whereas in fact it is a clause of much wider application. This chapter covers clauses 36 to 39 only.

The only change of note between the Sixth and Seventh editions in clauses 36 to 39 is that the text of clause 36(3) (cost of tests) is in part redrafted and split into new clauses 36(3) and 36(4). This may have been done partly in the interests of improving clarity but the redrafting does seem to have produced a significant change in respect of liability for costs not intended by or provided for in the contract. Liability is now allocated on a default basis whereas under the Sixth edition it was on a results basis.

14.2 Materials and workmanship generally

The contractor's responsibilities for providing materials and workmanship to a particular standard derive from the express terms of the contract and from terms implied by common law or statute.

Common law

The courts will imply two independent warranties into contracts for work and materials:

- that materials will be of good and merchantable quality
- that materials will be reasonably fit for the purpose for which they are used.

These warranties apply unless the circumstances of the contract can be shown to exclude them.

The general rule was laid down in the case *of Myers* v. *Brent Cross Service Co* (1934) where Mr Justice du Parcq said:

'a person contracting to do work and supply materials warrants that the materials which he uses will be of good quality and reasonably fit for the purpose for which he is using them, unless the circumstances of the contract are such as to exclude any such warranty.'

Two later decisions of the House of Lords show how the general rule may be modified by particular circumstances. In *Young & Marten* v. *McManus Childs Ltd* (1969) a roofing subcontractor was held to be responsible for a batch of defective tiles although both the brand and manufacturer of the tiles (Somerset 13) were specified. It was held that the specification of a particular brand excluded the warranty that the tiles would be fit for purpose but the circumstances of the case were not such as to exclude the warranty that the tiles would be of merchantable quality. Since the subcontractor had fixed tiles with latent defects he was in breach. Lord Pearce had this to say on the circumstances which would exclude a warranty:

'If it is known to both parties that the manufacturer gives no warranty to the contractor, that fact is a strong indication that no warranty is being given by the contractor. So, too, of course, if a contractor advises against a particular material. But the circumstances of contracts are so various that it must be a question of fact and degree whether the circumstances of a particular case suffice to exclude a warranty which the general rule implies.

In the present case the employer's choice of Somerset 13 tiles, which were manufactured by only one firm, is not in itself sufficient to exclude the warranty of quality.'

In *Gloucestershire County Council v. Richardson* (1969) concrete columns made by a nominated supplier were found to be defective after erection. It was held that the contractor was not responsible since the circumstances of the case indicated an intention to exclude from the main contract any implied terms that the columns would be of good quality or fit for their purpose. This is how Lord Wilberforce explained the matter:

'The situation thus created was one of a special and complex character, differing greatly from that which arose in *Young & Marten Ltd* v. *McManus Childs Ltd*. There the employer nominated a brand article to be supplied by the manufacturer with no limitation on the contractor's freedom to contract with the manufacturer as he thought fit. The contractor could, and it would be the expectation that he would, or at least it would be his responsibility if he did not, deal with the manufacturer on terms attracting the normal conditions or warranties as to quality or fitness.

But here, the design, materials, specification, quality and price were fixed between the employer and the sub-supplier without any reference to the contractor; and so far from being expected to secure conditions or warranties from the sub-supplier, he had imposed upon him special conditions which severely restricted the extent of his remedy. Moreover, as reference to the main contract shows, he had no right to object to the nominated supplier, though, by contrast, the contract does provide a right to object to a nominated sub-contractor if the latter does not agree to indemnify him against his liability under the contract.'

Statute

The common law rules are now partly codified in the Supply of Goods and Services Act 1982. This is titled as:

'An Act to amend the law with respect to the terms to be implied in certain contracts for the transfer of property in goods, in certain contracts for the hire of goods and in certain contracts for the supply of a service; and for connected purposes.'

Supply of Goods and Services Act 1982

Part I of the Act applies to the supply of goods and Part II of the Act applies to the supply of services. A construction contract with its composite provisions for goods and services attracts both parts of the Act.

Under Part I goods must conform with statutory obligations as to quality and fitness as implied by Sections 2 to 5 of the Act unless these obligations are displaced by particular circumstances or excluded by agreement:

- Section 2 covers implied terms about title
- Section 3 covers implied terms where transfer is by description.
  In such a case there is an implied condition that the goods will correspond with the description.
- Section 4 covers implied terms about quality or fitness.
  Where goods are supplied in the course of business there is an implied condition that they are of merchantable quality. Where the intended purpose of the goods is made known, there is a further implied term that they are reasonably fit for that purpose.
- Section 5 covers implied terms where transfer is by sample.
  Where goods are supplied by reference to a sample there are implied conditions that the bulk will correspond with the sample in quality and that the goods will be free from defects rendering them unmerchantable which would not be apparent from examination of the sample.

In Part II of the Act relating to the supply of services, Sections 13 to 16 deal with implied terms:

- Section 13 – Implied term about care and skill
  In a contract for the supply of a service where the supplier is acting in the course of a business, there is an implied term that the supplier will carry out the service with reasonable care and skill.

- Section 14 – Implied term about performance
  Where the time for the service to be carried out is not fixed by the contract, left to be fixed in a manner agreed by the contract or determined by the course of dealing between the parties, there is an implied term that the supplier will carry out the service within a reasonable time.
- Section 15 – Implied term about consideration
  Where the consideration for the service is not determined by the contract, left to be determined in a manner agreed by the contract or determined by the course of dealing between the parties, there is an implied term that the party contracting with the supplier will pay a reasonable charge.
- Section 16 – Exclusion of implied terms
  (1) Obligations under the Act may be negotiated or varied by express agreement, or by the course of dealing between the parties, or by such usage as binds both parties.
  (2) An express term does not negate a term implied by the Act unless consistent with it
  (3) Nothing in the Act prejudices any rule of law which imposes on the supplier a duty stricter than that imposed by Section 13 or 14 above.

It is worth noting that although Section 13 refers only to care and skill, any common law rule imposing the stricter obligation of fitness for purpose would remain effective by virtue of Section 16.

Specified materials

The cases of *Young & Marten* v. *McManus Childs* and *Gloucestershire County Council* v. *Richardson* illustrate some of the difficulties which can arise from the specification of particular materials. The more recent case of *Rotherham Metropolitan Borough Council* v. *Frank Haslam Milan & Co Ltd and M. J. Gleeson (Northern) Ltd* (1996) shows that a similar problem can occur with material specified by type rather than by name. In the *Rotherham* case Haslam was engaged to do advance site preparation work. This involved constructing foundations and supplying and placing imported hardcore to the underside of ground floor slab construction. The bills of quantities defined 'hardcore' as including 'slag'. Haslam used steel slag unaware that it was liable to expand in confined conditions. After the super-structure works had been completed by Gleeson the steel slag expanded, causing heaving and cracking in the buildings. Rotherham sued for breach of implied terms that the slag used should have been of merchantable quality and fit for purpose as fill material. During the trial it emerged that although it was known to specialist organisations such as the Building Research Establishment that steel slag was not a safe product for use as fill in confined conditions, that was not known to the parties or pro-fessionals involved in the project.

The judge at first instance found Haslam to be in breach but the Court of Appeal in a judgment of considerable importance in reviewing the law reversed the decision, finding that Rotherham did not rely on Haslam's skill in deciding the type of hardcore/slag to be used and holding (amongst other things):

- A term of fitness for purpose will not be implied without good reason; the implied term must be founded on the presumed intention of the parties and on reason. In this case the steel slag supplied arose from and was within the wording of the employer's specification, and the employer had the means of knowledge that the steel slag was not inert, and therefore it would be unreasonable and unjust to impose a term of fitness for purpose on the contractor.
- However, the warranty of merchantability was satisfied if the material was fit for some of the purposes within the description under which it was sold and saleable under that description without abatement of price. It did not have to be fit all purposes for which materials under that description were used. Here the steel slag was perfectly good steel slag for use as hardcore in road building or any other situation where it was not confined, and could have been sold as such without abatement of price; therefore it was merchantable.

Options and alternatives

To give the contractor choice, specifications sometimes name suppliers either with an option or with the phrase 'or other approved'. The two expressions give different legal rights.

In *Leedsford Ltd* v. *Bradford City Council* (1956) it was held that the conduct of an architect in rejecting alternatives put forward by the contractor under an item for 'Empire Stone Company or other approved' was not breach of contract by the employer. There was an absolute obligation on the contractor to supply Empire Stone unless the employer, through the architect, approved some other stone. The employer was neither obliged to approve some other nor give reasons for rejection of the alternatives.

In *Crosby & Sons Ltd* v. *Portland Urban District Council* (1967) however, it was held that the contractor was entitled to a variation where the engineer instructed that pipes should be 'Staveley' thereby depriving the contractor of the choice in the bill item which stated that pipes were to be 'Stanton or Staveley'.

14.3 Quality of materials and workmanship

Clause 36(1) requires firstly that all materials and workmanship shall be of the kinds described in the contract and in accordance with the engineer's instructions. It then stipulates that materials and workmanship shall be subject to such tests as the engineer may direct, on or off the site, and that the contractor shall provide:

- such assistance, instruments, labour and materials etc. as normally required for examining, measuring and testing, and
- samples of materials for testing before incorporation into the works as selected and required by the engineer.

The phrase 'and in accordance with the Engineer's instructions' is probably intended to be no more than a reminder that the engineer has the power to vary the works and, in certain situations, such as impossibility, an obligation to do so by clauses 13 and 51. It is doubted if clause 36(1) confers an additional free-standing power.

The engineer's power to order off-site testing is stated to be at the place of manufacture or fabrication or such other places 'as may be specified in the contract'. This can be taken to include for named testing laboratories or the engineer's or employer's premises. It may not be wide enough to cover all possible locations – for instance, off-site borrow pits or other sources of materials.

The requirement for the contractor to provide assistance, instruments, labour etc. for measuring and testing is qualified in clause 36(1) by the words 'as are normally required'. By itself, this could lead to dispute as to what is 'normally required' but in many contracts the specification and the bills detail what is required, leaving additional requirements to be dealt with by way of variations.

Cost of samples

Clause 36(2) provides that all samples shall be supplied by the contractor at his own cost if:

- the supply is clearly intended by the contract, or
- the supply is provided for in the contract.

Otherwise, the cost of samples is to be met by the employer.

Note the distinction between 'Contract' as used in clause 36(2) and 'Contract . . . and Engineer's instructions' in clause 36(1). It is clear that if the engineer orders samples beyond those required by the specification, bills of quantities or drawings, they are not in a narrow sense 'in the Contract' and must be paid for by the employer.

This is not always recognised by engineers who sometimes assume that the contractor is under a general duty to supply samples of all materials to prove that they meet the specification.

Costs of tests

Clauses 36(3) and 36(4) detail how the costs of testing as between the contractor and the employer are to fall. Clause 36(3) deals with the contractor's liabilities; clause 36(4) with the employer's liabilities. In the Sixth edition both sets of liabilities were given in a single clause, clause 36(3), but the wording was difficult to follow. It is much improved in the Seventh edition.

Clause 36(3) provides that the contractor shall bear the costs of tests which are:

- clearly intended by the contract, or
- provided for in the contract, or
- for load tests, particularised in the specification or bill of quantities in sufficient detail to enable the contractor to have allowed for the cost in his tender.

Clause 36(4) provides that in all other cases the cost of making any test shall be borne by the employer unless the need for such test results from the contractor's default or from failure on the part of the contractor to observe and perform his obligations under the contract.

To the extent that the contractor will normally allow in his tender for tests clearly intended by or provided for in the contract it might be said that it is the employer not the contractor who bears the costs of such tests and that the only circumstances where the contractor is truly to bear the costs of tests is where he has failed to make due provision in his tender or, as a result of some defect or failure, he is instructed to make additional tests.

Under the Sixth edition the costs of additional tests fell on the contractor if the tests failed, but on the employer if the tests passed. One consequence of that was that if additional testing was considered necessary by the engineer on grounds of prudence having regard to some identified defect in one part of the works, the employer was liable to pay for the additional testing unless further defects were found. The Seventh edition deals with this by changing the criteria for the employer's liability from the results of tests to the need for tests. The employer is not liable for the costs of tests, the need for which results from the contractor's default, whether the tests pass or fail. There is, however, another side to this. Where the need for additional testing does not arise from the contractor's default or failure, the contractor is entitled to recover his costs whether the tests pass or fail.

14.4 Access to site

Clause 37 provides:

- the engineer and any person authorised by him shall have access at all times to the works and the site and to all places where work is being prepared or materials etc. being manufactured.
- the contractor shall afford every facility and every assistance in obtaining such access or right to access.

The phrase 'The Engineer and any person authorised by him' differs from the phrase more commonly used in the Conditions – 'The Engineer or any person acting under him'. Presumably it is to ensure that rights of access are afforded to persons who the engineer considers best suited to the particular test – whether or not that person is on his staff or has delegated powers.

It can be implied that the requirement for the engineer to have access 'at all times' is subject to practical considerations such as safety, security and working hours. Similarly, it can be implied that the right to off-site access may be subject to restrictions for some manufacturing processes and to some manufacturing establishments.

The clause is silent on the costs which the contractor may incur in providing access. Where these are likely to be significant or when delay or disruption is likely to be caused, the contractor would be well advised to seek an instruction under clause 13 detailing the engineer's requirements. Recovery can then be sought under clause 13(3) for cost which could not have been foreseen at the time of tender. Alternatively where the engineer can foresee access requirements, such as the regular or full time provision of lifting platforms, these can be specified and included in the bill of quantities.

14.5 *Examination of work*

Clause 38 deals in 38(1) with the examination of work before covering up and in 38(2) with uncovering and making openings in work for inspection purposes.

Examination before covering up

The provisions of clause 38(1) can be summarised as follows:

- no work to be covered up without the consent of the engineer
- the contractor to afford full opportunity to the engineer to examine and measure
- the contractor to give due notice when work is ready for examination
- the engineer to attend without unreasonable delay, unless he considers it unnecessary and informs the contractor accordingly.

These apparently simple and practical rules are not without difficulty in their application. Firstly, there is the question of the importance or the volume of the work itself and then there is the question of timing. Both matters are related to the level of supervision available to the engineer and both have some bearing on what is unreasonable delay.

The contractor may claim that anything which impedes continuity of progress is unreasonable delay; the engineer may reply that he, or his staff, can only be in one place at a time and some measure of delay is inevitable and is envisaged by clause 38(1).

In practice common-sense usually prevails and the contractor and engineer establish a working relationship often based on a formal notice procedure with written notices from the contractor and return slips from the engineer. Clause 38 does not actually require written notices but on a large contract they are indispensable.

In the event of dispute, two questions arise:

- can the contractor proceed without waiting for the engineer?
- is the contractor entitled to an extension of time and recovery of cost for unreasonable delay?

Covering up without consent

If the contractor does put work out of view without the consent of the engineer that on the face of it is a breach of contract for which clause 38(2) provides a remedy in the power of the engineer to order uncovering. However, if the contractor has given notice to which the engineer has not responded, and this is accepted as a regular occurrence or pattern of behaviour on the particular contract, then it may be reasonable to imply consent to proceed from the engineer's conduct, and there is no breach by the contractor.

Recovery of cost for delay in attending

Note that strictly on the wording of clause 38(1) the contractor is prohibited from covering up without consent. It follows that he is obliged to stand and wait for the engineer to attend if notice has been given. It also follows that the contractor is entitled to his costs for unreasonable delay by the engineer in attending.

However, because there is no express provision for recovery of cost in the clause itself, the contractor may be obliged to rely on breach of contract – with reference, perhaps, to clause 2(1)(a) and the engineer's failure to carry out his duties.

Extension of time for delay in attending

In the Sixth edition the grounds for extension of time in clause 44 did not readily cover delays of the sort which can arise through the operation of clause 38(1). That left the employer vulnerable to the contention that time was 'at large' as a result of any delay caused by the engineer's delay in attending. In the Seventh edition the new grounds for extension in clause 44(1)(e) (delay/default by the employer) may have resolved the difficulty – always assuming that failure by the engineer to perform his duties can be treated as a default for which the employer is responsible.

Uncovering and making openings

Clause 38(2) requires firstly that the contractor shall, as the engineer may so direct, uncover or make openings in any part of the works and shall thereafter reinstate to the satisfaction of the engineer. The clause then provides that:

- if any parts have been covered up or put out of view after compliance with the requirements in clause 38(1) for examination before covering up, and
- are found to have been carried out in accordance with the contract, then
- the costs of uncovering, making openings and reinstatement shall be borne by the employer, but
- in any other case the costs shall be borne by the contractor.

That is to say, the contractor is only entitled to payment where work is found to be to the required standard and the engineer was invited to inspect before covering up.

To be certain of payment the contractor needs to show that he has been given consent to cover up and the work is satisfactory. Neither on its own is sufficient. However, it is suggested that the contractor has a good case for payment if he can show that he gave notice, allowed a reasonable time before covering up and the work was satisfactory.

Providing the contractor can prove his entitlement to payment he should have no difficulty obtaining an extension of time for any delay caused by the uncovering and reinstatement. This will follow the engineer's instruction to uncover, which can be attributed to clause 13.

14.6 *Removal of unsatisfactory work and materials*

Clause 39 deals through three sub-clauses with the important subject of removal of unsatisfactory work and materials:

- 39(1) sets out the powers of the engineer
- 39(2) states the consequences of default by the contractor
- 39(3) states that failure to disapprove does not prejudice action to require removal.

Removal

Under clause 39(1) the engineer has power during the progress of the works to instruct in writing:

- removal from the site of any materials which in the opinion of the engineer are not in accordance with the contract and substitution with materials which are in accordance, and

- removal and replacement of any work which in respect of materials, workmanship or design by the contractor or for which he is responsible is not in the opinion of the engineer in accordance with the contract.

These express conditions are necessary because it is doubtful if there is an implied term which empowers the engineer to take similar action by intruding into the contractor's performance prior to completion.

It has been suggested by some commentators on ICE Conditions that to be of full practical effect the provisions should go beyond their negative approach and should empower the engineer, with the employer's consent, to accept sub-standard work albeit at a reduced price, as does the New Engineering Contract and certain standard building forms.

'during the progress of the Works'

This phrase suggests that the engineer's powers under clause 39(1) are not exercisable during the defects correction period. At that stage the provisions of clauses 49 and 50 on outstanding works and defects become applicable.

'in the opinion of the Engineer'

Instructions to remove work and materials are frequently contentious. The question is what does the contractor do if he disagrees with the opinion of the engineer? Does he follow the instructions as he is required to do by clause 13 and seek to recover his costs by a claim, adjudication or arbitration or has he other alternatives?

Firstly the contractor should consider if it is the engineer himself or the engineer's representative with delegated powers who has given the instruction. If it is the latter the contractor can refer the matter to the engineer under clause 66(2). If no change is forthcoming the contractor has little option but to comply first and claim later. Any alternative course of action risks determination under clause 65 or direct action by the employer under clause 39(2).

'notwithstanding any previous test – or interim payment'

The engineer's power to order removal and reinstatement is stated in clause 39(1)(c) to be notwithstanding any previous test or interim payment.

This may seem hard on the contractor but the obligation is to construct the works in accordance with the contract; passing tests is incidental. If work is found not to be up to standard it is the standard and not the previous test which governs. As to interim payment, clause 60(8) gives the engineer express power to correct certificates.

'design by the Contractor'

The power of the engineer to order removal of work under clause 39(1)(c)(ii) which in his opinion is not in accordance with the contract in respect of 'design by the Contractor or for which he is responsible' is curiously worded. It treats equally 'design by the Contractor' and 'design for which he is responsible'. It might be expected that design by the contractor for which he is not responsible is design for which the engineer is responsible – in which case any requirement for the removal of work would be treated as a variation not as a contractor's default. Perhaps the 'or' in the clause should be 'and'.

Default in compliance

Clause 39(2) provides that if the contractor defaults in carrying out instructions of the engineer to remove unsatisfactory work and materials the employer shall be entitled to employ others to carry out the same and to recover the costs from the contractor.

Note also that under clause 65(1)(h), if the contractor fails to remove materials or pull down and replace work within 14 days of receiving an instruction from the engineer that the said materials and work have been condemned and rejected, the procedures for termination can be put into effect. This is clearly an extreme remedy and engineers/employers considering its application should look carefully at what may appear to be small differences in wording between clauses 39(2) and 65(1) but which might be held to be significant in legal terms.

Failure to disapprove

Clause 39(3) states that failure of the engineer or any person acting under him pursuant to clause 2 to disapprove any work or materials shall not prejudice the power of the engineer or any such person subsequently to take action under clause 39(1).

It is not unknown for contractors to complain that it is unfair that work which has once been approved can later be rejected. But this is not as unfair as it seems. The contractor's obligation is to construct the works in accordance with the contract and that is what the employer is entitled to receive. The engineer supervises on behalf of the employer, not the contractor, and he is not empowered to relieve the contractor of any of his obligations under the contract. If the engineer misses a fault or makes a mistake he is not only empowered to correct it, he has a duty to do so.

The contractor has no obvious redress against the effects of reversals of approvals which can be considerable in both cost and delay. He can refer the matter to the engineer under clause 66 but if that fails he must comply with instructions and then claim if he sees fit and has adequate grounds.

There have been instances where contractors have catalogued a series of late rejections and reversals of approvals on a particular contract and have claimed against the employer on the general ground that unreasonable behaviour by the engineer is a breach of contract. The difficulty is proving unreasonable behaviour but if it can be sustained the claim cannot lightly be dismissed.

'any person acting under him'

The wording of clause 39(3) is unusual in its reference to both the engineer and to 'any person acting under him pursuant to clause 2'. This suggests that persons other than the engineer have powers under clause 39. But that is clearly not so. The clause empowers only the engineer to act.

The engineer may of course delegate his powers under clause 39 as he can with most of the other clauses of the contract. But throughout the contract persons acting with delegated powers still come under the reference of the engineer.

What then is to be made of the reference to this other 'person'? Perhaps it is intended that the powers of delegation operate slightly differently under this clause from elsewhere. Thus if the resident engineer has delegated powers and fails to reject faulty work then the engineer can, without reversing that delegation, himself reject such work.

Chapter 15
Suspension of work

15.1 Introduction

The Seventh edition of ICE Conditions follows the format of earlier editions by including under the general heading 'Materials and Workmanship' clause 40 which carries the marginal note 'Suspension of work'.

But for the statement in clause 1(3) that headings are not to be taken into consideration in interpretation of the contract it might be assumed that clause 40 is concerned only with suspensions relating to materials and workmanship. However, on its wording, the clause is not so restricted and it is clearly intended to confer a general power on the engineer to order a suspension of work whenever the engineer considers it necessary and for reasons which go beyond concerns on materials and workmanship. Because of this, clause 40 would be better placed with clauses under the heading 'Commencement Time and Delays'.

Engineer's power to suspend

Clause 40 is the only clause of the Conditions which expressly provides for suspension of work on the order of the engineer. There is power under clause 12(4)(d) for the engineer to order a suspension in respect of problems arising from unforeseen physical conditions but that is stated in the clause to be a suspension under clause 40.

Instructions given by the engineer under clause 13 could arguably include orders to suspend but engineers should be cautious of giving such instructions since the contractor's right to recover costs arising from a suspension is not necessarily the same under clause 13 as under clause 40. Under clause 13 the basis of recovery is cost which could not reasonably have been foreseen at the time of tender; under clause 40 the basis is reasonable cost incurred – unless the suspension is ordered for one of the reasons specified in the clause.

Reasons for suspension

Clause 40 is not concerned with reasons for suspension except in connection with the contractor's right to payment of extra cost and extension of time for completion. The clause apparently gives the engineer discretionary power to suspend on whatever grounds he considers necessary. And the wording may be wide enough to allow the engineer to act on the instructions of the employer as well as on his own professional judgement.

Reasons for suspending progress will generally fall under one of the following broad headings:

- change of mind by the employer
- possession and access problems
- unexpected restrictions imposed on the works
- pre-planned closures
- unforeseen conditions encountered
- adverse weather
- safety
- unsatisfactory performance by the contractor.

Temporary lack of funds by the employer might also be added to the list, although there is a case for saying this may be an abuse of clause 40. From the location of the clause in the Conditions and its wording it appears to be intended principally for practical matters relating to when and how the works are constructed and not to financial matters.

15.2 *Contractor's right to suspend*

The question of whether the contractor has a right to suspend the works of his own volition needs to be considered in three parts:

- exercise of rights for practical or commercial reasons
- exercise of contractual rights in the event of employer's breach
- exercise of statutory rights in the event of employer's breach.

There are no express rights in the Seventh edition Conditions covering any of the above but there is a hint in clause 65(1)(g) that the contractor may have some rights to suspend. That clause, which is concerned with contractor's defaults uses the phrase 'has suspended the progress of the Works without due cause'. This seems to suggest that the contractor may, with due cause, suspend the works without being in default. However, it should be said that on its full wording – 'has suspended the progress of the Works without due cause for 14 days after receiving from the Engineer written notice to proceed' – the clause may only be intended to apply to continuing suspension after withdrawal of an ordered suspension.

But whatever the intention or application of clause 65(1)(g) and notwithstanding the absence of express rights, a contractor under ICE Seventh edition Conditions is not wholly devoid of any rights.

Suspension for practical or commercial reasons

By clause 41(2) of the Conditions the contractor is obliged to proceed with the works with due expedition and without delay. By clause 43 the contractor is required to complete the works within the time stipulated in the contract. Providing the contractor can avoid breach of these obligations he will generally have the freedom to plan and progress the works as he sees fit. If it makes good sense for practical or commercial reasons to suspend the progress of the works at certain times, such as terminating earthworks for the winter, both the contractor and the employer may be the beneficiaries.

Sometimes the contractor may have foreseen in the production of his clause 14 programme the need for certain periods of suspension; sometimes they may be imposed by the specification. In either case that may be sufficient to dispose of any suggestion that the contractor is in default by suspending work. Where the suspension is not so planned the contractor may have to show that the overall effect of a suspension is to improve prospects for timely completion or that there is sufficient time in hand, by way of float or advanced progress, that timely completion will not be jeopardised.

The risk for the contractor in suspending the progress of the works of his own volition is that his action might be taken as default under clause 65 of the Conditions, giving rise to the possibility of termination. For further comment on clause 65 see Chapter 27 but note that in *Hill* v. *Camden* (1980) it was held on the facts that ceasing work was neither abandonment nor repudiation; and in *Greater London Council* v. *Cleveland Bridge & Engineering Co* (1986) the Court of Appeal upheld the point that the contractor should be free to programme his work as he felt fit provided he planned to finish on time.

Contractual right to suspend

The Conditions do not provide any contractual rights for suspension by the contractor in the event of the employer's breach of contract by non-payment or other default. The only remedy given by clause 64 (employer's default) is the right of the contractor to terminate his own employment under the contract.

Prior to the implementation of the Housing Grants, Construction and Regeneration Act 1996 this was seen as something of an omission in traditional ICE Conditions, particularly in respect of non-payment by the employer of amounts certified by the engineer. The ICE Design and Construct Conditions of Contract published in 1992 rectified the omission in clause 64 by giving the contractor the right to suspend work when payment was 21 days or more late providing that 7 days' notice of intention to suspend was given. A similar clause might well have been included in the Seventh edition had the 1996 Act not made its need largely redundant.

Legal right to suspend

The Housing Grants, Construction and Regeneration Act 1996 has significantly changed the law on suspension by the contractor in response to non-payment by the employer so far as construction contracts in the UK are concerned.

The Act provides at Section 112, a statutory right to suspend performance for non-payment. Section 112 reads:

'(1)   Where a sum due under a construction contract is not paid in full by the final date for payment and no effective notice to withhold payment has been given, the person to whom the sum is due has the right (without prejudice to any other right or remedy) to suspend performance of his obligations under the contract to the party by whom payment ought to have been made ("the party in default").

(2)   The right may not be exercised without first giving to the party in default at least seven days notice of intention to suspend performance, stating the ground or grounds on which it is intended to suspend performance.

(3)   The right to suspend performance ceases when the party in default makes payment in full of the amount due.

(4)   Any period during which performance is suspended in pursuance of the right conferred by this section shall be disregarded in computing for the purposes of any contractual time limit the time taken, by the party exercising the right or by a third party, to complete any work directly or indirectly affected by the exercise of the right.

Where the contractual time limit is set by reference to a date rather than a period, the date shall be adjusted accordingly.'

It should, perhaps be explained that the 'final date for payment' mentioned in Section 112(1) is not the date for final payment but is the final date for payment of any amount due – and, in the Conditions, that is set in clause 60 as 28 days after certification.

The position at common law has for some years been clearly established as being that the contractor has no implied right to suspend for non-payment, see for example the cases of *Canterbury Pipelines Ltd* v. *Christchurch Drainage Board* (1979) and *Lubenham Fidelities Ltd* v. *South Pembroke District Council* (1986). Consequently contractors should be careful to ensure in contemplating suspension for non-payment that they are either operating in accordance with contractual provisions (if any) or statutory provisions (such as the 1996 Act).

15.3 Suspension under clause 40

Clause 40 of the Conditions is concerned only with suspensions ordered in writing by the engineer. The suspensions may apply to the whole or any part of the works and the engineer is empowered to order such suspensions as he may consider necessary.

The only change between clause 40 in the Sixth edition and clause 40 in the Seventh edition is that the wording of the contractor's entitlement to recovery of extra cost and extension of time for completion occasioned by suspensions (other than those for which such recovery is specifically excluded) is tidied up. In the process the right to recover 'extra cost' is changed to recovery of 'such extra cost as may be reasonable'; and, perhaps inadvertently, the grounds for extension of time are reduced.

Contractor's obligation to suspend

Clause 40(1) commences by requiring that the contractor shall:

- on the written order of the engineer
- suspend the progress of the whole or any part of the works
- for such time or times
- as the engineer may consider necessary
- and during such suspension shall protect and secure the work
- so far as is necessary in the opinion of the engineer.

On the wording of the clause it is not open to the contractor to challenge an order to suspend. However, through the dissatisfaction procedures in clause 66 or, in the case of a suspension ordered by an 'assistant' using delegated powers, through the procedure in clause 2(5)(c), the contractor can ultimately contest the validity of a suspension order.

Failure to suspend on receipt of a written order is not specifically listed as a contractor's default in clause 65 – although suspending progress without due cause after notice to proceed is a specified default under clause 65(1)(g). If failure to suspend when so ordered is a default within the scope of clause 65 it would seem to fall within clause 65(1)(j) – 'persistently or fundamentally in breach of his obligations under the Contract'.

'the written order of the Engineer'

For good project management the engineer should ensure that any order given under clause 40 is clearly recognisable as such and should avoid giving instructions under clause 13 which could lead to the contractor claiming receipt of a constructive order to suspend. Additionally, although clause 40 does not expressly require it, the engineer should state the reason for, or the purpose of, the suspension order.

Until the reason is established the contractor cannot be certain whether the suspension is at his own cost and time or is on a recoverable basis and this is such an unsatisfactory situation that it may justify the contractor refusing to comply with the order until the reason is given.

Not infrequently situations arise where the contractor is obliged to suspend progress, usually of parts of the works, for compliance with specification requirements and the like. Normally these will not require a further written order under clause 40, and clause 40(1)(a) expressly excludes such matters from recovery of extra cost and time. Problems can arise, however, as to whether the contract is clear on whether suspensions of work are stipulated or necessary requirements and it may be possible for the contractor to gain the benefits of clause 40 in some circumstances without a specific written order to suspend. For instance, if the contractor could show that compliance with a revised drawing or specification necessitated suspension, the issue of the relevant drawing or specification might, by itself, be the equivalent of a written order to suspend.

In the case of *Harrison & Co (Leeds) Ltd* v. *Leeds City Council* (1980) it was held that an architect's instruction under a building contract for the main contractor to use a particular nominated subcontractor who had timing conditions attached to his offer amounted, on the facts, to an instruction to postpone parts of the main contract works.

'for such time or times'

It appears from the wording of clause 40(1) – 'The Contractor shall . . . suspend . . . for such time or times . . . as the Engineer may consider necessary' – that it is for the engineer to state when a suspension should start and when it should end. However, in the case of *Crosby & Sons Ltd* v. *Portland UDC* (1967), best known for recognition of the global claim, it was held, under an ICE Fourth edition contract, that no term was to be implied that the engineer should lift a suspension order given on grounds for which there was no right to cost and time recovery once the conditions ceased to apply and accordingly no claim for damages for breach of such an obligation could be sustained.

Clearly the engineer has a duty to the employer to ensure that a suspension which is costing the employer time and money is lifted as soon as possible but the *Crosby* case suggests that the engineer has no comparable duty to the contractor. It would seem, and not without good reason, that where the contractor carries responsibility for the suspension, it is for the contractor to instigate its lifting.

'properly protect and secure the work'

In clause 40(1) these words are followed by 'so far as is necessary in the opinion of the Engineer'.

It might be expected that the burden of deciding what was necessary to properly protect and secure the works during suspension would fall wholly on the contractor and there are potential pitfalls for the employer if the engineer becomes involved. The clause opens the way to the contractor seeking instructions on what to do and then seeking recovery through clause 13 of costs not recoverable through clause 40.

Engineers should be careful not to convert their opinions to instructions and should remain mindful that under clause 20 the contractor carries full responsibility for care of the works.

15.4 Payment for suspension

The second part of clause 40(1) provides that if compliance with the engineer's instructions involves the contractor in delay or extra cost, the engineer shall take such delay into account in determining any extension of time due and the contractor shall be paid, subject to clause 53 (which relates to notice) the amount of such extra cost as may be reasonable, except to the extent the suspension is:

- provided for in the contract
- necessary by reason of weather conditions
- necessary because of contractor's default
- necessary for proper execution of the work
- necessary for safety.

Profit is to be added to cost in respect of any additional permanent or temporary work.

Note that the entitlement to extra cost and time is related to 'compliance with the Engineer's instructions under this clause'. By its reference to 'instructions' this is potentially wider in scope than compliance with the 'order' to suspend.

Shared responsibility

Clause 40(1)(c) recognises that a suspension for proper execution or safety could arise from some default of the engineer or employer or could be an excepted risk under clause 20(2). Of these, use by the employer is the most obvious. The exception to the contractor's right to payment is modified accordingly in such circumstances.

'provided for in the Contract'

The exception to the contractor's right to payment in clause 40(1)(a) is expressed as 'otherwise provided for in the Contract'.

It is suggested that this relates only to specific close-down periods which are mentioned in the contract documents. It is not thought to be a reference to the provisions for suspension in clause 12 (unforeseen physical conditions).

Weather conditions

Suspension for weather conditions is sometimes given at the contractor's request – usually in the expectation by the contractor that at the very least it will attract an extension of time. However, see the comment later in this chapter under the heading 'extension of time' on the revised position in the Seventh edition on this.

The position when the engineer suspends for weather conditions against the contractor's wishes is potentially contentious. The engineer will probably have in mind that temperatures are too low or too high for compliance with specifications or that working in wet weather or snow will render materials unsuitable. The contractor will often contend that he proceeds at his own risk and that the engineer has no right to interfere. However, by clause 40(1) the engineer has a right to interfere and it is suggested he has a duty to the employer to prevent needless destruction of materials where the costs of replacement fall on the employer or there is the prospect of the site being rendered unsafe by continued working.

Contractor's default

The exception to entitlement to payment and extension of time for contractor's default is wide enough to cover both practical and administrative defaults. Examples of the latter could include failure to provide a clause 14 programme; failure to provide permanent works designs under clause 7(6); or failure to provide methods of construction under clause 14(6) if any of these would produce a situation where it was unsafe, impractical or seriously unsatisfactory for work to continue.

Safety

Notwithstanding the contractor's responsibility for safety as expressed in clauses 8(3) and 19(1) of the Conditions, the engineer has a statutory duty to act when he sees danger in work in which he is involved. Clause 40 gives the engineer contractual power to act and he could be negligent in not using that power.

Extension of time

Under the Sixth edition the contractor's entitlements to extra cost and/or extension of time for suspensions were expressed in separate paragraphs and the exceptions in clauses 40(1)(a), (b) and (c) applied only to cost. For extension of time the exceptions were stated separately as 'except when such suspension is otherwise provided for in the Contract or is necessary by reason of some default on the part of the Contractor'.

Under the Seventh edition the exceptions in clauses 40(1)(a), (b) and (c) apply to both extra cost and extensions of time.

This has the odd and probably unintended effect, that so far as clause 40 is concerned no extension of time is due for a suspension necessary by reason of weather conditions. The contractor may still have the right to claim for exceptional adverse weather conditions under clause 44(1)(d) but this is a poor substitute for the right which previously existed in clause 40.

15.5 *Suspension lasting longer than 3 months*

Clause 40(2) deals with an ordered suspension where the engineer has not given permission for work to resume within 3 months of the date of the suspension.

In such a situation, unless the suspension is either provided for in the contract or continues to be necessary by reason of the contractor's default then:

- the contractor may serve written notice on the engineer requiring permission to proceed within 28 days, and
- if the engineer does grant such permission
- the contractor may by further written notice:
  - treat suspension of part of the work as an omission variation under clause 51, or
  - treat suspension of the whole of the works as abandonment of the contract by the employer.

Omission variation

In the event of the contractor giving notice that he intended to treat a prolonged suspension of part of the works as an omission variation, that would relieve the contractor of the obligation of constructing and completing the relevant part of the works.

It is not exactly clear how the valuation of such an omission variation fits into the Seventh edition scheme for valuation of ordered variations in clause 52. It is questionable if the rules of clause 52 apply to a deemed variation (as distinct from an ordered variation) and it is difficult to see how the contractor can be bound by the quotation obligations of clause 52 in circumstances where it is the contractor rather than the engineer who is entitled to state the operation of the variation.

There is no obvious barrier to the engineer reinstating work omitted by prolonged suspension by the issue of a later addition variation but such reinstated work would then arguably fall to be valued as a variation rather than as ordinary measured work.

Abandonment

The entitlement of the contractor to treat prolonged suspension of the whole of the works as abandonment of the contract by the employer is not carried through into clause 64 (default of the employer) as one of the grounds allowing the contractor to terminate his employment under the contract. Presumably it is intended to stand on a different legal footing.

Given the legal complications which frequently follow any charge of abandonment, the parties and the engineer would be well advised to consult their lawyers before taking any precipitous steps.

Is there a 28 day rule?

One final aspect of clause 40(2) which deserves some attention is the wording of that part of the clause which reads 'If within the said 28 days the Engineer does not grant such permission the Contractor by a further written notice so served may (but is not bound to) elect to treat the suspension . . .'.

There are at least two points of uncertainty here. One, whether the contractor is obliged to make his election immediately at the end of the 28 day period or, whether he can do so at any time thereafter. The other is whether the contractor loses his rights of election if the engineer acts to lift the suspension after expiry of the 28 day period – whether before or after the contractor's further notice. Again, these are questions which, if they arise, are best left to lawyers.

Chapter 16
Possession of the site, commencement and rate of progress

16.1 Introduction

This chapter covers all the clauses which appear in the Conditions under the heading 'Commencement Time and Delays' except clause 44 on extension of time for completion which is considered in the next chapter with the associated subject of liquidated damages for late completion.

In this chapter, therefore, the clauses considered are:

- Clause 41 – works commencement date
- Clause 42 – possession of site and access
- Clause 43 – time for completion
- Clause 45 – night and Sunday work
- Clause 46 – rate of progress.

The only changes of significance in this batch of clauses between the Sixth and Seventh editions are changes to clause 41(1)(b) and to clause 42 (possession of the site). These are changes of potentially major implication and are discussed in Section 16.4 below.

16.2 Works commencement date

The 'Works Commencement Date' is defined in clause 1(1)(q) as – 'as defined in Clause 41(1)'.

Clause 41(1) states that the works commencement date shall be:

(a)   the date specified in the appendix, or, if no date is specified
(b)   a date between 14 and 28 days of the award of the contract (to be notified to the contractor in writing by the engineer) or
(c)   such other date as may be agreed between the parties.

Thus by one route or another, clause 41(1) fixes the works commencement date as a particular date for any contract. However, that does not explain what the works commencement date is in legal or contractual terms. That is to be found in clause 43. There it states that the time for completion is calculated from the works commencement date.

What is also clear from clause 43 and the format of the appendix to the form of tender is that there is only a single works commencement date for any particular contract notwithstanding that the works may be divided into parts and sections with different stipulated commencement dates.

Effect of works commencement date

By clause 41(2) the contractor is to start the works on or as reasonably practicable after the works commencement date. The effect therefore of the date is twofold:

- it fixes the date on which the contractor's obligations to proceed commence, and
- it fixes the date for completion of the works (or dates where there are stipulated sections) by reference to the allowable times for completion.

It is, of course, essential that in contracts such as ICE Conditions where times for completion are specified rather than dates for completion (as in most building contracts), there is a firm reference point such as the works commencement date for fixing the due date for completion.

Fixing the works commencement date

Application of the rules for fixing the works commencement date set out in clause 41(1) means that if no date is specified in the appendix to the form of tender then the contractor is entitled to a start date 28 days after the award of the contract although he may, but is not obliged to, agree to some other date. And likewise, if no date is specified, the engineer is entitled to insist that the contractor should commence 28 days after the award. Note that because provision for specifying the commencement date appears only in part 1 of the appendix, and not in part 2 which is filled in by the contractor, only the employer or engineer can specify the commencement date. In practice, however, there may be sound reasons for seeking the views of tenderers on the commencement date. Such consultation could result in the fixing of a date which although not in the appendix could be made contractually binding by pre-award exchange of letters.

In clause 41(1)(b) there is a sensible change in wording between the Sixth and Seventh editions in that the previous wording 'a date within 28 days of the award' is now 'a date between 14 and 28 days of the award'. This prevents the engineer from fixing the start immediately upon award and it allows the contractor a modest time to mobilise resources.

'Such other date as may be agreed'

Clause 41(1)(c) allows for the works commencement date to be 'such other date as may be agreed between the parties'.

It is suggested that this is to be read, particularly by reference to the word 'or' at the end of clause 41(1)(b), as meaning a date agreed other than a date within the 14 to 28 day period mentioned in clause 41(1)(b).

It is doubted if it is intended to support the practice, which is not uncommon, of the employer/engineer writing in the appendix under 'works commencement date (if known)' the phrase 'to be agreed'. The law does not favour agreements to agree because of the uncertainty which follows from failure to agree. Either one party accepts the will of the other or there is deadlock.

The purpose of the provision for agreement in clause 41(1)(c) is probably no more than to allow advancement or deferment of the 28 day date by genuine post-award agreement.

Note that under the wording of the clause the agreement is between the parties not between the engineer and the contractor.

Note also that there is no express requirement for any agreement so reached to be in writing. This is something of an omission and it is suggested that words such as 'and confirmed in writing by the Engineer' should be added at the end of clause 41(1)(c) to avoid the problem in *Kemp* v. *Rose* (1858) where the date for commencement was omitted from a written contract and the court declined, in the face of conflicting oral evidence, to set a date.

Late works commencement date

If clause 41(1) is operated properly there should be no such thing as a late or delayed works commencement date. If no date is specified in the appendix then, except by agreement, the latest date is 28 days after the award of the contract.

However, it is not unknown for dates to be set unilaterally which are well outside the 28 day period – sometimes because the construction drawings are not ready, sometimes because the employer cannot give immediate possession of the site. In such circumstances the contractor probably has an option: he can either accept the late date and possibly claim damages for breach; or he can insist that the 28 day rule fixes the date and claim costs and extra time under various applicable clauses of the contract for late information, late possession or the like.

In extreme cases of delay in commencement, such as in the old case of *Bush* v. *Whitehaven* (1888) where the delay moved a drainage contract from summer working to winter working, and the contractor was held to be entitled to recovery on a quantum merit basis, a contractor under ICE Seventh Conditions might well be able to argue for a variation within the scope of clause 51 (change in timing of construction) and thereby escape from the rigidity of the bill rates to a more appropriate method of valuation.

16.3 Start of works

Clause 41(2) requires the contractor to start the works on, or as soon as reasonably practicable after, the commencement date and thereafter to proceed with due expedition and without delay.

These are obligations the contractor must take seriously because, apart from any liability for liquidated damages for late completion, there are provisions in clause 65 for determination of the contractor's employment for failing without reasonable excuse to commence the works in accordance with clause 41 or failing to proceed with due diligence.

The 'Works' referred to in clause 41(2) are the 'Works' as defined in clause 1(1)(p), i.e. 'the Permanent Works together with the Temporary Works'. Although it is not expressly stated in the definitions of the permanent works and the temporary works that these include for design it is arguable that for timing purposes they should include for such design as is the contractor's obligation. Commencement of the works, therefore, may not necessarily mean commencement on site and where the contractor has taken on obligations and responsibilities in respect of the design of the permanent works there may well be a period of off-site activity prior to any start on site.

'reasonably practicable'

The question of what is a 'reasonably practicable' time in clause 41(2) is a matter of fact to be determined in the light of all circumstances. Factors to be taken into account could include:

- the notice given by the engineer for commencement
- the complexity of the works and the need for pre-planning
- delivery times on plant and materials
- the contractor's programme
- the time for completion
- the extent of contractor's design.

'due expedition'

Although failure by the contractor to proceed with due expedition and without delay is a breach of clause 41(2), it is not fully clear whether or not the employer has a remedy other than damages for late completion. Clause 65 includes the default of failing to proceed with due diligence as ground for determination but it is arguable that diligence relates to correctness and that it does not extend to expedition.

'in accordance with the Contract'

The closing phrase of clause 41(2) 'in accordance with the Contract' is sometimes taken as placing an obligation on the contractor to comply with his programme.

Whilst it is doubtful if there can be any such obligation in respect of the clause 14 programme the position may be different for a tender programme which is listed as a contract document.

The phrase may also relate to prescriptions on possession of the site and access given under clause 42 and it may link in with clause 46 which deals with the expedition of progress and accelerated completion. However, an alternative explanation is that the phrase is intended to place assessment of 'due expedition and without delay' in the context of the time allowed for completion in the contract. This would be consistent with provisions for damages which apply only to late completion.

It is possible, therefore, that the phrase 'in accordance with the Contract' does not add to the contractor's obligation to proceed with due diligence and without delay but it simply qualifies that obligation so that instead of being absolute the obligation is secondary to finishing on time.

16.4 Possession of the site and access

Clause 42 deals with possession and access in four parts:

- 42(1) – prescriptions on access and order of progress
- 42(2) – employer's obligation to give possession of the site
- 42(3) – contractor's entitlement for employer's failure to give possession
- 42(4) – access and facilities to be provided by the contractor.

Clause 42(1) is identical to the corresponding clause in the Sixth edition, as is clause 42(4).

Clause 42(3) is reworded but more by way of a tidying up exercise than any change of policy.

Clause 42(2) is significantly changed. The employer's primary obligation is now to give possession of the whole of the site on the works commencement date, subject to any prescriptions in clause 41(1)(a), whereas in the Sixth edition the obligation was to give possession of so much of the site as was required by the contractor.

Meaning of possession

Possession of the site under the contract does not mean that the contractor is literally the party in possession. The word 'possession' is used for convenience and it is not intended to confer on the contractor the full range of legal rights and liabilities which accompany possession by ownership.

Possession given to the contractor is more in the nature of a licence to occupy the site. Thus in *H. W. Neville (Sunblest) Ltd* v. *William Press & Sons Ltd* (1981) it was said:

'Although [the contract] uses the word "possession" what it really conferred on William Press was the licence to occupy the site up to the date of completion.'

And in *Surrey Heath Borough Council* v. *Lovell Construction* (1988) the following passage from *Hudson on Building Contracts* (tenth edition) was accepted as an accurate statement of the law:

> 'In the absence of express provisions to the contrary, the contractor in ordinary building contracts for the execution of the works upon the land of another has merely a licence to enter upon the land to carry out the work. Notwithstanding that contractually he may be entitled to a considerable degree of exclusive possession of the site for the purpose of carrying out the works, such a licence may be revoked by his employer at any time, and thereafter the contractor's right to re-enter upon the site of the works would be lost. The revocation, however, if not legally justified, will render the employer liable to the contractor for damages for breach of contract, but subject to this the contractor has no legally enforceable right to remain in possession of the site against the wishes of the employer.'

Prescriptions

Clause 42(1) states the contractor may prescribe:

(a) the extent of portions of the site which the contractor is to be given possession of from time to time
(b) the order in which such portions shall be made available to the contractor
(c) the availability and nature of the access to be provided by the employer
(d) the order in which the works shall be constructed.

There is no entry in the appendix to the form of tender for these prescriptions and in Seventh edition contracts they will usually be found in the drawings, specification or special conditions. But they will, of course, generally be prescriptions imposed by the employer rather than by the contractor.

Order of construction

There are clear dangers in fixing the order of construction by contractual requirements because any directions on how the contractor is to operate can rebound on the employer as claims. The employer starts with the basic obligation not to impede the contractor in the performance of his obligation to complete the works on time. But by prescriptions, or directions, the employer can end up with subsidiary obligations.

The case of *Yorkshire Water Authority* v. *Sir Alfred McAlpine and Son (Northern) Ltd* (1985) illustrates the point. In that case the contractor submitted with his tender, as instructed, a method statement showing he had taken note of certain specified phasing requirements providing for the construction of the works upstream. The formal contract agreement incorporated the method statement.

The contractor maintained that in the event it was impossible to work upstream and after some delay work proceeded downstream. The contractor then sought a variation order under clause 51(1). The court held:

- the incorporation of the method statement into the contract imposed an obligation on the contractor to follow it so far as it was legal or physically possible to do so
- the method statement, therefore, became a specified method of construction and the contractor was entitled to a variation order and payment accordingly.

See also the case of *Havant Borough Council* v. *South Coast Shipping Co Ltd* (1996) as a further warning to employers on stipulating how the works are to be carried out.

Obligations to give possession

Clause 42(2)(a) states that subject to clause 42(1)(a) the employer shall give the contractor possession of the whole of the site on the works commencement date together with such access as necessary to enable the contractor to commence and proceed. Clause 42(1)(a) states that the contract may prescribe the extent of portions of the site of which the contractor is to be given possession from time to time.

It is clear, therefore, that unless the contract contains prescriptions of the type permitted by clause 42(1)(a) the employer is under a firm obligation to give possession of the whole of the site at commencement. This is a significant change from the position in the Sixth edition where the employer's obligation under clause 42(2) was to give possession at commencement of only so much of the site as required to enable the contractor to commence and proceed.

In many civil engineering projects, pipelines, roadworks, land reclamation and similar, the contractor frequently occupies the site in stages taking over parts of the site when the works on other parts are completed. The Seventh edition Conditions can cope with this but only if the contract documents disclose that possession of the site will be staged.

Difficulties may arise when possession is to be in stages linked to the contractor's programme since the extent of the stages may not be known until after the contract is awarded. Clause 14(1)(a) requires the contractor's programme to be prepared having regard to the provisions of clause 42(1) but in some cases this could put the cart before the horse because the prescriptions in clause 42(1) could follow and not precede the programme. The problem did not arise in the Sixth edition where clause 42(2)(b) expressly linked possession of parts of the site to the contractor's clause 14 programme but in the Seventh edition clause 42(2)(b) refers only to access (not possession) and, in any event, drops all reference to the contractor's programme.

One way of overcoming the difficulty is that where possession of the site is to be given in stages linked to the contractor's programme and/or progress, then the prescriptions within the contract for clause 42(1)(a) should be expressed in terms relating to the programme and/or progress.

Failure to give possession

Clause 42(3) provides that if the contractor suffers delay and/or extra cost from failure by the employer to give possession (or access) as required by clause 42, the engineer shall take such delay into account in determining any extension of time due under clause 44 and, subject to clause 53 (notice provisions), the contractor shall be paid 'any cost to which he may be entitled' – with profit added in respect of any additional or temporary work.

The wording of clause 42(3) differs from that in the Sixth edition but not to any significant practical effect except that the unusual requirement in the Sixth edition – 'The Engineer shall notify the Contractor accordingly with a copy to the Employer' – no longer applies.

One unsatisfactory aspect of the wording does remain, however, and that is the expression of the contractor's financial entitlement as 'any extra cost to which he may be entitled'. Elsewhere in the Conditions phrases are used such as 'cost as may be reasonable' and 'costs which may reasonably have been incurred'. The clause 42(3) phrase leaves open the question as to what is the entitlement.

Obligations on access

The employer's obligations on access are stated in two parts:

- clause 42(2)(a) – employer to give such access as necessary to enable the contractor to commence and proceed
- clause 42(2)(b) – thereafter, the employer to give during the course of the works such further access 'in accordance with the Contract' as necessary to enable the contractor to proceed with due despatch.

Additionally the employer has the extra cost and extra time liabilities arising from clause 42(3) in the event of failure to give access as required under the contract – liabilities which were surprisingly missing from the Sixth edition, where clause 42(3) applied only to 'possession'.

The employer's obligation on access in clause 42(2)(a) is expressly made subject of clause 42(1) which states at 42(1)(c) – the contract may prescribe 'the availability and nature of the access which is to be provided by the Employer'. There is no similar express link between clause 42(2)(b) and clause 42(1) but if the contract does contain prescriptions on the employer's obligations they probably over-rule the apparent generality of the clause

42(2)(b) obligations. Indeed this may be the intention of the phrase 'further access in accordance with the Contract' which appears in clause 42(2)(b).

The contractor's obligations on access are stated in clause 42(4) which provides:

- the contractor shall bear all costs and charges for any access required additional to those provided by the employer, and
- the contractor shall provide at his own cost any additional facilities outside the site required for the purpose of the works.

The first of these obligations is consistent with the contractor's obligation under clause 11(2)(c) to satisfy himself as to means of access, but the second does not fit easily with the provision in clause 26(2) that the employer shall pay the rates and taxes on structures outside the site used for the purposes of the works. It must be that the cost of 'providing' as meant by clause 42(4) is not to be taken as paying rates and taxes.

Giving of access/provision of access

Amongst the problems potentially arising from clause 42 is whether any distinction is to be made between the requirements in parts of the clause for access to be 'given' and in other parts for access to be 'provided'.

In practical terms access can be 'given' without anything physical being 'provided'. There is nothing in ICE Conditions comparable to the express obligation in the IChemE model forms of contract for process plants which requires the purchaser (employer) to provide access from a convenient point on a road, railway or dock. However, where a civil engineering site does not have direct access to transport facilities the employer may wish to control, contribute to, or even provide the access. Explanation of what is meant by giving and providing access may in some contracts be desirable if not essential.

16.5 Time for completion

Clause 43 states the contractor's obligation to substantially complete the whole of the works and any sections within the times stated in the appendix to the form of tender or such extended times as allowed under clause 44 or as revised by agreement under clause 46(3), with all such times calculated from the works commencement date.

A similar obligation can be deduced from the form of tender where it is stated, 'We undertake to complete and deliver the whole of the Permanent Works comprised in the Contract within the time stated in the Appendix hereto'; but clause 43 is the more specific by the use of its phrase 'substantially completed'. Without that phrase the contractor's obligations would be subject to a more rigid test of completion.

Sectional completion

The obligation of completion in clause 43 extends to any section for which a particular time is stated in the appendix. The scheme is that there should be stated in the appendix either:

- a time for completion of the whole of the works, or
- times for various sections with the 'remainder' of the works forming a final section.

All times for completion of sections run from the single commencement date and are not intermediate periods.

Contractor's own times

The Conditions make provision for the practice of allowing the contractor to state his own times for completion. They do this by a two-part appendix to the form of tender. The employer can either insert required times for completion in part 1 or allow the contractor to insert his own times in part 2.

The advantages of giving the contractor freedom are firstly, tenders can be compared on time as well as price, and secondly, disputes on shortened programmes should largely be avoided.

Completion of phases (or parts)

Clause 43 places no obligation on the contractor to complete phases or parts of the works by times which may be specified in the specification, bills or drawings, unless those phases or parts are designated as sections and are properly included in the appendix.

Correspondingly, there is no obligation on the employer to take possession of a phase or part not so designated unless it has been occupied or put into use. By clause 48(2) the employer is however obliged to accept any section for which the engineer has issued a certificate of substantial completion.

Damages for late completion of phases

Because the Conditions do not oblige the contractor to complete phases or parts not designated as sections within specified times the provisions for liquidated damages do not apply to late completion of such phases or parts.

In *Bruno Zornow (Builders) Ltd* v. *Beechcroft Developments Ltd* (1990) a contract was negotiated for a housing development on the basis of a first tier tender which showed a detailed programme to complete in 16 months and second stage agreements for completion of the work in two overlapping phases. The architect calculated liquidated damages of £40,000 based on the stipulated rate of £200 per week per block from the date shown on the original works programme and the contractor sued for the return of this amount. It was held:

- the contract did not incorporate documents which specified dates for sectional completion but only phased provisions for the transfer of possession
- a claim for liquidated damages could only be made in respect of failure to meet specified completion dates and not failure to meet transfer of possession dates – which operated on a consent basis
- no term would be implied for any sectional dates for completion.

A similar situation arose in *Turner* v. *Mathind* (1986) where there was a clear requirement in the bills for phased completion but the sectional completion supplement was not used and the appendix contained only a rate for liquidated damages for late completion of the whole of the works. It was held that it was not appropriate that the employer should either pro-rata the stipulated damages to the number of phases or apply the stipulated rate to each phase.

Completion within the time allowed

Both clause 43 and the appendix to the form of tender state the contractor's obligation to complete 'within' the time allowed. The contractor is therefore entitled to finish early and, unless a special clause is introduced, as is sometimes the case, is entitled to programme to finish early.

16.6 Night and Sunday work

Clause 45 provides that none of the works shall be carried out during the night or on Sundays without the permission in writing of the engineer unless it is:

- subject to any provisions in the contract to the contrary
- unavoidable
- necessary for saving life or property
- necessary for the safety of the works
- work which it is customary to carry out outside normal working hours by rotary or by double shifts.

It is clear from the wording of the clause that it is intended to apply principally to construction activities on the site. On strict interpretation it may encompass design activities but any practical impact of this is likely to be slight.

'provisions to the contrary'

The proviso which opens clause 45 is probably intended to apply to pre-
planned night or Sunday work relating to road closures, rail track posses-
sions and the like – details of which are stipulated in the drawings, specifications or special conditions. It has no obvious application to other clauses
of the Conditions.

Work which is unavoidable

The wording of the clause is not absolutely clear with regard to work which
is unavoidable. Two interpretations are possible. One is that unavoidable
relates to any qualifying work; the other is that unavoidable attaches only to
the words relating to the saving of life or property or for safety of the works.
If unavoidable has the wider application the question then becomes who
decides what is, and what is not, unavoidable.

Steps to expedite progress

Note that under clause 46(2) if the engineer has served notice that the contractor shall take steps to expedite progress, permission to work at nights or
on Sundays shall not be unreasonably refused.

16.7 *Rate of progress and acceleration*

Clause 46 has two major parts: clause 46(1) concerning the contractor's rate
of progress and clause 46(3) which provides for accelerated completion by
agreement of the parties. The third part – clause 46(2) – is, as noted above,
simply concerned with permission for night or Sunday work.

Rate of progress

Clause 46(1) places a duty on the engineer to notify the contractor in writing
when at any time, in his opinion, the rate of progress is too slow to ensure
substantial completion by the due date. The contractor is then obliged to take
such steps as are necessary at his own cost and to which the engineer may
consent, so as to complete by the due date. The provisions apply equally to
sections and the whole of the works.

Clause 46(1) has remained largely unchanged through the Fifth and Sixth
editions to the Seventh edition despite concerns that, as drafted, it has little
or no teeth in pressurising the contractor to remedy any shortcomings in his
rate of progress but it does have the potential to trip up the engineer by
providing the contractor with opportunities for claims.

The main problem with the clause is that it is expressed in mandatory fashion – 'the Engineer shall notify the Contractor'. Strictly, therefore, the engineer is obliged to constantly monitor progress and to issue 'expedition' notices whenever there is the possibility of late completion. Failure to do so is a breach and although it might be difficult for the contractor to prove damages for such a breach, the contractor may be on better ground in drawing inferences from the engineer's failure to act under clause 46(1) – for example, that late completion is to be covered by an extension of time. If the engineer does act he has to be particularly careful of the opening words of the clause – 'If for any reason which does not entitle the Contractor to an extension of time'.

Should it later transpire or be recognised after the engineer has given notice to the contractor to expedite progress that an extension of time was due, the contractor may have grounds for claiming constructive acceleration or reimbursement through clause 13(3) of costs of complying with an instruction which could not have been foreseen at the time of the tender. The position is particularly sensitive for the engineer when there is already a disputed extension of time claim on the table. That claim may well be the subject of review in the engineer's final determination or under the dispute resolution procedures of the contract. But whatever the engineer does he should not issue any notice under clause 46(1) before all applications for extensions of time have been considered and the interim provisions of clause 44(3) have been operated.

Another problem with clause 46(1) is that there is no sanction if the contractor ignores the expedition notice – except in extreme cases where the default provisions of clause 65 might apply, i.e. failing to proceed with due diligence. Usually the contractor who is running late has a commercial choice between incurring extra production costs or paying liquidated damages for late completion.

One well used approach to the problems of clause 46(1) is to amend the clause so that the phrase 'the Engineer shall notify the Contractor' becomes 'the Engineer may notify the Contractor'. Use of the clause then becomes discretionary and the engineer can assess the value and the potential effects of serving an expedition notice. The engineer might, for example, restrict his use of the clause to serious delay situations where the possibility of having to use the termination provisions of clause 65 is in contemplation and there is need and value in having warning notices on record.

Permission to work at night or on Sundays

Clause 46(2) links back to clause 45 so that if the engineer has served a clause 46 notice he cannot then unreasonably refuse the contractor permission to work at night or on Sundays. The phrase 'on site' clarifies the need for permission.

Accelerated completion

Clause 46(3) has nothing to do with constructive acceleration as mentioned above. It relates only to 'requested' acceleration where the employer or the engineer requests the contractor to complete within a time less than that in the appendix to the form of tender or as extended.

The clause does no more than provide that if the contractor agrees to accelerate, any special terms and conditions of payment shall be agreed between the parties before action is taken. Clearly this cannot be binding on the parties if they choose by agreement to ignore it but it does have some practical effect in that the contractor can ask whatever price he likes.

This latter point offers a possible explanation for the inclusion of what appears otherwise to be a somewhat unnecessary clause. There is a view that the powers of the engineer under clauses 13 and 51 are so great that he can order acceleration and evaluate its cost under clause 52. If that is in any way correct, then its operation is severely diminished by clause 46(3).

Liquidated damages

Note that the reference to clause 46(3) in clause 47(1) indicates that agreed acceleration could involve the payment of liquidated damages from an earlier date than would otherwise apply. But again this is clearly a point for agreement/negotiation between the parties.

Chapter 17
Extensions of time and liquidated damages for delay

17.1 Introduction

Clause 44 of the Conditions deals with extension of time for completion; clause 47 deals with liquidated damages for delay.

Both are lengthy clauses and for the most part are word for word as in the Sixth edition. Clause 44, however, has gained an additional ground for extension of time – 'any delay impediment prevention or default by the Employer'. This has long been missing from ICE Conditions and its inclusion in the Seventh edition is as necessary as it is late. It closes the door on the argument that ICE Conditions are defective in not fully covering delays for which the employer is responsible – a defect with the consequence that any such delay not expressly covered in the Conditions puts time at large thereby preventing the employer from recovering liquidated damages.

One side effect of the correction now made to clause 44 is likely to be renewed attention to delay situations in contracts still running under ICE Fifth and Sixth editions. Engineers have been disposed to cover employer's delays by granting extensions of time under 'other special circumstances' but with the weakness of that practice now exposed, contractors under the Fifth and Sixth editions may feel more confident in claiming time to be at large.

17.2 Extensions of time generally

The contractor is under a strict duty to complete on time except to the extent that he is prevented from doing so by the employer or is given relief by the express provisions of the contract. The effect of extending time is to maintain the contractor's obligation to complete within a defined time and failure by the contractor to do so leaves him liable to damages, either liquidated or general, according to the terms of the contract.

In the absence of extension provisions, time is put at large by prevention and the contractor's obligation is to complete within a reasonable time. See *Dodd* v. *Churton* (1897) and *Peak Construction* v. *McKinney Foundations* (1970).

When time is put at large the contractor's liability is only for general damages; but first it must be proved that the contractor has failed to complete within a reasonable time. See *Rapid Building Group Ltd* v. *Ealing Family Housing Association* (1984).

Extension of time clauses, therefore, have various purposes:

- to retain a defined time for completion
- to preserve the employer's right to liquidated damages against acts of prevention
- to give the contractor relief from his strict duty to complete on time in respect of delays caused by specified neutral events.

Relevant events

In building contracts the events which entitle the contractor to extensions of time are commonly called 'relevant events'. ICE Conditions do not use the term but the grounds for extension listed in clause 44 amount to the same thing.

Relevant events vary in scope and description from contract to contract according to the policy of the particular contract on risk sharing in respect of delays. For example, some contracts allow extensions of time for strikes whereas others do not. However, as discussed above, all contracts should include for delays caused by the employer or his engineer to ensure that time can be extended in respect of such delays.

'Neutral events' is a term generally used to describe events over which neither party has direct control such as weather and strikes. Where neutral events are included in a contract as relevant events it is sometimes said that the loss lies where it falls. That is, the contractor carries the costs of delay in completing the works whilst the employer carries the cost (or losses) he suffers from late completion. For legal comment on this see the case of *Henry Boot Construction Ltd* v. *Central Lancashire Development Corporation* (1980).

Reimbursable/non-reimbursable delays

Some standard forms of contract expressly link extension of time clauses to clauses on the contractor's entitlements to be paid delay costs. To the extent that that happens in ICE Conditions it is simply that the two entitlements, extra time and extra money, are generally stated together in clauses covering entitlements. There is no direct link from clause 44 to rights of payment although in practice the well established maxim 'get time first and money will follow' frequently operates.

However, of itself, an extension of time under ICE Conditions gives no entitlement to payment to the contractor and the question of whether a particular delay is reimbursable or non-reimbursable is properly determined from the cause of delay (and proof of cost arising) rather than whether or not an extension of time has been granted.

17.3 *Liquidated damages generally*

Damages for breach of contract are either assessed after the breach on proof of loss – in which case they are called general or unliquidated damages – or

they are fixed in advance of the breach in which case they are called liquidated damages.

As to which type of damages applies in any particular circumstances the
position is:

- liquidated damages only apply when expressly stipulated in the contract
- liquidated damages only apply when they meet certain legal tests
- unliquidated damages apply when the contract does not stipulate liquidated damages or when stipulated liquidated damages cannot be
  enforced due to some legal fault or other irregularity.

Contracts can provide for limitation of either liquidated or unliquidated
damages to particular amounts or percentages of the contract value and can
similarly provide in respect of certain breaches that nil damages apply. ICE
Conditions, Sixth and Seventh editions, for example, provide in clause
47(4)(a) for limitation of liquidated damages for late completion and in
clause 47(4)(b) that where no stipulated rate for liquidated damages is
provided in the contract, then no damages are payable.

Liquidated damages

The essence of liquidated damages is a genuine covenanted pre-estimate of
loss. The characteristic of liquidated damages is that loss need not be proved.

Most standard forms of contract are drafted to permit the parties to fix in
advance the damages payable for late completion. When these damages are
a genuine pre-estimate of the loss likely to be suffered, or a lesser sum, they
can rightly be termed liquidated damages. In short, liquidated damages are
fixed in advance of the breach whereas general, or unliquidated damages,
are assessed after the breach.

Not all breaches of contract can be satisfactorily addressed by liquidated
damages and in construction it is usually only damages for late completion
and low performance which are liquidated. Even then there are differences
of application. For instance liquidated damages for late completion,
although common in main contracts, are less common in subcontracts. The
reason is the difficulty of pre-estimating the full loss to the contractor of the
subcontractor's breach.

Reasons for using liquidated damages

There are sound commercial reasons for using liquidated damages. They
bring certainty to the consequences of breach and they avoid the expense
and dispute involved in proving loss.

Liquidated damages provisions are not solely for the benefit of the
employer. They are also beneficial to contractors for they not only limit the
contractor's liability for late completion to the sums stipulated, but they also

indicate to the contractor at the time of his tender the extent of his risk. Thus, if a contractor believes that he cannot complete within the time allowed he can always build into his tender price his estimated liability for liquidated damages. All that the employer gets out of liquidated damages is relief from the burden of proving his loss and usually the right to deduct liquidated damages from sums due to the contractor.

Exhaustive remedy

It needs to be emphasised that if liquidated damages provisions are valid they provide an exhaustive and exclusive remedy for the specified breach.

This was one of the points confirmed in the case of *Temloc* v. *Errill Properties Ltd* (1987) where it was said:

'I think it clear, both as a matter of construction and as one of common sense, that if . . . the parties complete the relevant parts of the appendix . . . then that constitutes an exhaustive agreement as to the damages which are . . . payable by the contractor in the event of his failure to complete the works on time.'

The effect of this is that the employer cannot choose to ignore the liquidated damages provisions and sue for general damages. Nor can he recover any other damages for late completion beyond those specified.

Limitation of liability

One important aspect of liquidated damages is that they operate as a limitation of the contractor's liability.

In the case of *Widnes Foundry Ltd* v. *Cellulose Acetate Silk Co Ltd* (1933) the contractor was not prepared to take the financial risk of any greater sum than £20 per week for late completion and this was inserted in the contract. When a dispute arose on final payment after a 30 week delay in delivery and erection the purchaser counter-claimed not the £600 due as liquidated damages but £5850 as unliquidated damages. It was held that the contractor's liability was limited to £600.

In the *Temloc* case mentioned above the limitation effect was even more apparent. In that case, liquidated damages had been entered into the contract at £nil. And it was held that £nil was the effective sum for the calculation of the liquidated damages due.

If the provisions fail

It is well settled that when a liquidated damages clause fails to operate because it is successfully challenged as a penalty, or fails because of some defect in legal construction, act of prevention or other obstacle, then general damages can be sought as a substitute. Thus it was said in *Peak* v. *McKinney* (1970):

'If the employer is in any way responsible for the failure to achieve the completion date, he can recover no liquidated damages at all and is left to prove such general damages as he may have suffered.'

There is no firm ruling in English law that liquidated damages invariably act as a limit on any general damages which may be awarded in their place, although this is generally thought to be the position.

Liquidated damages and penalties

There is a good deal of misunderstanding on the relationship between liquidated damages and penalties. It is sometimes thought that if the employer cannot prove his loss then any sum taken as liquidated damages must be a penalty. This is certainly not the case.

Providing the sum stipulated as liquidated damages is a genuine pre-estimate of loss it is immaterial whether or not the loss can be proved or even suffered – a point well illustrated by the case of *BFI Ltd v. DCB Integration Systems Ltd* (1987).

However, if the stipulated sum is extravagant relative to any likely loss it may be held by the courts to be a penalty. English law does not allow the recovery of penalties and the courts will look at the stipulated sum irrespective of whether it is called liquidated damages or a penalty and if it is found to be a penalty will limit damages to the proven amount flowing from the breach. It is this which encourages contractors and their lawyers to find ways of challenging liquidated damages as penalties.

The rules on distinguishing penalties from liquidated damages laid down by Lord Dunedin in *Dunlop Pneumatic Tyre Co Ltd* v. *New Garage and Motor Co Ltd* (1915) are founded on the principle that liquidated damages must be a genuine pre-estimate of loss. Although an extravagant sum is the most obvious target for challenge many cases have succeeded before the courts on technical arguments on whether or not the stipulated sum could, in all circumstances, be a genuine pre-estimate of loss.

Such arguments no longer carry the weight they once did. The Privy Council held in the case of *Phillips Hong Kong Ltd* v. *The Attorney General of Hong Kong* (1993) that:

- the time when the clause should be judged was at the time of making of the contract not the time of the breach
- so long as the sum payable was not extravagant having regard to the range of losses which could be reasonably anticipated, it can be a genuine pre-estimate of the loss
- the argument that, in unlikely and hypothetical situations, the clause

might provide damages greater than the loss actually suffered, was not valid
• what the parties have agreed should normally be upheld.

The decision in the above case was followed in *J.F. Finnegan Ltd* v. *Community Housing Association Ltd* (1993) where it was also held that the use of a formula to calculate the amount of liquidated damages was justified.

17.4 Grounds for extension of time under ICE Seventh Conditions

Clause 44(1) states the grounds for extension of time under the Conditions as follows:

(a) any variation ordered under clause 51(1) or
(b) increased quantities referred to in clause 51(4) or
(c) any cause of delay referred to in the Conditions or
(d) exceptional adverse weather conditions or
(e) any delay, impediment, prevention or default by the employer or
(f) other special circumstances of any kind whatsoever which may occur.

As noted in Section 17.1 above, ground (e) – 'any delay impediment prevention or default by the Employer' – is new to ICE Conditions and its addition is a change of some significance.

Ground (a) – variations

Clause 44(1)(a) refers specifically to ordered variations but it probably applies to any variation which falls within the scope of clause 51(1).

Note that under the new provisions on quotations for variations in clause 52 the contractor can submit his estimate of any delay and that may either be accepted by the engineer or may form the basis for negotiations.

Ground (b) – increased quantities

The reference in clause 44(1)(b) to increased quantities 'referred to in clause 51(4)' is sometimes taken as suggesting that all increases on re-measurement might qualify as grounds for extension of time.

It is doubted if this is the correct position. Increased quantities under clause 51(4) do not necessarily add to the work content of the contract; it may simply be that certain work shown on the drawings has been omitted or understated in the bill of quantities. Entitlement to an extension then becomes a matter of judgement on what is fair.

Ground (c) – delays referred to in the Conditions

The delays referred to in the Conditions are most obviously those in clauses with specific reference to clause 44. There can be delays under other clauses without such reference – e.g. clause 32 (fossils); clause 20 (excepted risks) – and here, unless an instruction under clause 13 or a variation under clause 51 is involved, the extension would have to be given under delays by the employer or other special circumstances. Clauses which do refer to clause 44 include:

- 7(4)   – late drawings and instructions
- 12(2)   – adverse physical conditions or artificial obstructions
- 13(3)   – instructions causing delay
- 14(8)   – delay in engineer's consent to the contractor's methods or because of the engineer's requirements
- 27(6)   – variations relating to streetworks
- 31(2)   – facilities for other contractors
- 40(1)   – suspension of work
- 42(3)   – failure to give possession
- 59(4)(f) – nominated subcontractor's default

Ground (d) – exceptional adverse weather

Adverse weather of itself does not give grounds for non-performance of contractual obligations. Unless there are provisions in the contract offering relief, the contractor is deemed to have taken all risks from weather. However, many, but not all, standard forms of construction contract make some provision for extension of time in respect of delays caused by adverse weather.

The extent to which adverse weather applies as a relevant event depends on the wording of the contract. ICE Conditions use the phrase 'exceptional adverse weather conditions' but it is not just the phrase which has to be considered but also its context. There has to be delay, not just exceptional adverse weather. This may seem theoretical but it is common for contractors to apply for extensions of time on the grounds that the weather has been worse than average and sight can become lost of the need for proof of delay. The practice of obtaining local weather records and comparing them on a year to year basis, or of a particular year against average, may show that the weather has been exceptional but it says nothing about delay.

The point came up in the case of *Walter Lawrence and Son Ltd* v. *Commercial Union Properties (UK) Ltd* (1984) where a contractor was suing for return of amounts deducted as liquidated damages. It was held that:

'. . . When considering an extension of time under clause 23(b) of JCT 63, on the ground of "exceptionally inclement weather" the correct test for the architect to apply is whether the weather itself was "exceptionally inclement" so as to give rise to delay and not whether the amount of time lost by the inclement weather was exceptional...'

Another matter of interest arose in the *Walter Lawrence* case in respect of the time at which the weather should be assessed. The architect in correspondence had said: 'It is our view that we can only take into account weather conditions prevailing when the works were programmed to be put in hand, not when the works were actually carried out.'

The contractor refuted this and claimed that his progress relative to programme was not relevant to his entitlement to an extension. It was held that the effect of the exceptionally inclement weather is to be assessed at the time when the works are actually carried out and not when they were programmed to be carried out even if the contractor is in delay.

Perhaps it should be added however that the judge in the *Walter Lawrence* case drew a distinction between delays which occur during the original or extended time for completion and delays after the due date when the contractor is in culpable delay.

Ground (e) – delay attributable to the employer

The new ground for extension at clause 44(1)(e) – 'any delay impediment prevention or default by the Employer' – is discussed in Section 17.1 above.

On its wording it should be sufficient to cover most types of delays where the contractor blames the employer and where the other specified grounds for extension are not applicable. However, if there is any question on its sufficiency it is whether the wording covers delays caused by the engineer and which are not within the other specified grounds – for example, delay by the engineer under clause 4 in giving consent to subcontractors or failure by the engineer to attend 'without unreasonable delay' the examination of work for covering up under clause 38(1).

If it can be assumed that under the Conditions the employer takes responsibility for the performance of the engineer and effectively warrants that the engineer will carry out his duties under the contract without causing the contractor delay, there is no problem as to the sufficiency of clause 44(1)(e). However, the Conditions do not go so far as to state any such warranty and that puts reliance on an implied general term or implied particular terms on the employer's responsibility. And, in the absence of applicable court rulings, such reliance is inevitably prone to uncertainty.

Ground (f) – other special circumstances

Prior to the introduction in the Seventh edition of ground (e) (employer's default) there were differing views on what could be included within the scope of 'other special circumstances'.

On one view the wording of clause 44(1)(f) is so wide – 'other special circumstances of any kind whatsoever which may occur' – that there are no exclusions. The other, and it is suggested better view, is that catch-all phrases cannot include for unspecified breaches of contract by the employer. Thus, in *Fernbrook Trading Co Ltd* v. *Taggart* (1979) it was held that the words 'any special circumstances of any kind whatsoever' were not wide enough to empower the engineer to extend time for delays caused by the employer's breaches of contract in making late payments on interim certificates.

However, leaving aside the above argument on employer's default there is little consensus on what other situations can properly be included as 'special circumstances'. Such matters as delay caused by the unexpected discovery of antiquities (clause 32) are unlikely to be controversial but there is argument about whether delays caused by industrial disputes, non-availability of materials, shortage of labour and the like qualify under the clause. There is something to be said for accepting the test 'beyond the contractor's control' as a valid test but it may be too restrictive to say that it is the only test.

17.5 Procedures for extensions of time

Contractor's application

Clause 44(1)(a) provides that, should the contractor consider that any of the relevant events entitles him to an extension of time, he shall:

> 'within 28 days after the cause of the delay has arisen or as soon thereafter as is reasonable'

deliver to the engineer full and detailed particulars in justification of the extension claimed in order that the claim may be investigated at the time.

The clause does not require the contractor to give notice of delay unless he is seeking an extension, which is a curious omission in the light of clause 44(2) – assessment of delay. It appears that it is open to the contractor to claim an extension of time whether or not he expects to overrun the specified time for completion.

Failure by the contractor to make an application under clause 44(1) does not automatically disentitle him to an extension. The engineer is required to assess the contractor's entitlement at the date for completion and on the issue of the certificate of substantial completion whether or not the contractor has made an application.

Where the contractor will be at risk is in making a late application, or no application at all, in respect of neutral events such as adverse weather. He will similarly be at risk in gaining his full entitlement for other relevant events where, by failing to claim, he has prevented the engineer from investigating the delay at the time.

Assessment of delay

Clause 44(2)(a) states that the engineer shall upon receipt of particulars submitted by the contractor consider all the circumstances known to him at the time and make an assessment of the delay (if any) suffered by the contractor as a result of the alleged cause. He shall notify the contractor in writing of his assessment.

Clause 44(2)(b) states that the engineer may, in the absence of any claim by the contractor, make an assessment of the delay he considers has been suffered by the contractor as a result of any of the circumstances listed in clause 44(1). He shall notify the contractor in writing of his assessment.

The procedure in clause 44(2) for assessment of delay was first introduced into ICE Conditions in the Sixth edition. It requires the engineer to respond to the contractor's application for extension of time by assessing the delay as a result of the alleged cause before going on to consider whether or not any extension is due.

The intention of this is understood to be:

- to establish and place on record details of delays for availability in subsequent disputes, and
- to avoid the granting of extensions of time when they are not strictly necessary in relation to the time for completion.

If these are truly the intentions the first would have been improved by a requirement for the contractor to give notice of any delay, whether or not it involves extension of time; the second, by a definitive statement that any extension of time would apply only to delay beyond the date for completion.

An odd aspect of the provisions in clause 44(2), in so far as they may be intended to eliminate the granting of extensions when completion can still be achieved within the original time, is that the contractor's application under clause 44(1) is firmly stated as an application for extension of time. It would have been better stated as either an application for assessment of delay or for extension of time if that was the intention.

The delay assessment procedure of clause 44(2) does not appear to act as a condition precedent to granting extensions of time and although it is clearly intended to operate prior to the granting of interim extensions under clause 44(3) it is doubtful if it has any application to assessments under clauses 44(4) and 44(5). It is most unlikely that extensions granted without operation of the procedure would be invalidated and perhaps it should be seen as supplementary rather than integral to the rules for extension of time.

The practical detail of clause 44(2) is less than precise. No time limits are imposed for action, unlike clause 44(3) which requires extensions to be granted forthwith and clauses 44(4) and 44(5) which have 14 day and 28 day limits respectively.

Moreover, although the engineer can make an assessment in the absence of a claim he can only make an assessment of the delay caused by matters defined as relevant events for extensions. So he is not entitled, although it might be useful, to issue a notice under clause 44(2) where it is apparent that the contractor has been delayed by circumstances within his control which give no entitlement to extension.

Interim grant of extension of time

Clause 44(3) provides firstly that should the engineer consider that the delay suffered fairly entitles the contractor to an extension of the time for substantial completion of the works or any section, an interim extension shall be granted forthwith and shall be notified to the contractor in writing with a copy to the employer. It secondly provides that in the event that the contractor has made a claim for an extension of time but the engineer does not consider the contractor entitled to an extension, he shall so inform the contractor without delay.

There is much to be said for contractual provisions such as clause 44(3) which require extensions of time to be granted when there is still time to use them – rather than granting them after completion when costs may have been incurred by the contractor on acceleration measures endeavouring to fulfil his obligations under clause 43 to complete within stated times.

This was recognised in the redrafting of clause 44 in the Fifth and Sixth editions and the policy of the Conditions is clearly directed towards revision of the time for completion as the works progress. Clause 44(3) emphasises the point by requiring interim extensions to be granted 'forthwith'.

In *Fernbrook Trading Co Ltd* v. *Taggart* (1979) and *Perini Corporation* v. *Commonwealth of Australia* (1969), both cases under civil engineering contracts, the judgments support the proposition that there is breach of contract if the engineer fails to grant extensions in reasonable time.

Under clause 44(3) where there is a specified duty to grant interim extensions 'forthwith', the need for prompt action by the engineer is even more acute. And even if the employer does not face claims for breach of contract as a result of any failings by the engineer he may well face claims for constructive acceleration.

Assessment at the due date for completion

Clause 44(4) requires that the engineer shall, not later than 14 days after the due date or extended date for completion of the works or any section, consider all the circumstances known to him at the time and take action similar to that provided for in clause 44(3) (i.e. assess the contractor's entitlement to extension). Such action is to be taken whether or not the contractor has made any claim for an extension. Should the engineer consider that the contractor is not entitled to an extension of time he is to notify the employer and the contractor.

The procedure under clause 44(4) is clearly intended to operate on a repeat basis and is to be put into effect after each extended completion date has expired.

Note that, if the engineer does not consider the contractor entitled to an extension of time, he is to notify both the employer and the contractor. This is presumably so that the employer can consider deducting liquidated damages under clause 47.

Final determination of extension

Clause 44(5) requires that the engineer shall:

- within 28 days of the issue of the certificate of completion for the whole of the works or for any section
- review all the circumstances of the kind referred to in clause 44(1) – i.e. the relevant events, and
- finally determine and certify to the contractor with a copy to the employer
- the overall extension of time (if any) he considers due to the contractor.

The clause concludes with the statement that no such final review shall result in a decrease in any extension of time already granted under clauses 44(3) or 44(4).

A small point to note about clause 44(5) is that the time allowed for the final review in the Seventh edition is 28 days whereas in the Sixth edition it was only 14 days. This was a necessary change since under clause 44(1) the contractor has 28 days from an event to make an application for an extension. The 14 day time limit could result in the final review preceding notification of delays occurring in the fortnight before completion.

With regard to the 28 day time limit, and the 14 day limit on clause 44(3), it is unlikely that any failure by the engineer to act within the prescribed times puts time at large. It is probable that these time limits should be regarded as 'directory' only following the decision on a similar point in a building contract in the case of *Temloc Ltd v. Errill Properties Ltd* (1987).

Another small point to note on clause 44(4) is that the 28 day period runs from the date of issue of the relevant completion certificate and not from the date of completion stated in the certificate.

A further point to note, but one which is anything but small, is that the provisions of clause 44(5) apply with equal effect to each and every section as well as to the whole of the works – as indeed do the provisions of clauses 44(1), 44(2), 44(3) and 44(4). Therefore each stage of procedure must be followed for each and every section. A tiresome task sometimes for engineers, but a duty nevertheless.

Extent of the engineer's duty

The engineer's final review under clause 44(5) is to consider the 'overall' extension due. It is not therefore a matter of granting incremental extensions as in clauses 44(3) and 44(4) but rather a matter of fixing a final date for completion. This is reflected in the requirement for the employer to receive a copy of the certificate issued to the contractor.

The change of wording, from the engineer considering 'all the circumstances known to him' in clause 44(4) to 'all the circumstances of the kind referred to in sub-clause (1) in clause 44(5)', is interesting. It may not indicate a great deal but perhaps it illustrates that the clause 44(5) review should be undertaken on an analytical basis rather than on a response basis.

But whether or not this is deliberately intended any engineer making a final review of the contractor's entitlement to time under ICE Seventh edition or otherwise, should be mindful of the decision in the case of *John Barker Construction Ltd* v. *London Portman Hotel Ltd* (1996). The case related to an architect's final assessment under a building contract but its principles are likely to be of general application. The following summary in *Building Law Reports* (83 BLR) of the relevant parts of the decision speaks for itself:

'In the circumstances of the case, the architect's assessment of the extension of time due to the plaintiffs was fundamentally flawed because he did not carry out a logical analysis in a methodical way of the impact which the relevant matters had or were likely to have on the plaintiffs' planned programme; he made an impressionistic rather than calculated assessment of the extensions; he misapplied specific contractual provisions; where he allowed time for relevant events, the allowance made in important instances bore no logical or reasonable relation to the delay caused. Therefore, although there was no bad faith or excess of jurisdiction on the part of the architect, his determination of the extension of time due to the plaintiffs was not a fair determination nor was it based on a proper appreciation of the provisions of the contract and it was accordingly invalid.'

The consequence of an invalid assessment is loss by the employer of his entitlement to liquidated damages, and the potential consequence of that on the engineer is a claim for negligence.

Reduction of extensions granted

Clause 44(5) states that the final review shall not result in any decrease in any extension granted at the clause 44(3) and 44(4) assessments; that is at the interim and due date stages. This inevitably raises the question, can the earlier assessments under 44(3) and 44(4) produce reductions? Commentators are divided on this but logically, if an extension has been granted for an additional variation under 44(3) and there is subsequently a significant omission variation, it should be possible to recognise this at a later clause 44(3) assessment if not at a clause 44(4) assessment. However, see the comment in section 17.6 below on the recovery and reimbursement of liquidated damages under clause 47(5) of the Conditions.

Final review

Although clause 44(5) describes the engineer's review after the issue of a certificate of substantial completion as his 'final review', it is questionable whether it is necessarily his last review. If either the contractor or employer is dissatisfied with the extension granted either can call for an engineer's decision under clause 66, which would seem to oblige the engineer to carry out a further review.

Should the engineer decide that he has no power to carry out a further review – which might be the correct approach – that leads in turn to the question of whether an adjudicator, arbitrator or the courts have the power to review what has been stated in the contract at clause 44(5) as finally determined.

The courts probably have no such power except in cases when the final determination is ruled invalid – see for example *Balfour Beatty Civil Engineering Ltd* v. *Docklands Light Railway Ltd* (1996) and the *John Barker* case referred to above. As for adjudicators and arbitrators, they are given powers under clause 66 of the Conditions and the ICE arbitration and adjudication procedures, to open up and review any certificates. There remains, however, the question of whether in doing so they are bound by the same rules as the engineer – in particular, whether by clause 44(5) they are precluded from decreasing any extension of time granted under clauses 44(3) and 44(4).

There is no obvious legal reason why they should not be so bound and the practical effect of reducing, after completion, an extension already granted would be not far different in practical terms from reducing the original time for completion – a reduction which can only be made by agreement of the parties.

17.6 *Liquidated damages for late completion*

The provisions for liquidated damages for delay are the same in the Seventh edition as in the Sixth edition, subject to minor wording changes such as 'fails to achieve substantial completion' in place of 'fails to complete' and the omission from the end of clause 47(6) of a redundant phrase on interest.

The basis of the scheme in the Seventh edition is that liquidated damages for delay shall be stipulated on a daily or weekly basis albeit that the stipulated rate may be nil. If the parties wish to have unliquidated damages for delay, clause 47 should be deleted in its entirety along with the corresponding entries in the appendix to the form of tender and other references to liquidated damages in the Conditions. The necessity for this can be gathered from the statement in clause 47(4) that where there are nil liquidated damages, then damages shall not be payable – as in the decision in *Temloc Ltd* v. *Errill Properties Ltd* (1987).

Liquidated damages for the whole of the works

Clause 47(1) deals with liquidated damages where the whole of the works is not divided into sections. The clause has three major parts:

- clause 47(1)(a) which requires (in mandatory fashion) that the appendix to the form of tender shall include a sum which represents the employer's genuine pre-estimate (expressed per week or per day) of the damages likely to be suffered by him if the whole of the works is not substantially completed within the time prescribed by clause 43 or by any extension thereof granted under clause 44 or by any revision thereof agreed under clause 46(3)
- clause 47(1)(b) which states that if the contractor fails to achieve substantial completion of the whole of the works within the prescribed time he shall pay the employer for every week or day (as applicable) which elapses between the prescribed time and the date of substantial completion, the sum stipulated as liquidated damages
- a concluding part which provides that if any part of the works is certified as substantially complete before the completion of the whole of the works, the stipulated sum shall be reduced by the proportion which the value of the part so completed bears to the value of the whole of the works.

Pre-estimate of loss/proportioning down

The mandatory requirement in clause 47(1)(a) for the employer to state in the appendix to the form of tender his genuine pre-estimate of loss is slightly unusual in so far as any lesser sum will be equally valid as liquidated damages.

Both clause 47(1)(a) and the appendix allow for liquidated damages to be expressed per week or per day. It is usually better to use 'per day' since damages 'per week' cannot be proportioned down for a part week. Thus the contractor is only liable for each full week of delay if weeks are used. The not uncommon alteration of 'week' to 'week or any part thereof' is dangerous since that wording is open to challenge as a penalty. It is not necessary for liquidated damages to be expressed in the same time periods as the time for completion.

The final paragraph of clause 47(1) is a provision for proportioning down the specified figure for liquidated damages when parts (as distinct from sections) of the works have been completed prior to the whole. The general position is that unless the contract provides a mechanism for the issue of partial completion certificates and a corresponding mechanism for proportioning down liquidated damages, the contractor remains liable for full damages up to the date of total completion. However, where a contract does make provision for the issue of partial completion certificates, albeit on a consent basis, then it is essential to have corresponding provisions for proportioning down liquidated damages. The courts will not imply such a term if it is missing and the liquidated damages clause may then fail for uncertainty or as a penalty.

Advancing the date for completion

The reference to clause 46(3) at the end of clause 47(1)(a) allows the parties, by agreement, to advance the date from which liquidated damages apply.

Liquidated damages for sections

Clause 47(2) repeats the provisions of clause 47(1) so that they apply when the whole of the works is divided into sections.

Thus the provisions in clause 47(1) for the whole of the works as a unity are effectively omitted and each section is given its own time and liquidated damages with the 'remainder of the works' regarded as a further section.

Clause 47(2)(c) confirms a point fundamental to the scheme – that liquidated damages in respect of two or more sections may run concurrently.

Damages not a penalty

Clause 47(3) states, presumably for the avoidance of doubt rather than by way of legal assertion, that all sums payable by the contractor pursuant to clause 47 are paid as liquidated damages and not as a penalty.

If such sums were a penalty, proof of loss would be required to make them enforceable but as liquidated damages they are due without proof of loss.

However, the terminology itself is not decisive and if the courts find, as a matter of construction, that liquidated damages are penalties or vice versa, they will award accordingly. Thus, Lord Dunedin in *Dunlop Pneumatic Tyre Co* v. *New Garage and Motor Co Ltd* (1915) said:

> 'Though the parties to a contract who use the words "penalty" or "liquidated damages" may prime facie be supposed to mean what they say, yet the expression is not conclusive.
>   The court must find out whether the payment stipulated is in truth a penalty or liquidated damages.'

Limitation of liquidated damages

Clause 47(4) contains two significantly different provisions:

- 47(4)(a) states that the total amount of liquidated damages in respect of the whole of the works or any section shall be limited to the appropriate sum

stated in the appendix to the form of tender. If no such limit is stated then liquidated damages without limit apply
- 47(4)(b) states that should there be omitted from the appendix to the form of tender any sum required to be inserted by way of liquidated damages or if any such sum is stated to be 'nil' then to that extent damages shall not be payable.

It is common in some contracts – process and plant contracts for example – for the total amount payable for late completion to be limited on commercial grounds to 5% or 10% of the contract sum. The practice of limitation is less common in civil engineering contracts and when it is seen, the limitation figure is likely to be considerably more than 5% or 10%.

The concluding statement in clause 47(4)(a) that if no limit is stated then 'liquidated damages without limit shall apply' means only that the contractor's liability has no ceiling at the stipulated rates.

The effect of clause 47(4)(b) is that the liquidated damages provisions do not become inoperative because of a blank or a nil entry. They remain in force but nothing can be recovered. The employer cannot sue for general damages for late completion because he has exhausted his remedy for the contractor's breach. This was one of the matters confirmed in the case of *Temloc Ltd* v. *Errill Properties Ltd* (1987) where the entry in the appendix to a JCT 80 contract was stated as £nil liquidated damages. The contract was finished late and the employer/developer, who was himself liable to the property purchaser for damages, unsuccessfully sought to recover them as general damages from the contractor.

Clause 47(4)(b) goes further than the decision in the *Temloc* case by equating a blank entry with a nil entry. The effect of this is that if the employer genuinely wants the right to general damages as opposed to liquidated damages for late completion, the whole of clause 47 needs to be deleted from the contract with corresponding amendments made to other clauses.

Recovery and reimbursement of liquidated damages

Clause 47(5) provides:

- the employer may deduct the amount of any liquidated damages due from any other sums due or which become due under the contract
- the employer may require the contractor to pay any sums due as liquidated damages forthwith
- if the engineer on review of the circumstances causing delay grants a further extension the employer shall 'no longer be entitled to liquidated damages in respect of the period of such extension'
- any sums deducted or paid as liquidated damages for such an extended period shall be reimbursed forthwith to the contractor with interest
- interest shall be at the rate provided for in clause 60(7) compounded

monthly from the date on which the sums were deducted from, or paid by, the contractor.

Clause 47(5) commences significantly with the phrase 'The Employer may' – a clear indication of discretion. It is not absolutely clear from the wording of the clause whether the discretion is intended to apply only to the method of collection of any liquidated damages or whether it applies more broadly to whether they should be imposed at all. It probably applies to both but whichever is assumed it is evident that deduction of liquidated damages is a matter for the employer not for the engineer.

It is worth noting that clause 47(5)(a) entitling the employer to deduct or require payment of liquidated damages comes into effect without any condition precedent other than failure by the contractor to complete by the due date. Unlike certain other standard forms (including ICE Fifth edition) there is no requirement for a certificate of non-completion or a certificate that damages are due.

It is possible, therefore, for liquidated damages to be deducted when the engineer is still considering the contractor's entitlement to extension of time, although if both the contractor and the engineer strictly follow the notice and assessment requirements of the Conditions, the potential consequences are minimised.

As to operation of the employer's discretion, employers need to consider the extent of their accountability to others (rate payers and the like) in deciding whether they properly have the discretion to forego recovery from the contractor of amounts effectively due as damages for breach of contract.

'shall no longer be entitled to liquidated damages'

The part of clause 47(5) containing the above phrase requires some comment. The part reads:

> 'If upon a subsequent or final review of the circumstances causing delay the Engineer grants a relevant extension or further extension of time the Employer shall no longer be entitled to liquidated damages in respect of the period of such extension.'

It is quite possible that nothing more is intended by the phrase 'shall no longer be entitled to liquidated damages' than that the granting of a further extension cancels the employer's rights in respect of what was previously a period of delay. However it raises questions as to whether an adjudicator or arbitrator later called upon to open up and review the engineer's final assessment can reinstate the employer's entitlement to liquidated damages. The wording of the clause suggests that they cannot.

Such an interpretation may appear somewhat surprising in its consequences but taken in conjunction with clause 44(5) which provides that no final review by the engineer shall result in a decrease of an extension of time already granted, it may be correct. Much depends on what is to be taken from the words 'shall no longer'.

Intervention of variations

Clause 47(6) deals with the contractor's liability for delays which occur after the due date for completion but when the works are still progressing and unfinished and the contractor is operating in what is sometimes called culpable delay.

In general where such delays occur and they are the contractor's responsibility they simply add to his problems. But when the delays are not the contractor's responsibility because they are due to variations, prevention, unforeseen conditions or the like, there needs to be some mechanism in the contract for either extending time or otherwise relieving the contractor of any added liability for liquidated damages.

Clause 47(6) originates from a time when there was uncertainty on what legal rules applied to extensions for delays in periods of culpable delay – i.e. was the contractor entitled to:

- a gross extension from the due date to the date the delay ended
- a net extension for the period of the delay
- claim time to be at large.

The uncertainty was removed by the decision in the case of *Balfour Beatty Building Ltd* v. *Chestermount Properties* (1993) where it was held that the contractor was entitled only to a net extension.

Clause 47(6) had by then appeared in the ICE Sixth edition tackling the uncertainty not by way of further extension but by way of suspension of liquidated damages.

Clause 47(6) provides:

- if after liquidated damages becomes payable
- the engineer orders a variation, or adverse physical conditions/obstructions within the meaning of clause 12 are encountered or any other situation outside the contractor's control arises
- any of which in the opinion of the engineer results in further delay to 'that part of the Works'
- the engineer shall notify the contractor and the employer in writing, and
- the employer's further entitlement to liquidated damages in respect of 'that part of the Works' shall be suspended until the engineer notifies the contractor and the employer in writing that the further delay has come to an end.

The clause concludes by stating that such suspension shall not invalidate any entitlement to liquidated damages which accrued before the period of further delay started to run, and subject to any subsequent or final review of the circumstances causing delay any monies already deducted or paid as liquidated damages under the provisions of the clause may be retained by the employer.

The problems with the operation of clause 47(6) are immediately obvious from its wording. The grounds for suspension of liquidated damages are far too wide and go well beyond the events which would entitle the contractor to extension of time under clause 44. Of particular concern is the phrase 'beyond the Contractor's control'. This could bring in all time lost due to weather and even, following the decision in *Scott Lithgow* v. *Secretary of State for Defence* (1989), supplier and/or subcontractor delays. The reference to clause 12 circumstances presumably means circumstances which could not reasonably have been foreseen but it does not say so. Worse still are the references in the clause to liquidated damages in respect of 'part of the Works'. The Conditions make no provision for such damages and it is by no means clear how if part of the works is delayed the calculations for suspension of damages are to be made.

Some users of the Seventh edition may try to improve clause 47(6) by amendment. Others may take the view that it has been rendered superfluous by the decision in *Balfour Beatty* v. *Chestermount* and that it can safely be deleted.

Chapter 18
Completion

18.1 Introduction

The provisions in the Seventh edition on completion of the works and the procedures for the issue of certificates of substantial completion are identical to those in the Sixth edition. For the most part they are contained in clause 48 of the Conditions but the obligation to complete within a particular time is stated in clause 43 and damages for late completion are covered in clause 47.

The Seventh edition retains the curious and somewhat unsatisfactory arrangements that whereas the test for the issue of a certificate of substantial completion is when the works (or any section) are substantially complete in the opinion of the engineer, the engineer has no general obligation, and arguably no power, to issue a certificate of substantial completion until the contractor has given notice that he considers the works (or any section) are substantially complete. And to complicate matters still further the contractor has no obligation to give such notice.

The Conditions do not define what is meant by substantial completion and the definition of a 'Certificate of Substantial Completion' in clause 1(1)(r) goes no further than to state that it means a certificate issued under clause 48. In effect, unless and until the formal disputes procedures of the Conditions are invoked and an objective analysis is made of when the works were substantially complete, it is solely the opinion of the engineer which is determinative – but that opinion can apparently be withheld until sought.

18.2 General meaning of completion

Various phrases are used in construction contracts to define completion:

- completion
- practical completion
- substantial completion.

ICE Conditions have traditionally used the term 'substantial completion' although the ICE Fifth edition adopted 'completion' and the ICE Minor Works Conditions refer to practical completion.

The significance of completion in construction contracts, however expressed, is generally that it marks:

- the transfer of risks for care of the works from the contractor to the employer
- the commencement of the defects liability period
- the end of the employer's entitlement to damages for late completion
- the employer's entitlement to repossess the site.

Disputes on completion are commonplace. Contractors may want early completion to reduce liabilities for liquidated damages and insurances and perhaps to secure part payment of retention monies. Employers may want later completion to ensure that the works are better finished or because they wish to delay occupation.

Completion

In its precise legal sense 'completion' means strict fulfilment of obligations under the contract, and when used in the context of 'entire contracts' which attract the doctrine of substantial performance failure to complete produces the apparently harsh result that no payment is due. Thus in the case of *Cutter* v. *Powell* (1795), when the second mate on a ship bound for Liverpool from Jamaica died before the ship reached Liverpool, his widow was unsuccessful in a claim for a proportion of his lump sum wages of 30 guineas.

Fortunately for contractors, construction contracts rarely fall into the category of 'entire contracts'. Indeed if they did the risks for contractors would be immense since an employer unwilling to pay anything would only have to point to a modest default or item of unfinished work to escape the obligation of making payment.

The courts take a practical view of construction contracts as illustrated by this extract from the judgment of Lord Justice Denning in the case of *Hoenig* v. *Isaacs* (1952) which concerned the decorating and fitting-out of a one room flat:

'In determining this issue the first question is whether, on the true construction of the contract, entire performance was a condition precedent to payment. It was a lump sum contract, but that does not mean that the entire performance was a condition precedent to payment. When a contract provides for a specific sum to be paid on completion of specified work, the courts leap against a construction of the contract which would deprive the contractor of any payment at all simply because there are some defects or omissions. The promise to complete the work is, therefore, construed as a term of contract, but not as a condition. It is not every breach of that term which absolves the employer from his promise to pay the price, but only a breach which goes to the root of the contract, such as abandonment of the work when it is only half done. Unless the breach does go to the root of the matter, the employer cannot resist payment of the price. He must pay it and bring a cross-claim for the defects and omissions, or alternatively, set them up in diminution of the price. The measure is the amount which the work is worth less by reason of the defects and omissions, and is usually calculated by the cost of making them good.'

In cases concerned with substantial performance the issue is generally the employer's payment obligation and little else. Under most standard forms of construction contracts the issues are likely to be wider – liability for liquidated damages, release of retention etc. Generally, therefore, substantial performance is not relevant to the meaning of 'completion' as mentioned in construction contracts or determination of terms such as 'practical completion' and 'substantial completion' used in such contracts.

Practical completion

Practical completion is the phrase commonly used in building contracts to define the point at which the works are fit to be taken over by the employer.
It is also used in the ICE Minor Works Conditions where it is stated:

'Practical completion of the whole of the Works shall occur when the Works reach a state when notwithstanding any defect or outstanding items therein they are taken or are fit to be taken into use or possession by the Employer.'

In the case of *Emson Eastern Ltd* v. *EME Developments Ltd* (1991) the court had to decide whether the issue of a certificate of practical completion under a JCT 80 contract constituted 'completion of the works' as mentioned in the determination clause of the contract. Judge Newey QC held that it did. He said:

'In my opinion there is no room for "completion" as distinct from "practical completion". Because a building can seldom if ever be built precisely as required by drawings and specification, the contract realistically refers to "practical completion", and not "completion" but they mean the same.'

From the cases of *H. W. Neville (Sunblest) Ltd* v. *William Press & Sons Ltd* (1981) and *City of Westminster* v. *J. Jarvis & Sons Ltd* (1970) the following rules to determine practical completion have been developed:

- practical completion means the completion of all the construction work to be done
- the contract administrator may have discretion to certify practical completion where there are minor items of work to complete on a *de minimis* basis

- a certificate of practical completion cannot be issued if there are patent defects
- the works can be practically complete notwithstanding latent defects.

Substantial completion

The term 'substantial completion' is a more flexible concept than 'practical completion' and the provisions for dealing with outstanding works under ICE Conditions suggest that it is not the *de minimis* principle which applies to such works but whatever is acceptable to the engineer.

One consequence of this lack of precision is increased scope for dispute. Another is uncertainty for the parties on how the achievement of the particularly important milestone of substantial completion is to be assessed.

Occupation/use

The Seventh edition expressly provides in clause 48(3) that occupation or use by the employer entitles the contractor to a certificate of substantial completion.

But for this provision it is doubtful if occupation or use would be a decisive test. It is clear from two building cases, *BFI Ltd* v. *DCB Ltd* (1987) and *Big Island Contracting (HK) Ltd* v. *Skink Ltd* (1990) that occupation/use by the employer does not by itself establish practical completion.

18.3 *Notification of substantial completion*

Clause 48(1) of the Conditions provides that the contractor may give notice to the engineer when he considers that:

- the whole of the works, or
- any section with its own time in the appendix

has been:

- substantially completed, and
- has passed any final test prescribed in the contract.

Such notice is to be accompanied by an undertaking to finish any outstanding work in accordance with the provisions of clause 49(1).

Note that strictly the contractor does not apply for a certificate of substantial completion under clause 48(1) although he does so under clause 48(3) where there is premature use by the employer. Under clause 48(1) the contractor merely gives notice that he considers the works to be substantially complete and undertakes to finish the outstanding work.

Contractor 'may give notice'

Although the burden of giving notice falls on the contractor, it is not expressed in the mandatory term of 'shall give notice' but in the permissive term of 'may give notice'. This raises various questions:

- is the engineer empowered or required to issue a certificate notwithstanding lack of notice?
- is the issue of a certificate of substantial completion deferred indefinitely if the contractor fails to give notice?
- is the date of substantial completion itself affected by delay or failure on the part of the contractor to give notice?

The answers, it is suggested, are as follows:

- since the engineer is empowered to issue certificates for parts of the works under clause 48(4) without any request from the contractor, the engineer could effectively certify completion of the whole of the works without notice – albeit in parts
- the engineer cannot be required by the contractor to certify completion of the whole of the works without notice under clause 48(1) but there could be circumstances where the engineer might have a duty to the employer to certify such completion
- certificates of completion are intended principally for the contractor's benefit and if the contractor fails to activate the contractual procedures for their issue they may be deferred
- it is clear from clauses 48(2) and 48(4) that it is the engineer's opinion which fixes dates of substantial completion, not the date of any notification by the contractor
- as neither clause 43 (time for completion) nor clause 47 (liquidated damages) refers, in respect of the whole of the works or sections, to substantial completion as certified but only to substantial completion it is arguable that it is the actual state of affairs which regulates those clauses, not certificates
- however, unless there is something on record equivalent to a certificate indicating the actual state of affairs, certificates are necessary.

Notice to engineer's representative

Although the contractor is entitled under clause 48(1) to give notice to the engineer's representative, no powers are conferred in clause 48 on the engineer's representative. Clause 2(4)(c) prohibits delegation of the engineer's powers under clause 48 so the engineer's representative can strictly do no more than pass on the notice to the engineer.

In reality, named engineers to contracts are frequently detached from detailed knowledge of the state of completion and rely on reports from subordinates. Every engineer should recognise that it is his opinion not someone else's which validates certificates given under clause 48. If there is any hint of potential dispute over dates of substantial completion, the engineer needs to see for himself how the works are progressing.

18.4 Certification of substantial completion

Clause 48(2) requires the engineer to respond to the contractor's notification under clause 48(1) within 21 days of the 'date of delivery' of such notice by either:

(a) issuing a certificate of substantial completion to the contractor (and copied to the employer) stating the date on which in his opinion the works, or section, were substantially completed 'in accordance with the Contract', or

(b) giving instructions in writing to the contractor specifying the work which in the engineer's opinion requires to be done by the contractor before the issue of the certificate.

Clause 48(2) concludes by stating that if the engineer gives instructions on further work, the contractor shall be entitled to receive the certificate of substantial completion within 21 days of completion of the specified work 'to the satisfaction of the Engineer'.

Date of certificate

For the purpose of clause 20 (care of the works) and clause 60(6) (payment of retention), it is the date 'of issue' of the certificate which is important. For other contractual purposes, clause 47 (liquidated damages) and calculation of the defects correction period under clause 1(1)(s), it is the date of completion 'on' the certificate which is important.

Engineer's failure to respond

Clause 48 is silent on the position which applies if the engineer fails to respond to the contractor's notice, in time or at all.

The contractor has a strong case for arguing that once the 21 days have passed it is too late for the engineer to give instructions on works to be completed; and he has an equally strong case for arguing that his notice should be effective in fixing the date of substantial completion.

Instructions on work to be done

A frequent source of dispute on whether or not the works are complete is disagreement between the contractor and the engineer on the amount of work necessary to permit the issue of a certificate of substantial completion.

The provision in clause 48(2)(b) requiring the engineer to give instructions in writing specifying the work which in 'his opinion' requires to be done suggests that subject to the test of reasonableness the opinion of the engineer prevails. However, the test which applies in clause 48(2)(a) of 'substantially completed in accordance with the Contract' must be highly relevant to what is a reasonable opinion in clause 48(2)(b). So if it comes to a formal dispute before an adjudicator or an arbitrator the issue is likely to be at what date were the works substantially completed in accordance with the contract.

There are, of course, some disputes where it is not a matter of the contractor getting on with the instructed works under protest and arguing later about the proper date for completion, but more a matter of whether or not the instructed works can or should be undertaken at all. In such cases the issue then at adjudication or arbitration is more of a technical nature.

The wording of clause 48(2)(b) requiring the engineer to give 'instructions' on work to be done before the issue of a certificate is not ideal. The engineer should be careful in giving any instructions which might give rise to claims under clause 13. Instructions telling the contractor how to do the work should certainly be avoided.

Even instructions doing no more than scheduling the work to be done could possibly give rise to a claim under clause 13 if the contractor could show that it had caused him to incur cost beyond that which could reasonably have been foreseen at the time of tender.

The contractor must, of course, comply with the instructions of the engineer and then pursue his remedies under the contract.

There is nothing in clause 48 itself to suggest that the engineer can relate his 'instructions' to an undertaking by the contractor to complete outstanding work within a specified time, as envisaged by clause 49(1). In practice, however, the two are likely to be related.

Completion of work to be done

The final part of clause 48(2) states that the contractor is entitled to receive a certificate of substantial completion within 21 days of completion of the specified work to the satisfaction of the engineer.

There is no express requirement for the contractor to give notice of completion or make application for the certificate but it is obviously in the contractor's interest that he should do one or the other.

The reference to 21 days in the final sentence of clause 48(2) presumably has the same application as the reference to 21 days in the opening sentence of the clause. That is, the engineer is required to issue a certificate within 21 days stating the date of completion. Taken on its own the final sentence can be read as potentially adding 21 days to the date of completion.

Contractor's design

The engineer should be aware when considering action under clause 48(2) that somewhat unusually one clause of the contract, clause 7(6)(b), overrides his opinion and by that clause he is prevented from issuing a certificate covering any part of the permanent works designed by the contractor until all drawings and manuals have been submitted and accepted.

18.5 Premature use by the employer

Clause 48(3) provides that:

- if any substantial part of the works has been occupied or used by the employer, other than as provided for in the contract
- the contractor may request in writing and the engineer shall issue a certificate of substantial completion in respect thereof
- any such certificate shall take effect from the date of the contractor's request
- upon the issue of such a certificate the contractor is deemed to have undertaken to complete any outstanding work during the defects correction period.

At face value the procedure under clause 48(3) is simple. The contractor may request and the engineer shall issue. There is no reference in the clause to the opinion of the engineer; there is no requirement on the contractor to give an undertaking on outstanding work; and there is no provision for the engineer to refuse to issue the certificate until further work has been completed. In reality there is ample scope for dispute on such matters as:

- what constitutes a 'substantial part'
- the meaning of 'occupied or used'
- the meaning of 'other than as provided in the Contract'.

It is not clear if the requirement on the provision of operating and maintenance manuals in clause 7(6)(b) applies to certificates issued under clause 48(3). It may be that provision applies only to certificates given at the engineer's discretion and that occupation or use by the employer nullifies the clause 7(6)(b) provision.

Substantial part

Disputes on what constitutes a 'substantial part' are common. If resolved formally such disputes require a finding of fact given having regard to all the circumstances. It is not clear from the wording of clause 48(3) whether at a less formal level it is the opinion of the contractor or the opinion of the engineer which is intended to prevail but the engineer is in the stronger position to impose his will.

Occupied or used

In some types of civil engineering projects, occupation or use of parts of the works as the contract proceeds is inevitable and unavoidable. To avoid problems with clause 48(3) this needs to be provided for in the contract.

Note the use in clause 48(3) of the past tense 'has been'. It may be arguable from this that continuing use is not a determining factor.

'provided in the Contract'

It is doubted if provisions in the contract relating to sections have any relevance to clause 48(3). It is more likely that the phrase 'provided in the Contract' intends reference to requirements or explanations on drawings, in specifications, or in special clauses on practical aspects of use of the site.

18.6 Substantial completion of parts of the works

Clause 48(4) gives the engineer discretionary power to issue a certificate of substantial completion for any part of the works he considers has:

- been substantially completed and
- passed any final test prescribed in the contract.

Upon the issue of such a certificate the contractor is deemed to have undertaken to complete any outstanding work during the defects correction period.

Clause 48(4) is unusual to the extent that it does not depend upon any of the criteria expressed elsewhere in clause 48 for the issue of a certificate of completion. There is no mention of:

- notice or application from the contractor
- an undertaking to complete outstanding work (except that this is deemed to be given)
- any obligation to complete outstanding work other than during the defects correction period
- provision of operating and maintenance manuals
- occupation or use by the employer
- the part being a substantial part
- the contractor being in agreement
- the certificate being copied to the employer.

The apparently wide discretion given to the engineer under clause 48(4) is not, it is suggested, a discretion to be exercised lightly. The issue of any certificate of substantial completion has repercussions for the employer under clause 20 (care of the works); clause 47 (liquidated damages); and clause 60 (release of retention). It is questionable whether the engineer should issue a certificate under clause 48(4) without consulting or at least notifying the employer and it would not be inappropriate for the employer to stipulate in part one of the appendix to the form of tender (at part 19) that the approval of the employer is needed before the engineer can act under clause 48(4).

18.7 Reinstatement of ground

Clause 48(5) provides only that a certificate of substantial completion for a section or a part shall not be deemed to certify completion of any ground or surfaces requiring reinstatement unless the certificate expressly so states.

This is a practical provision having regard to the contractor's continued occupation of the site.

Chapter 19
Outstanding works and defects

19.1 Introduction

The clauses considered in this chapter are:

- clause 49 which deals with works outstanding at the date of issue of a certificate of substantial completion and defects occurring during the defects correction period
- clause 50 which states the contractor's obligation to carry out searches and trials to determine the causes of defects
- clause 61 which provides for the issue of a defects correction certificate at the end of the defects correction period and when all outstanding works have been completed and all defects remedied
- clause 62 which empowers the engineer to give instructions on urgent repairs and entitles the employer to carry out the work himself or using others if the contractor is unwilling to do so.

There are no changes of substance in any of the clauses between the Sixth and Seventh editions although clause 62 is completely reworded in the interest of clarity.

Defects correction period

Clause 1(1)(s) defines the defects correction period as that period stated in the appendix to the form of tender calculated from the date on which the contractor becomes entitled to a certificate of substantial completion for the works, a section, or any part.

For any one contract there can therefore be a number of defects correction periods. The length of each will depend on the entry made by the employer in part one of the appendix to the form of tender. More often than not this is 52 weeks but for some works it can be as short as 12 weeks and for others as long as 5 years. In some contracts more than one length is entered into the appendix to allow different length correction periods for different parts or sections of the works. But whether there be a single length or differing lengths it is not unusual for a contract to have concurrent defects correction periods.

Making good defects generally

Provisions in contracts requiring the contractor to make good defects for a specified period after taking-over or completion usually have various effects:

- they oblige the contractor to make good defects for which he is responsible, and
- they entitle the contractor to make good such defects
- they oblige the employer to allow the contractor entry to the works to remedy defects.

The entitlement can be seen as a substantial benefit to the contractor in that if defects do occur during the period of liability the contractor will normally be able to rectify them at less cost than he would have to pay as damages to the employer.

If the employer takes it on himself to rectify defects within the specified defects liability period thereby depriving the contractor of his right to make good his own defects, the employer will be able to recover as damages not the full amount of his expenditure, but only the amount of cost that the contractor would have incurred. See, for example, the case of *Tomkinson* v. *The Church Council of St Michael* (1990).

Latent defects

Defects which are not apparent at the date of completion of the works are categorised as latent defects (as distinct from patent defects).

The legal position on latent defects is complex but to the extent they are treated as breaches of contract the contractor's liability is governed by the applicable legal limitation periods. Under English law and the Limitation Act 1980 those periods are 6 years for simple contracts and 12 years for contracts executed as a deed (or under seal). Where the defects are treated as arising from negligence, legal actions are subject to the Latent Damage Act 1986 which provides for extended limitation periods – with a 15 year longstop.

Standard forms of construction contract rarely have anything to say on liability for latent defects except for those defects which appear during the defects correction period/defects liability period/period for making good defects/maintenance period or whatever else it is called. However, some process and plant contracts do expressly exclude the contractor's liability for latent defects appearing after the end of the contractual defects correction period. And some building contracts have clauses on final certificates which are so worded by phrases such as 'the issue of the certificate shall be conclusive evidence of fulfilment of the contractor's obligations' that, intentionally or not, they exclude the contractor's further liability once the final certificate is issued.

ICE Conditions have neither an express exclusion of liability nor final certificates which are stated to be conclusive evidence (or similar). Clause 61(2) expressly states that the defects correction certificate does not relieve the parties from liabilities and obligations. Accordingly, under the Conditions, liability for latent defects is a matter of general law once the final defects correction period has expired.

19.2 *Outstanding works and defects*

Clause 49 is in four parts:

- 49(1) which deals with the times within which outstanding works are to be completed
- 49(2) which states the contractor's obligation to rectify defects notified during the defects correction period
- 49(3) which explains which party is to bear the cost of rectifying defects
- 49(4) which entitles the employer to rectify defects if the contractor fails to do so.

Outstanding works

Clause 48(1) requires the contractor, when giving notice that he considers the works or any section to be substantially complete, to provide an accompanying undertaking to finish any outstanding work in accordance with clause 49(1).

Clause 49(1) provides that the undertaking to be given under clause 48(1) may, after agreement between the engineer and the contractor, specify a time or times within which the outstanding works shall be completed. It then states that if no such times are specified any outstanding work shall be completed as soon as practicable during the defects correction period.

The provision for agreement of a time in which to complete outstanding work was new to the Sixth edition and, although perhaps not so intended, in practice it may have some influence in decision on whether or not to allow contractors their certificates of substantial completion. It has long been a point of discontent with engineers and employers that contractors, having been given certificates of completion, leave the site and put off finishing outstanding works until the very end of the defects correction period.

Contractors may argue that the provision amounts to unfair pressure – either agree to complete all outstanding work within a set time or no certificate will be issued until it is complete. Employers are more likely to say that this is how things should be. But in any event, the procedure for agreement applies only to clause 48(1) and not to certificates issued under clauses 48(3) and 48(4).

A point which is none too clear is what remedy, if any, does the employer have if the contractor defaults on his agreement to complete outstanding work in a set time?

It is possible that clause 49(4) may address the issue by allowing the employer to engage others to complete the work but, as shown below, there is a strong case for saying that clause 49(4) applies only to defects. If that is the case the employer has no obvious contractual remedy but he could sue for damages for breach if the outstanding works impeded his use of occupation of the works. That apart, the contractor's agreement may be no better than his word.

Works of repair

The first part of clause 49(2) requires the contractor to 'deliver up to the Employer' the works, or any section or part for which a certificate has been issued, as soon as practicable after the end of the relevant defects correction period, in the condition:

- required by the contract (fair wear and tear excepted), and
- to the satisfaction of the engineer.

The second part of clause 49(2) continues:

'To this end the Contractor shall as soon as practicable carry out all work of repair amendment reconstruction rectification and making good defects of whatever nature as may be required of him in writing by the Engineer during the relevant Defects Correction Period or within 14 days after its expiry as a result of an inspection made by or on behalf of the Engineer prior to its expiry.'

It is far from clear from the wording of Clause 49(2) what exactly is the contractor's obligation in respect of the time for completion for repair works. The first part of the clause says 'as soon as practicable after the end of the relevant Defects Correction Period'. The second part of the clause can be taken as an obligation to complete repair works either during the relevant period or within 14 days of its expiry.

However, a better interpretation of the second part of the clause is that the engineer 'may' notify defects during the defects correction period but 'must' notify them, in any event, within 14 days of its expiry. And the inspection made by the engineer must be made before expiry of the defects correction period.

On this basis the contractor's obligation remains no more than to undertake repair works as soon as practicable after the end of the defects correction period.

Inspection by the engineer

As to the situation when the engineer fails to give notice of defects within 14 days after expiry of the defects correction period, or fails to make his inspection prior to its expiry, contractors may argue that they are then relieved of any obligation to deal with defects. It is unlikely that this argument is wholly effective since the obligation in the first part of clause 49(2), and the contractor's general obligation as expressed in clause 8 and elsewhere to complete the works in accordance with the contract may take precedence. Nevertheless this is not a matter which any engineer should risk putting to the test.

'defects of whatever nature'

The phrase 'defects of whatever nature' in clause 49(2) indicates that the clause is not confined to repairs and defects for which the contractor is responsible. This is supported by the wording of clause 49(3) which includes for repairs which may be valued as additional works and the wording of clause 51(1) which states that variations may be ordered during the defects correction period.

Cost of repair works

Clause 49(3) provides that all work required under clause 49(2) shall be carried out at the contractor's expense if 'in the Engineer's opinion it is necessary due to the use of materials or workmanship not in accordance with the Contract or to neglect or failure by the Contractor to comply with any of his obligations under the Contract'. In any other event the work is to be valued and paid for as if it were additional work.

When disputes arise, as they frequently do, on whether the contractor is to bear the cost of, or is entitled to be paid for, repair works, one point at issue is often which party has the burden of proof. Is it for the contractor to prove that his work was up to standard; or is it for the employer to prove that the work was below standard?

The test for payment as expressed in clause 49(2) is the 'Engineer's opinion' and from this it would seem to follow that it is for the employer to show that the engineer's opinion was properly founded.

Contractor's failure to carry out repairs

Clause 49(4) entitles the employer, in the event of the contractor's failure to do 'any such work as aforesaid', to carry out such work himself. The employer can recover the cost from the contractor if the work should have been carried out at the contractor's own expense.

It is not clear whether the phrase 'any such work as aforesaid' covers both outstanding works under clause 49(1) and remedial works under clause 49(2) or whether it applies only to the latter.

There is an argument for saying that because the phrase 'such work' appears in clause 49(3) relating to repairs, then the phrase only applies to repairs in clause 49(4). However, that would leave the employer without a practical remedy in the event of uncompleted work.

The difficulty with including outstanding work within the scope of clause 49(4) is that it then entitles the employer to act within the defects correction period once the period described as 'as soon as practicable' in clause 49(1) is adjudged to have elapsed.

Whatever the scope of the employer's remedy it apparently operates before the expiry of the defects correction period. This limits the general proposition that the contractor has the benefit of the defects correction period to enter the site and put right his own defects. It would seem that if the contractor fails to undertake the required work 'as soon as practicable' the benefit can be extinguished.

However, before taking action with the intention of charging the contractor, the employer needs to consider carefully how he is going to justify his action if a formal dispute arises. By acting, the employer will probably destroy the practical evidence for his case and he will need to rely on records of the defects he has put right.

19.3 Searches and trials

Clause 50 commences by obliging the contractor to carry out at the direction of the engineer:

- searches
- tests, or
- trials

as may be necessary to determine the cause of any:

- defect
- imperfection, or
- fault

if so required, in writing, by the engineer.

The clause continues by stating that unless the defect is one for which the contractor is liable under the contract, the cost of searches etc. are to be borne by the employer. If the defect is one for which the contractor is liable, the costs are to be borne by the contractor and he shall repair and make good the defect at his own expense 'in accordance with Clause 49'.

There is nothing in clause 50 expressly restricting its application to the defects correction period. However, the reference in its last line to 'expense in accordance with Clause 49' implies that it is intended for use in the defects correction period. It can also be said that clause 36(3) (checks and tests) and clause 38(2) (uncovering and making openings) deal with tests during the construction period.

Against this there is the point that clauses 36, 38 and 50 serve slightly different purposes:

- clause 36(3) deals with routine testing for quality
- clause 38(2) deals with uncovering work put out of view
- clause 50 deals with searches in respect of patent defects.

Costs of searching

The provision in clause 50 that the contractor is to bear the costs of searching if the defects are found to be his liability, otherwise the employer is to bear the costs, is not without difficulties.

As with clauses 36 and 38 there can be problems in establishing how far the employer's liability extends when the searches and tests show some of his work to be acceptable, but those searches and tests are necessitated by other defects for which the contractor is liable.

Works of repair

Where the contractor is liable for any defect, clause 50 obliges the contractor to repair and rectify at his own expense 'in accordance with Clause 49'. This latter phrase presumably involves the timescale of clause 49.

Clause 50 says nothing on the ordering or execution of repairs where the contractor is not liable, but the engineer has powers under clauses 13, 51 and 62 to instruct repairs.

19.4 Defects correction certificate

Clause 61(1) provides that:

- at the end of the defects correction period, or last such period if more than one, and
- when all outstanding works under clause 48 and all repairs under clause 49 and 59 have been completed, then
- the engineer shall issue to the employer (with a copy to the contractor) a defects correction certificate, stating
- the date on which the contractor completed his obligations to construct and complete the works to the satisfaction of the engineer.

Clause 61(2) states that the issue of the defects correction certificate shall not be taken as relieving either the contractor or the employer from any liability towards one another arising out of, or in any way connected with, the performance of their respective obligations under the contract.

19.4 *Defects correction certificate* 251

Status of the certificate

The effect of clause 61(2), combined with the wording at the end of clause 61(1) that the defects correction certificate states the date on which the contractor completed his obligations 'to construct and complete the Works to the Engineer's satisfaction', is that the defects correction certificate does not have the status of a binding final certificate. All that the certificate does is:

- terminate the contractor's entitlement to enter the site
- terminate the contractor's obligations to the extent they are concerned with performance to the engineer's satisfaction
- terminate the engineer's powers to give instructions.
- establish the last date for the submission of the contractor's final account under clause 60(4).

Note that the issue of the defects correction certificate does not directly influence the release of the final tranche of retention money under clause 60(6). That is triggered by the date which marks the end of the defects correction period – although where there are outstanding works or repairs to be done under clause 49 or 50 then monies relating to those can still be retained.

19.5 *Urgent repairs*

Clause 62 in the Seventh edition is a redraft of the Sixth edition clause but without any change of substance.

The clause which is now in three distinct parts deals with remedial work or other work or repairs which, in the opinion of the engineer, are urgently necessary. The clause refers specifically to accident or failure but it is comprehensive in its scope by its reference to 'or other event'. And it expressly says 'either during the carrying out of the Works or during the Defects Correction Period'.

The clause provides that, firstly, the engineer shall inform the contractor with confirmation in writing. Thereafter, if the contractor is unwilling or unable to carry out the work or repair, the employer can make other arrangements to undertake the work. Furthermore the cost of the work finally so done is charged to the contractor if it is his liability, otherwise the cost falls on the employer.

Chapter 20
Variations

20.1 Introduction

The Seventh edition retains unchanged from the Sixth edition clause 51 (ordered variations). It also retains, largely unchanged, clause 52(4) (notice of claims), although this is renumbered and renamed in the Seventh edition as clause 53 (additional payments).

The major changes from the Sixth to the Seventh edition are in clauses 52(1) and 52(2) on the valuation of variations. These introduce for the first time in traditional ICE Conditions quotations from the contractor as the first stage of the valuation process. This is potentially the most significant change from the Sixth to the Seventh edition.

Quotations for variations

Many standard forms of contract make provision for quotations to be submitted prior to the ordering and/or implementation of variations. Sir Michael Latham in his 1994 report *Constructing the Team* recommended that the practice in effective modern contracts should be:

'Taking all reasonable steps to avoid changes to pre-planned works information. But, where variations do occur, they should be priced in advance, with provision for independent adjudication if agreement cannot be reached.'

It is obviously essential when decisions are to be made as to whether or not to proceed with variations depending on their cost, that quotations or estimates be given in advance; similarly for options on alternative variations. It is also desirable, particularly in lump sum contracts, that the employer and the project manager should know as the works progress how the price is rising. Additionally there is the expectation that pre-pricing of variations eliminates or reduces the scope for disputes on variations. And there is the possibility, supported by anecdotal evidence to the effect that contractors frequently do well financially out of pre-pricing, that there may be a reduction in the incidence of other claims.

It is not surprising therefore, given the points in favour of pre-pricing and the general trend towards its use, that it has been introduced into the Seventh edition. However there are aspects of the Conditions and civil engineering works which need to be kept in mind as points which can undermine or counter the anticipated benefits.

The first is that civil engineering variations usually arise from necessity rather than simple change of mind by the engineer or the employer. The definition of a variation under ICE Conditions has long been something which is necessary or desirable for completion or desirable for improved functioning. There is far less commercial logic for the pre-pricing of essential variations than there is for the pre-pricing of non-essential change proposals.

A second point is that under re-measurement contracts such as ICE Conditions, certainty of the final contract price cannot be assured in the same way as for lump sum contracts. There is even the danger that the imposition of a quotation system for variations might lead some employers towards a false understanding of how the Conditions operate in establishing the final price.

Finally there is the point that where quotation systems have been introduced into civil engineering contracts, for example, the New Engineering Contract, there seems to be a relatively high incidence of breakdown of operation of the contractual mechanisms. This may be because too much is expected or attempted. It is one thing to have a quotation system which operates on a take-it or leave-it basis and which operates only occasionally. It is something else when the system applies to essential changes with high frequency.

The problems of the latter are particularly obvious where it is the intention that quotations shall include not only the cost of work changes but also delay and disruption costs. This is the intention in the New Engineering Contract and similarly the intention in the Seventh edition – although in the Seventh edition disruption costs are not covered as clearly as delay costs. It can become impractical if not impossible to accurately estimate in advance the consequential effects of a multitude of inter-related changes. Sometimes this leads to end-of-contract global claims; sometimes it leads to formalised abandonment of the contractual provisions for valuation and ad hoc adoption of a cost-plus approach; and frequently it leads to dispute.

The Seventh edition quotation system

The quotation system in the Seventh edition is grafted onto the traditional ICE rules for the valuation of variations. The old three tier approach of using bill rates for similar work, bill rates as the basis for other work so far as reasonable, and a fair valuation where not so reasonable, remains in the Conditions and applies when there is disagreement between the engineer and the contractor on the acceptability of a quotation.

This emphasises a point which should not be overlooked, that contracts under traditional ICE Conditions (including the Seventh edition) are not far short of schedules of rates contracts and the policy of the Conditions has always been to maintain the link between the bill rates and the contract price so far as reasonable.

It is unlikely that the draftsmen of the Seventh edition have intentionally set out to break the link in respect of variations – indeed when the contractor submits a quotation of his own accord he is required under the new clause 52(2) to have due regard to any rates and prices in the contract. Nevertheless it is probable that two of the most contentious aspects of the new quotation system will be firstly, whether all quotations should be rates based so far as reasonable as a matter of contractual obligation, and secondly, whether as matters of fact particular quotations are rates based. The requirement to include for delay consequences in quotations is bound to complicate this latter issue.

One of the more interesting aspects of the Seventh edition quotation system is the obligation on the engineer to either accept a quotation or to negotiate with the contractor. The engineer under ICE Conditions has not traditionally possessed powers to negotiate and no express power is conferred in the Seventh edition except by the reference to negotiation in the revised clause 52.

Legal cases on variations

The courts were particularly busy in the 1990s with cases relating to or touching on variations under ICE Conditions and other standard forms. Cases of particular interest and/or importance include:

- *McAlpine Humberoak* v. *McDermott International* (1992)
    - revised drawings did not transform the contract into a different contract or distort its substance and identity
- *Laserbore* v. *Morrison Biggs Wall* (1993)
    - a fair and reasonable valuation can be on a dayworks basis rather than a costs basis
- *Shanks & McEwan* v. *Strathclyde Regional Council* (1994)
    - engineer's letter accepting contractor's proposals for repair work held to be a variation
- *Tinghamgrange* v. *Dew Group* (1996)
    - contractor entitled to recover loss of profit charged by a subcontractor on cancellation of order arising from a main contract variation
- *Havant Borough Council* v. *South Coast Shipping* (1996)
    - contractor entitled to a variation if the specified method of construction is legally or physically impossible
- *AMEC Building* v. *Cadmus Investments* (1996)
    - contractor entitled to loss of profit on work omitted by variation but subsequently awarded to another contractor
- *Henry Boot Construction* v. *Alstom Combined Cycles* (2000)
    - not unreasonable to use bill rates in valuing variations because, by mistake, they are too high or too low.
- *Weldon Plant* v. *Commission for New Towns* (2000)
    - a fair valuation should include an element for profit and a contribution towards fixed or running overheads.

Comment on these cases is made in later sections of this chapter.

20.2 *Variations generally*

A broad definition of a variation is work which is not expressly or implicitly included in the original contract price.

In the absence of an express provision in the contract giving the employer the power to order variations, the contractor is not obliged to undertake them. The contractor's general obligation is merely to complete the work specified in the contract and such work as can reasonably be inferred for completion.

Reasons for variation clauses

Variation clauses are included in contracts principally to permit the employer to make changes and to order extras but they may have other purposes:

- to authorise the engineer/contract administrator to order variations
- to entitle the contractor to a variation if the specified work proves impossible to construct
- to define the scheme for the valuation of variations
- to prescribe procedures to be followed to recover payment
- to preclude payment for unauthorised variations
- to set financial limits on authorised variations.

Not infrequently variation clauses also seek to define what is permitted as a variation and to indicate the scope of the contractual provisions.

Scope of variation clauses

The limits of the variation provisions in any particular contract can usually be deduced from the nature of the contract work and the wording of the variation clause. Thus in the ICE Seventh edition a variation is either necessary for completion of the works or desirable for completion and improved functioning. Clearly the varied works need to have some relationship to the original work. But it does not follow that they will be a variation simply because they are carried out at the same time as the original works and on the same site. In a case under the ICE Fifth edition, *Blue Circle Industries plc* v. *Holland Dredging Co (UK) Ltd* (1987), it was held on-site that work in attempting to form an island bird sanctuary with dredged material was not within the scope of the variation clause.

The *Blue Circle* case was somewhat unusual in that it was the contractor who was seeking to extend the scope of the variation clause. Usually on a dispute over scope the contractor will seek to nullify a variation in order to be paid on a more favourable basis. This extract from the judgment of Lord Cairns in the old case of *Thorn* v. *Corporation of London* (1876) illustrates the point:

'Either the additional and varied work which was thus occasioned is the kind of additional and varied work contemplated by the contractor, or it is not. If it is the kind of additional or varied work contemplated by the contractor, he must be paid for it, and will be paid for it, according to the prices regulated by the contract. If, on the other hand, it was additional or varied work, so peculiar, so unexpected, and so different from what any person reckoned or calculated upon, that it is not within the contract at all; then, it appears to me, one of two courses might have been open to him; he might have said: I entirely refuse to go on with the contract - *non haec in foedera veni*: I never intended to construct this work upon this new and unexpected footing. Or he might have said, I will go on with this, but this is not the kind of extra work contemplated by the contract, and if I do it, I must be paid a *quantum meruit* for it.'

The logic of the Thorn decision was applied recently in the Scottish case of *ERDC Construction* v. *H. M. Love & Co* (1994) where it was held that a contractor's claim for payment for changed work on a *quantum meruit* basis was inconsistent with the contractor's continued performance of the contract and the continuing existence of the contract.

Disputes on scope

Issues which frequently arise as disputes on the scope of variation clauses are:

- is a variation necessary for a particular item of work or change? Is it the contractor's responsibility or is it already allowed for in the contract rates and prices?
- is a particular variation within the scope of the variation clause or is it something which should be valued outside the contract?

Disputes which have reached the courts on cases under ICE Conditions or similar on whether or not particular works were varied works under the contract include:

- *Yorkshire Water Authority* v. *Sir Alfred McAlpine & Son (Northern) Ltd* (1985) where it was held that the incorporation of a method statement into the contract made it a specified method of construction and the contractor was entitled to a variation order for an alternative method of working
- *Holland Dredging (UK) Ltd* v. *Dredging & Construction Co Ltd* (1987) where it was held that the contractor was entitled to a variation for measures in rectifying a shortfall in backfill to a sea outfall pipe arising from loss of dredged material at the dumping ground
- *English Industrial Estates* v. *Kier Construction Ltd* (1991) where it was held that an engineer's letter instructing the contractor to crush all suitable hard arisings deprived him of the option of importing suitable fill and was therefore a variation

- *Havant Borough Council* v. *South Coast Shipping* (1994) where it was held that a method of mechanical screening of materials in the contractor's method statement became a specified method by incorporation into the contract and if it proved impossible the contractor was entitled to a variation.

The most important modern case on whether variations can be valued outside the provisions of the contract is *McAlpine Humberoak* v. *McDermott International* (1992). The contractor contended that the number of revised drawings he had been given had effectively frustrated the contract and there had come into existence a substitute contract which entitled him to a reasonable time to complete the works and a reasonable price for performing them. The Court of Appeal held, over-ruling the judge of first instance, that the contract could not be frustrated by the operation of matters provided for in the contract (the ordering of variations) and that the contractor had failed to prove any breach of contract which entitled him to damages.

Omission variations

Variation clauses commonly include the power to make omissions – and clause 51 of ICE Conditions states that variations may include omissions.

However, that does not entitle the employer to take work out of the contract and give it to others. To do so amounts to a breach of contract for which the employer is liable in damages.

The leading legal case on this, *Carr* v. *J.A. Berriman Pty Ltd* (1953), was followed in another Australian case - *Commissioner for Roads* v. *Reed & Stuart Pty Ltd* (1974) – where it was held:

- it was a concept basic to the contract that the contractor should have the opportunity of performing the whole of the contract work
- the variation clause entitled the employer only to omit work from the contract works altogether. It did not permit the taking away of a portion of the contract work for it to be performed by another contractor.

Until recently there was some uncertainty on how these rules applied to provisional sums. Much would seem to depend on the wording of the contract in relation to the expenditure of such sums. Where they can be deleted from the contract without a variation order it was thought this could be done without creating any breach of contract. However in the case of *AMEC Building* v. *Cadmus Investments* (1996), the Court of Appeal held that the contractor was entitled to be compensated for an architect's instruction omitting work covered by provisional sums. The logic of the decision was explained by the court as follows:

'There is no dispute that the power is given to the architect in his sole discretion to withdraw any work from provisional sums for whatever reason if he considers it in the best interests of the contract or the employer to do so. The difficulty that arises in this case is that which arose in the Australian case [*Carr* v. *J. A. Berriman Pty Ltd* (1953)], namely that it would appear that the purpose was to remove it from the existing contractor and award the work to a new contractor. Without a finding that the architect was entitled to withdraw the work for reasons put forward by [counsel], and in view of the fact that the specific reasons he advances were expressly rejected by the arbitrator, it seems to me that the only conclusion I can come to is that the arbitrator had concluded that it was an arbitrary withdrawal of the work by the architect in order to give it to a third party other than AMEC. In those circumstances ... it seems to me that the arbitrator was perfectly correct in deciding that such an arbitrary withdrawal of work from the provisional sums and the giving of it to the third party was something for which AMEC were entitled to be compensated.'

In the case of *Tinghamgrange* v. *Dew Group* (1996) a change of specification led the engineer to instruct the contractor to cancel an order with a subcontractor for the supply and manufacture of pre-cast concrete blocks. The employer paid for the costs of moulds already made but refused to pay anything on the supplier's loss of profit claim. The Court of Appeal found the employer liable stating:

'The employer was, or should have been, well aware that the contractor would not itself be the supplier of the concrete blocks, but that these would be purchased under a subcontract. The employer was, or should have been, equally aware that the result of the variation which was made – the cancellation of the order for 209,874 concrete blocks – would necessarily result in a loss of profit for the subcontractor, Gryphonn, since the contractor would be contractually bound under the head contract to pass on to the subcontractor the variation consisting of the cancellation.'

The question of whether the omission of work because of the employer's financial position can validly be a variation, again depends on the terms of the contract and, in particular, the wording of the variation clauses. It is questionable whether under ICE Conditions, where the test for a valid variation is necessary for completion or desirable for completion and/or improved functioning, the employer's financial position or other wholly commercial considerations have any relevance to the test.

Limitations on the value of variations

Some contracts place a financial limit, expressed in terms of a percentage of the contract price, on the value of variations which can be ordered without the consent of the contractor. In the model forms for plant contracts this limit is 15%. In the IChemE model forms the limit is 25%.

The principal effect of these limits is that the contractor is not obliged to undertake variations which would take the revised contract price outside the stated percentages. Without such limits the contractor under the wording of most standard forms would be in breach of contract in refusing to perform any properly ordered variations whatever their value. With limits, however, the contractor has a choice. He can either accept variations which go beyond the limits and undertake them as part of the contract works or he can exercise his right of refusal.

Reasonable as this may be from a contractual viewpoint it does raise some practical problems:

- what is the position if there is a disagreement on the value of variations?
- how are the works to be completed if variations exceeding the limits are necessary for completion?
- what happens when an omission variation outside the limits is ordered?
- do variations proposed by the contractor fall within the scope of the limits?

ICE Conditions do not contain limitations on the value of variations in the sense that they define limits of the contractor's obligations. The nearest the Conditions come to any financial limitation is if a figure is stated in the appendix to the form of tender on the engineer's power to order variations without first obtaining the employer's specific approval. This is obviously intended to bind the engineer without impacting on the contractor but it is not unknown for contractors to argue that any stipulated figure has some relevance to the amount of varied work which could have been expected under the contract.

Time for ordering variations

Contractors sometimes allege that it is the time at which variations are ordered which leads to delay and disruption. From this they try to develop claims for breach of contract. The legal rule would seem to be, however, that providing variations are ordered at a time which is reasonable in the circumstances there is no breach of contract.

Thus in *Neodox Ltd* v. *Swinton & Pendlebury BC* (1958) the contractor alleged an implied term that instructions would be given to enable him to complete in an economic and expeditious manner. It was held by Mr Justice Diplock that under the terms of the contract it was clear that instructions would be given from time to time and what was reasonable did not depend solely on the convenience and financial interests of the contractor. He said:

'To give business efficacy to the contract, details and instructions necessary for the execution of the Works must be given by the Engineer from time to time in the course of the contract. If he fails to give such instructions within a reasonable time the Corporation are liable in damages for breach of contract.'

And in *A. McAlpine & Son* v. *Transvaal Provincial Administration* (1974), a South African case, a motorway contractor asked the court to define an implied term on the time for supplying information and giving instructions on variations as either:

- a time convenient and profitable to himself
- a time not causing loss and expense, or
- a time so that the works could be executed efficiently and economically.

The court declined on the grounds that under the contract variations could be ordered at any time irrespective of the progress of the works and that drawings and instructions should be given within a reasonable time after the obligation arose.

Delay and disruption effects of variations

The question of how the delay and disruption effects of variations should be valued is frequently contentious.

Amongst the issues which arise are: Has the contractor got a contractual claim other than in the variation clause or can damages be sought for breach of contract? To what extent can the contractor roll together the delay and disruption effects of variations and present them as a global claim? Can the contractor claim that variations have so distorted the contract that the original price no longer applies and the whole of the works should be valued on a cost-plus or *quantum meruit* basis?

The starting points for answers to these questions are usually simple if there is a variation clause in the contract with rules for valuation. Generally it will follow that:

- the contractor has undertaken to perform variations
- the ordering of variations is not, of itself therefore, a breach of contract
- claims relying on breach of contract are unlikely to succeed
- where the contract provides the scheme for the valuation for variations there is no place for the introduction of extra-contractual schemes
- rolled-up (or global) claims made outside the provisions of the contractual scheme for valuing variations usually suffer the difficulty of having no obvious contractual or legal basis
- only in exceptional circumstances where the scheme for valuing variations on an individual basis has broken down can rolled-up (or global claims) be made within the provisions of the contract
- claims that the contract has been distorted (or even frustrated) by the ordering of variations so as to justify abandonment of the contract price in favour of cost-plus or *quantum meruit* are unlikely to succeed.

Many of the basic principles listed above were examined in the case of *McAlpine Humberoak* v. *McDermott International* (1992) mentioned earlier in this chapter. In that case the contract for the deck structure of an off-shore drilling rig was let on a lump sum basis. The contractor claimed extra costs arising from variations. The judge, at first instance, held that the contract had been frustrated by the extent of the variations and awarded the contractor *quantum meruit*. The Court of Appeal overturned this decision and held:

- the variations did not transform the contract or distort its substance or identity
- the contract provided for variations
- the contractual machinery for valuing variations had not been displaced
- an award of *quantum meruit* could not be supported.

Duty to order variations

Some standard forms such as the model forms for plant contracts use the phrase 'the Engineer shall have the power' in relation to the making of variations. That wording seems to imply the exercise of discretion on the part of the engineer. It contrasts, for example, with the wording in the ICE Conditions 'the Engineer shall order any variation ... necessary for the completion of the Works'.

Clearly under the ICE Conditions, contractors are entitled to a variation order if the specified work or methods prove impossible to perform. Thus in *Yorkshire Water Authority* v. *Sir Alfred McAlpine & Son (Northern) Ltd* (1985) it was held that incorporation of a method statement into a contract made it a specified method of construction and when that method proved impossible the contractor was entitled to a variation for an alternative method of working. And in *Holland Dredging (UK) Ltd* v. *Dredging & Construction Co Ltd* (1987) it was held that the contractor was entitled to a variation order for measures in rectifying a shortfall in backfill to a sea outfall pipe arising from loss of dredged material. See also the *Havant Borough Council* case mentioned in section 20.1 above.

However, even where the wording of the contract seems to suggest discretion, for example by use of the phrase 'The Engineer may order', the proper interpretation of the contract is usually that there is a duty on the engineer to order variations which are necessary for completion of the works. But that, of course, is subject to consideration of the contractor's responsibility for design and impossibility of the contractor's own making or default.

20.3 *Variations in the Seventh edition*

Clause 51 of the Seventh edition is identical to clause 51 of the Sixth edition. The clause is in four parts:

- 51(1) – ordered variations
- 51(2) – ordered variations to be in writing

- 51(3) – variations not to affect the contract
- 51(4) – changes in quantities.

Ordered variations

Clause 51(1) commences by providing that the engineer:

(a) shall order any variation to any part of the works which in his opinion is necessary for the completion of the works, and

(b) may order any variation that for any other reason is in his opinion desirable for the completion and/or improved functioning of the works.

The second part of clause 51(1) states that:

'Such variations may include additions omissions substitutions alterations changes in quality form character kind position dimension level or line and changes in any specified sequence method or timing of construction required by the Contract and may be ordered during the Defects Correction Period.'

Necessary variations

The provision that the engineer 'shall', in the imperative, order any variation necessary for the completion of the works is of obvious benefit to the contractor but less so to the employer. Amongst other things it transfers the burden of dealing with impossibility from the contractor to the engineer, thereby eliminating from the Conditions the proposition that the employer does not warrant that the works can be completed. It also puts the engineer under pressure to intervene when the contractor is in difficulties. As discussed in Chapter 9 on impossibility, there is a fine dividing line in construction between impossibility and difficulty.

There is also a certain lack of economic realism in the provision that the engineer 'shall' order variations in so far as there is no point in the engineer being obliged to issue variations which are beyond the financial resources of the employer or which entail continuing with wasted expenditure on a project already rendered financially non viable.

Clauses of the contract which have a direct bearing on the engineer's duty to issue variations include:

- clause 5   – documents mutually explanatory
- clause 7   – further drawings and instructions
- clause 12 – adverse physical conditions
- clause 13 – engineer's instructions
- clause 20 – excepted risks

- clause 27 – street works
- clause 62 – urgent repairs

Other clauses can impose a duty in appropriate circumstances.

Desirable variations

The limitations on the power of the engineer to order 'desirable' variations may come from financial controls imposed by the employer – as covered by clause 2(1)(b) of the Conditions and the corresponding appendix entry – or from lack of proximity of function as described in the *Blue Circle* case.

Lack of proximity in the geographical sense may also be relevant if the varied work extends beyond the site boundaries, although the power to extend the site under clause 1(1)(v) could be helpful to the engineer in appropriate circumstances.

It is not thought that the words 'for any other reason' in clause 51(1)(b) are intended to confer a general power on the engineer to order variations. They probably mean no more than the words 'in his opinion necessary' do in relation to clause 51(1)(a). It is doubtful therefore if the engineer can, for example, order omission variations purely as cost saving measures – although in reality it has to be said that such practice is commonplace.

Variation by agreement

Where additional work with recovery of cost is involved contractors do not always oppose variations of dubious validity but sometimes welcome them as an opportunity of increasing turnover and profitability. So a contractor will often accept a variation which is outside the scope of clause 51, as either necessary or desirable, and to the extent that he is prepared to do so and the employer is of the same mind such a variation serves as a substitute for a separate works agreement. However, when this is done it is best that the variation is labelled a 'variation by agreement' to avoid any possibility of later dispute that it was unauthorised, unacceptable or void.

Range of variation items

The range of items listed in clause 51(1) as permissible variations is unusually wide compared with other construction contracts and the reference to changes in 'sequence method or timing' appears to give the engineer extensive powers to interfere in, or control, the contractor's arrangements.

Taken to the extreme there is a view that the engineer has power to order the contractor to accelerate or complete within a shorter time than that allowed in the contract, although the provisions in clause 46(3) for accelerated completion suggest this is a matter to be addressed by agreement rather than by instruction.

Variations to be in writing

Clause 51(2) states that all variations shall be ordered in writing but the provisions of clause 2(6) in respect of oral instructions shall apply.

Clause 2(6) is to the effect that if the contractor is given an oral instruction by the engineer or by any authorised person acting under him and the contractor confirms such oral instruction in writing, then, if such confirmation is not contradicted in writing forthwith, the instruction is deemed to be an instruction in writing.

Engineer's duty to order

It does not wholly follow from clause 51(2) that for there to be a variation there must either be an order in writing or written confirmation of an oral instruction. Such an interpretation would effectively define a variation to the exclusion of matters where the engineer fails to give an instruction, written or oral, notwithstanding that he might have a duty to do so. It is clear from many legal decisions – *Brodie* v. *Cardiff Corporation* (1918) and the various cases discussed earlier in this chapter – that the courts will look at the substance of whether or not there has been a variation within the meaning of the contract and that the contractor does not lose his entitlement to the valuation of a valid variation simply because it has not been formally ordered. The employer is not allowed to benefit from the engineer's failure to perform his duties.

However, the contractor needs to be aware that if he has no written instruction from the engineer or written confirmation of his own of an oral instruction then he will have to prove not only that he performed varied work but that he was ordered to perform it or that it was otherwise necessary as a variation within the terms of the contract. The contractor is not entitled to be paid for extra or varied work simply because he has done it. Consequently whilst the principal function of clause 51(2) is to place a duty on the engineer to ensure that variations are ordered in writing, it can be seen as having a secondary function in protecting the employer against claims for unauthorised variations.

Variations to be valued

Clause 51(3) which has the marginal note 'Variation not to affect Contract' provides:

- no variation ordered in accordance with clause 51 shall in any way vitiate or invalidate the contract

- the value (if any) of all variations shall be taken into account in the contract price except to the extent that the variation is necessitated by the contractor's default.

The marginal note hardly does justice to the content of the clause. Its importance lies as much in its statements on the valuation of variations as on its statement that variations shall not vitiate or invalidate the contract.

The intention of the statement on non-vitiation/non-invalidation may simply be to put beyond doubt that the performance of individual or multiple variations has to be accommodated by the contractor within the terms of the contract. Since the Conditions expressly provide for variations the argument that any particular valid variation could vitiate or invalidate the contract would be difficult to sustain even without the clause 51(3) provision. But it is possible to envisage a variation of such significance that it might be beyond the expectation or expertise of the contractor – for instance, a change from a bolted structure to a welded structure. And in such circumstances, in the absence of the clause 51(3) provision the contractor might have a case for arguing vitiation or invalidation.

More commonly, the claim for vitiation/invalidation is made when there has been a multitude of variations and the contractor seeks to move away from the contractual basis of pricing the contract works to cost-plus or *quantum meruit*. For the difficulty in advancing that see the *McAlpine Humberoak* case mentioned earlier in this chapter.

Contractor's default

It would seem to go without saying that the contractor would not be entitled to payment for a variation which was necessitated by his own default. But contracts are taken to mean what they say and there would have been a danger in leaving the clause unqualified on the default point after the decision in the case of *Simplex Concrete Piles* v. *St Pancras Borough Council* (1958) where an architect who assented to a contractor's proposals for overcoming defective work by his piling subcontractor was held to have issued a variation which entitled the contractor to payment.

Acceptance of contractor's proposals

The *Simplex* case and others since have shown how careful engineers or architects must be in accepting contractor's proposals for alternatives to the designed works or the rectification of defaults. Generally, if the engineer accepts an alternative to his design, that becomes the specified method or requirement. If the method proves impossible or the requirement unsatisfactory the engineer is then bound to issue instructions under clause 13 or a variation under clause 51(1). The contractor is then entitled to be paid because he is not contractually 'in default'.

Changes in quantities

Clause 51(4) states that no order in writing is required for increase or decrease in the quantity of any work where such increase or decrease is not the result of an order given under clause 51 but is the result of the quantities exceeding or being less than those stated in the bill of quantities.

This is a somewhat ambiguous clause which may be saying that changes in quantities are not to be regarded as variations or, alternatively, that changes in quantities are variations but they do not require an order in writing.

The most likely intention, having regard to other clauses in the Conditions, is that changes in quantities are not to be taken as variations. This would seem to follow from the separate provision in clause 44 (extension of time) for increased quantities and the provisions in clause 56(2) for increases or decreases of rates for changed quantities. Additionally, it is difficult to see how changes of quantities, if variations, would fit into the scheme for the valuation of variations in clause 52.

20.4 *Valuation of variations by quotations*

Clauses 52(1) and 52(2) of the Seventh edition are the newly introduced clauses dealing with the valuation of variations by quotation. In their entirety they read as follows:

'52 (1) If requested by the Engineer the Contractor shall submit his quotation for any proposed variation and his estimate of any consequential delay. Wherever possible the value and delay consequences (if any) of each variation shall be agreed before the order is issued or before work starts.

52 (2) Where a request is not made or agreement is not reached under sub-clause (1) the valuation of variations ordered by the Engineer in accordance with Clause 51 shall be ascertained as follows:
(a)   As soon as possible after receipt of the variation the Contractor shall submit to the Engineer
(i)   his quotation for any extra or substituted works necessitated by the variation having due regard to any rates or prices included in the Contract and
(ii)   his estimate of any delay occasioned thereby and
(iii) his estimate of the cost of any such delay
(b)   Within 14 days of receiving the said submissions the Engineer shall
(i)   accept those submissions or
(ii)   negotiate with the Contractor thereon.
(c)   Upon reaching agreement with the Contractor the Contract Price shall be amended accordingly.'

Structure of the quotation scheme

By clause 52(1) the contractor is required to submit quotations for proposed variations – but only if requested to do so by the engineer. By clause 52(2) the contractor is required to submit quotations for ordered variations – but only where a request has not been made or agreement reached under clause 52(1).

Looking at the clauses another way it can be said that clause 52(1) applies when the engineer requests a quotation and clause 52(2) applies when there is no such request.

Both viewpoints are too simplistic, however, to provide a definitive and consistent structure. Problems lie in many areas of interpretation of the clauses. They start with the relationship between the first and second sentences of clause 52(1); the relationship between that second sentence and the first sentence of clause 52(2); the meaning of that first sentence; and the relationship between clause 52(1) and the opening sentence of clause 52(3).

The opening sentence of clause 52(3) reads:

'(3) Failing agreement between the Engineer and the Contractor under either sub-clause (1) or (2) the value of variations ordered by the Engineer in accordance with Clause 51 shall be ascertained by the Engineer in accordance with the following principles and be notified to the Contractor.'

At first sight it appears that the reference to agreement in the second sentence of clause 52(1) is to agreement on the quotation requested in the first sentence. That would be entirely logical and proper. However, if the agreement is a reference to such a quotation then when there is no agreement, two apparently inconsistent but mandatory obligations follow: the contractor is required to submit a further quotation under clause 52(2) and the engineer is to ascertain the value under clause 52(3).

It may be that the second sentence of clause 52(1) is merely an exhortation to agree the value of all variations before they are ordered or commenced (and the Guidance Notes support this) and is not intended to apply only to quotations provided under the first sentence. It may be that it contemplates agreement between the contractor and the engineer which is not based on a quotation. If either is correct the provisions of the second sentence could have been better located.

As to the opening sentence of clause 52(2) the request which is referred to seems to be a request by the engineer for a quotation for a proposed variation. The contractor's obligation under clause 52(2), however, extends only to giving quotations for ordered variations. Possibly, therefore, although it may offend the rules of construction, the words 'Where a request is not made' at the start of the sentence should be detached from the reference to clause 52(1) and should be seen as free-standing in respect of ordered variations under clause 52(2).

Various possibilities emerge from the above. The following are considered the most likely:

- the engineer has the power to request a quotation for a proposed variation. If the quotation is not acceptable the engineer can either:
  - decide not to proceed, or
  - order the variation and accept or negotiate on a quotation provided under clause 52(2) or
  - order the variation and value the work under clause 52(3)
- the engineer can seek to agree the value of any variation without requesting a quotation but if no agreement is reached can order the variation and proceed as above under clauses 52(2) or 52(3)
- the contractor is obliged to submit a quotation for a proposed variation if requested to do so
- the contractor is obliged to submit a quotation for an ordered variation if there is no prior agreement on its valuation.

Details of the quotation scheme

Clause 52(1) is conspicuously light on details. The contractor is to submit his quotation and his estimate of any consequential delay. It is not wholly clear if the estimate of delay is to be in terms of time and money or just time.

Clause 52(2) by contrast has considerable detail. The contractor is to submit as soon as possible after receipt of the variation:

- a quotation for extra or substituted work (having regard to bill rates and prices)
- an estimate of delay
- an estimate of the cost of any delay.

Then within 14 days the engineer is to accept the submissions or negotiate with the contractor.

It may have been the intention of the draftsmen of the Conditions that some, if not all, of the detail in clause 52(2) should be applicable to quotations under clause 52(1), but that does not follow naturally from the manner in which the clauses are set out and drafted.

Nevertheless, it is reasonably obvious that the provision in clause 52(2)(c) for the contract price to be adjusted to cover any agreed valuation should apply to clause 52(1) as well as to clause 52(2). It is less obvious whether the requirement in clause 52(2)(a)(i) for quotations to have due regard for contract rates and prices and the obligation in clause 52(2)(b)(ii) for the engineer to negotiate, should also apply to clause 52(1).

One logical explanation for some of the differences between clause 52(1) and clause 52(2) details might be that clause 52(1) is intended for 'proposed' variations which might or might not be implemented – depending on the acceptability of the quoted price – whereas clause 52(2) is intended for 'ordered' variations to be implemented in any event. But that explanation is undermined by the wording of clause 52(3) which applies rates-based fall back valuations to both clauses 52(1) and 52(2).

Regard to contract rates and prices

The requirement in clause 52(2)(a)(i) for quotations for ordered variations to have regard to contract rates and prices cannot be intended as full application of the rules in clause 52(3); such application, by clause 52(3), only comes into play when there is no eventual agreement. Presumably it means some looser association – perhaps with inclusions for risk which would not normally be allowed in the operation of clause 52(3).

Delay and disruption

Both clauses 52(1) and 52(2) refer to delay and its consequences – although, for reasons which are not apparent, not in the same terms. Neither clause refers to disruption and it may be intended that the costs of delay are to include the costs of disruption. If this is the case it would be better if stated openly.

Engineer to accept or negotiate

The obligation under clause 52(2)(b) for the engineer to accept or negotiate within 14 days of receipt of the contractor's quotation and delay estimates raises some interesting questions:

- are the contractor's submissions deemed to be accepted if the engineer does not respond within 14 days?
- are the contractor's submissions deemed to be accepted if the engineer fails to engage in meaningful negotiations?
- can the engineer negotiate on a purely commercial basis (without regard to contract rates and prices) and if so from where does he get this power?
- can the employer challenge the outcome of the engineer's acceptance and/ or negotiations in dispute resolution procedures?

On the first two questions it is probably the case that the contractor's submission can be deemed to be accepted. It is difficult to see how the engineer can properly move the valuation to the next stage in clause 52(3) if he has not established disagreement within the stipulated 14 days.

As to negotiations, it is suggested that engineers should clear in advance the extent of their powers to negotiate and if such powers are subject to restriction the appropriate entries should be made in the appendix to the form of tender under clause 2(1)(b).

On the employer's challenge, it is worth noting that neither the list of matters for dissatisfaction under clause 66(2)(b) nor the list of matters within an arbitrator's powers under clause 66(11)(a) include opening up and reviewing acceptance of quotations or negotiations.

Omission variations

It is doubtful if the quotation schemes of either clause 52(1) or clause 52(2) apply to omission variations. Both schemes seem to contemplate delay and clause 52(2)(a)(i) refers expressly to extra or substituted works.

Contract price to be amended

The point is made above that it seems odd that the provisions in clause 52(2)(c) for the contract price to be amended to reflect any agreement on the value of a variation do not expressly extend to clause 52(1) agreements.

Another odd point is that clause 52(2)(c) refers to price but not to extension of time. If anything, it is the latter which is the more necessary. Clause 51(3) already requires the value of variations to be taken into account in the contract price, whereas clause 44 only provides for extension of time on an application and review basis.

Perhaps a distinction is intended and is to be made between delay and extension of time for the purposes of clauses 52(1) and 52(2).

Concluding comment

It is disappointing that the introduction of quotations for variations into standard ICE Conditions should be by way of provisions which raise more questions than they answer. Inevitably amendments will be made on an ad hoc basis with the aim of improvement. The sooner the clauses are formally amended and re-issued the better.

20.5 *Valuation of variations by rates/fair valuation*

Clause 52(3) of the Conditions brings the valuation of variation rules contained in clause 52(1) of the Sixth edition into the Seventh edition as fall-back provisions when the quotation scheme fails to produce agreement. Clause 52(3) reads:

'(3)   Failing agreement between the Engineer and the Contractor under either sub-clause (1) or (2) the value of variations ordered by the Engineer in accordance with Clause 51 shall be ascertained by the Engineer in accordance with the following principles and be notified to the Contractor.

(a)   Where work is of similar character and carried out under similar conditions to work priced in the Bill of Quantities it shall be valued at such rates and prices contained therein as may be applicable.

(b)   Where work is not of a similar character or is not carried out under similar conditions or is ordered during the Defects Correction Period the rates and prices in the Bill of Quantities shall be used as the basis for valuation so far as may be reasonable failing which a fair valuation shall be made.'

The 3 tier rules

Clause 52(3), or clause 52(1) as it was in earlier editions of the Conditions, is often described as a 3 tier rule system of valuation, i.e.:

* rule 1 – valuation at bill rates and prices
* rule 2 – valuation using bill rates and prices so far as is reasonable
* rule 3 – fair valuation.

The question, for any particular valuation, as to which tier of the system is applicable is usually how closely the varied work matches in description and method/timing of construction the billed items of work. The answer to the question is dependent on the facts. This is particularly the case for the transition from rule 1 (valuation at bill rates) to rule 2 (valuation using bill rates as the basis).

Operation of rule 3 – fair valuation

The transition to rule 3 (fair valuation) brings in a point of legal interpretation – the meaning of the phrase in clause 52(3)(b) 'so far as may be reasonable'. This has long been a point of debate. Does it refer to the process of calculation; or does it refer to the outcome of the calculation?

The definitive legal answer was provided in the case of *Henry Boot Construction Ltd* v. *Alstom Combined Cycles* (2000) in a dispute under the ICE Sixth edition Conditions. The contractor had by mistake entered particularly good rates for certain items of work in the bills. Similar work was ordered by way of variation. The engineer, and later the arbitrator, decided that it was not reasonable in the circumstances to use the bill rates as the basis of valuation. The matter ended up in the courts where the Court of Appeal held as follows.

Per Lord Lloyd:

'Clause 52(2) on which Lord Neill relied, provides an exception. But it is an exception which proves the rule. It applies "if the nature or amount of the variation relative to the nature or amount of the contract work" is such as to make it unreasonable to apply the contract rates to the variation. In such a case, the engineer may fix a reasonable rate. Thus, clause 52(2) creates a limited exception to Rules 1 and 2 where the scale or nature of the variation makes it unreasonable to use the contract rates. It certainly does not justify displacing the rates themselves because they were inserted by mistake or are too high or too low or otherwise unreasonable.

The same applies to another provision on which Lord Neill relied, namely, clause 56(2). It enables the engineer to increase or decrease the rates where "the actual quantities executed in respect of any item [are] greater or less than those stated in the Bill of Quantities". These limited exceptions underline the basic rule that the rates themselves are not subject to correction.

Any other view would have far reaching consequences. If the engineer were free to open up the rates at the request of one party or the other because they were inserted in the bill of quantities by mistake, it would not only unsettle the basis of competitive tendering, but also create the sort of uncertainty in the administration of building contracts which should be avoided at all costs.'

Per Lord Justice Bedlam:

'Under Clause 52(1)(b), in the case of work not of a similar character or not executed under similar conditions, the rates and prices in the Bill of Quantities are to be used as the basis for valuation "so far as may be reasonable". Thus the clause contemplates cases in which it may be reasonable to use rates and prices in the bill of quantities and in others not but it is the reasonableness of using the rates and prices, and not the reasonableness of the prices or rates, which has to be considered. I agree with the judge and Lord Lloyd that the fact that the price stated in the bill of quantities referred by mistake only to the works in the turbine hall was not a material consideration in deciding whether it was reasonable to use the price as the basis for valuing the other additional works. Insofar as the arbitrator rejected the price of £250,880 as a basis for valuation because it was not reasonable to enlarge the ambit of the mistake, in my view he took into account an irrelevant consideration and his decision cannot stand.'

Fair valuation

In the event that a fair valuation does become due, the question arises – what is a fair valuation?

More often that not it is taken as cost plus – and if that satisfies the parties then that is appropriate. However, cost plus will not always produce a fair valuation or one that is acceptable to the contractor. The problem is that cost, in relation to contractor owned plant, machinery, cabins etc., is usually no more than depreciation plus running costs rather than, as contractors would have it, notional or in-house hire cost. See the discussion on cost in Chapter 21.

The point came up in the case of *Laserbore Ltd* v. *Morrison Biggs Wall Ltd* (1993). In that case the work was carried out under the terms of a letter which stated that Laserbore would receive fair and reasonable payment for all work executed. Laserbore claimed on a dayworks rate basis using the FCEC (Federation of Civil Engineering Contractors) Schedules. Morrison Biggs

Wall argued that payment should be on a cost-plus basis. The judge came down in favour of a rates-based valuation and he rejected the cost-plus approach with this analysis:

'I am in no doubt that the costs-plus basis in the form in which it was applied by the defendants' quantum experts (though perhaps not in other forms) is wrong in principle even though in some instances it may produce the right result. One can test it by examples. If a company's directors are sufficiently canny to buy materials for stock at knockdown prices from a liquidator, must they pass on the benefit of their canniness to their customers? If a contractor provides two cranes of equal capacity and equal efficiency to do an equal amount of work, should one be charged at a lower rate than the other because one crane is only one year old but the other is three years old? If an expensive item of equipment has been depreciated to nothing in the company's accounts but by careful maintenance the company continues to use it, must the equipment be provided free of charge apart from running expenses (fuel and labour)? On the defendants' argument, the answer to those questions is, "Yes". I cannot accept that that begins to be right.

One problem in this case is that the plaintiffs' expert has in some instances relied not on general market rates but on institutional rates which are said to be too high in the circumstances. The plaintiffs' expert relies in some instances on FCEC daywork rates. The defendants' expert objects that use of those rates involves an element of double charging in that FCEC daywork rates include charges for insurance and head office charges which are separately made in Mr Spence's calculations. The experts have agreed a reduction in respect of those charges. Once the agreed reduction is made, I see no objection to the use of FCEC daywork rates in the instances where they have been used.

Despite those findings of principle, it is necessary to go through the points of difference between the experts and consider in respect of each point of difference what is fair and reasonable payment.'

Overheads and profit in a fair valuation

Where cost does form the basis of a fair valuation there can be questions on the extent to which head office overheads and profit should be included. Those questions were resolved in the case of *Weldon Plant* v. *Commission for New Towns* (2000) where it was held that in the absence of special circumstances a fair valuation under clause 52(1) of ICE Conditions Sixth edition has to include an element of profit and an element to cover a contribution towards fixed and running overheads. The judge said:

'Indeed in my judgement a fair valuation must, in the absence of special circumstances (none of which have been identified by the arbitrator), include an element on account of profit. First, a contractor is in business to make a profit on the costs of deploying its resources, and accordingly an employer must under clause 52(1) pay profit in a valuation made under any rule (via the rates or otherwise on a fair valuation) on costs for a valuation under clause 52 would not otherwise be a fair valuation within the meaning of those words. Secondly, a valuation which did not include profit would not contain an element which is an integral part of a valuation under rules 1 and 2. A fair valuation under rule 3 would not be in accordance with the principles of clause 52 if it did not include all relevant elements to be valued or represented in some significant manner in a valuation under that clause.

Overheads require separate consideration. It is important to be clear which elements of overheads are involved. Some overheads, e.g. site overheads, may be constant and not normally related to base costs (unless they are brought into an assessment of the costs of prolongation by reference to base costs); others are related to them. Overheads may be directly related to the value of work. Thus some will only be recovered if there is proof that they were in fact incurred or increased as they will be or will have been recovered from valuations of the work executed. The arbitrator correctly recognised that "the ICE Conditions of Contract undeniably allows for the addition of an overhead addition (in some form or another) to nett costs". The arbitrator took the view that time related overheads required to be established. To that extent, as a matter of fact, no exception could be taken to his approach. However, the arbitrator did not deal with the addition which in my judgement has to be made in order to ensure that the contractor obtains a contribution from the costs of the business it undertakes towards its fixed or running overheads. For reasons that I have already given it would not be fair if the valuation did not include an element on account of such contribution. It would mean that such a contribution would have to be found elsewhere, presumably from the contractor's margin for profit or risk. In my view a valuation which in effect required the contractor to bear that contribution itself would not be a fair valuation, in accordance with the principles of clause 52(1) which are intended to secure that the contractor shall not lose as a result of having to execute a variation (except, as I have stated, to the extent its costs etc. are of its making). Unlike overheads such as time-related overheads, it is not necessary to prove that they were actually incurred for the purposes of a fair valuation (although their approximate amount must of course be established, e.g. by deriving a percentage from the accounts of the contractor including where appropriate associated companies that provide services or the like that qualify as overheads).'

20.6 *Engineer to fix rates*

Clause 52(4) of the Seventh edition is a renumbered version of clause 52(2) of the Sixth and previous editions. The clause reads:

'(4) If in the opinion of the Engineer or the Contractor any rate or price contained in the Contract for any item of work (not being the subject of any variation) is by reason of any variation rendered unreasonable or inapplicable either the Engineer shall give to the Contractor or the Contractor shall give to the Engineer notice before the varied work is commenced or as soon thereafter as is reasonable in all the circumstances that such rate or price should be increased or decreased and the Engineer shall fix such rate or price as in the circumstances he shall think reasonable and proper and shall so notify the contractor.'

Operation of clause 52(4)

Clause 52(4) has the potential to operate at two levels. The first is the straightforward adjustment of a directly affected rate; for instance if the original contract work involves off-site disposal of excavated material and importation of fill and a variation is ordered that excavated material is to be used on site for landscaping mounds or the like, the contractor may have a case for seeking a change in the rate for imported fill on the basis that it was priced on a return load basis with off-site disposal of excavated material.

The second potential use of clause 52(4) is more complex. It derives from the proposition that clauses 52(3) and 52(4) between them provide a complete code for the valuation of variations (other than by quotation). That is to say the contractual scheme for the valuation of variations covers not only the valuation of the varied work itself but also the knock-on effects of delay and disruption. It is difficult to fault the logic of this proposition. If, as seems to be the case from various legal decisions, most notably *McAlpine Humberoak Ltd* v. *McDermott International* (1992), the contractor cannot claim for the delay and disruption effects of variations on a global basis as breach or on a *quantum meruit* basis, the contractor is forced back to his contractual remedies – i.e. the operation of clauses 52(3) and 52(4).

That leads to the conclusion that if clause 52(3) is taken to be effective for valuing the varied work itself, then clause 52(4) provides the remedy for delay and disruption.

Clause 52(4) adjustments

Clause 52(4) provides for the revaluation of bill rates and prices rendered unreasonable or inapplicable by the nature or amount of any variation. The revaluation can apply to any rates or prices so affected and can be in either direction; that is an increase or a decrease.

The test to be used is expressed in the most general terms 'if the nature or amount of any variation relative to the nature or amount of the whole of the contract or any part thereof shall be such ... that any rate or price in the contract ... is by reason of such variation rendered unreasonable or inapplicable'. Some employers attempt to regulate the use of the clause by issuing guidelines to their engineers suggesting that only variations above a certain percentage value, 10% or 15% or so, should trigger revaluation of rates but this cannot replace the operation of the clause under its proper construction.

Giving notice

Clause 52(2) requires either the engineer to give the contractor, or the contractor to give the engineer, notice requiring a revaluation. Notice is to be given before the varied work is commenced, or as soon thereafter as is reasonable in all the circumstances.

The provision for notice from either the engineer or the contractor will normally operate such that the engineer looks for reductions and the contractor looks for increases but if a bundle of rates is altered it may well be that some go up and some go down.

The question of timing of notices under near identical provisions to clause 52(4) in the ICE Second edition was considered in the case of *Tersons Ltd* v. *Stevenage Development Corporation* (1963). In that case the phrase 'as soon thereafter as is practicable' was given wide interpretation although the decision of the Court of Appeal made clear that compliance with the notice condition was a condition precedent to payment. This question of timing was considered again in the case of *Hersent Offshore* v. *Burmah Oil* (1978) under conditions similar to the ICE Fourth edition and there it was held that notice given after the varied work had been completed had not been given 'as soon thereafter as is practicable'.

20.7 *Dayworks*

Clause 52(5) provides that:

- the engineer may if in his opinion it is necessary or desirable
- order in writing
- any additional or substituted work to be carried out on a daywork basis
- in accordance with clause 56(4).

Clause 52(5) does no more than give the engineer power to order 'additional or substituted' work to be executed on a daywork basis. The engineer has no power to allow any of the original work to be executed on a daywork basis – that would amount to relieving the contractor of his obligation to construct the works at the rates and prices in his tender. And, as stated in clause 2(1)(c), the engineer has no authority to amend the terms of the contract or relieve the contractor of his obligations.

The detailed procedures for records and valuations for dayworks are in clause 56(4).

Chapter 21
Claims and additional payments

21.1 Introduction

In the Sixth and earlier editions of the Conditions the procedures for claiming additional payments were detailed in clause 52(4) which carried the marginal note 'notice of claims'. The Seventh edition retains identical procedures but the old clause 52(4) is renumbered as clause 53 and the marginal note changed to 'additional payments'. The marginal note change is appropriate because the old clause 52(4) was never a notice of claims clause in a general sense – nor is the renumbered clause in the Seventh edition.

Clause 53 deals expressly with claims for additional payments pursuant to clauses of the Conditions and it is doubtful if it has any application to claims for damages for breach of contract or misrepresentation. And even if the scope of the clause can be stretched to include claims for breach of express terms it is unlikely that it goes so far as to encompass claims for breach of implied terms.

Clause 53 – additional payments

The provisions of clause 53 can be briefly summarised as follows:

- the contractor is to give notice of claims within a defined time
- the contractor is to keep such contemporary records as necessary
- the engineer may instruct the contractor to keep further records
- the contractor is to permit the engineer to inspect his records
- the contractor is to submit a first interim account
- the engineer may require the contractor to submit up-to-date accounts
- the contractor is to submit such accounts with accumulated totals
- if the contractor fails to comply with the provisions of clause 53 the contractor is entitled to payment only to the extent that the engineer is not prejudiced from investigating the claims
- the contractor is entitled to interim payments of amounts considered due by the engineer
- where the contractor has not substantiated the whole of the amount claimed he remains entitled to payment of any amount which is substantiated.

21.2 Legal basis of claims

Claims for additional payments will either be claims under the terms of the contract or claims at common law. They are sometimes known as contractual claims and extra-contractual claims.

Contractual claims

The legal basis of contractual claims is the contract itself. The claiming party, usually the contractor, simply has to establish that the conditions for additional payment have occurred in order to acquire an entitlement to payment.

Subject to considerations of legality the parties are free to contract on whatever terms they wish. They cannot contract out of statutory obligations or restrictions but they can, and frequently do, depart from the rules of common law in apportioning risk and responsibilities.

So it will often be the case that contractual claims can be made where there would be no common law entitlement. For example, under ICE Conditions the contractor can claim for the additional costs of unforeseen ground conditions and for additional costs arising from labour tax and landfill tax fluctuations – neither of which could be the basis of a common law claim for breach.

Contrary to popular belief contracts need not be fair. The Unfair Contracts Terms Act 1977 sometimes mentioned in this context has more to do with preventing a party from excusing itself from its own negligence than with any general concept of fairness. But generally ICE Conditions are fair in the sense that they give contractors more opportunities to claim additional payments than does the common law.

Contractual claims also have other advantages for contractors in that interim payments may be permitted and the basis of recovery may be clearly defined. Against this, and leaning towards the advantage of the employer, is that contractual entitlements to payment may have strict notice and other administrative requirements.

Common law claims

In the main, common law claims depend upon establishing a breach of contract. The breach will usually be related to an express term of the contract which places on one party an obligation to the other; for example, to give possession of the site or to provide information necessary for construction to proceed.

But the breach may also be related to a term implied into the contract to give it business efficacy; for example, that one party will not hinder the other in the performance of the contract. However, it is not the purpose of implied terms to improve the contract that the parties have made and a term will not be implied when there are express terms in the contract dealing with the point at issue.

Contracts rarely provide provisions for the submission, valuation or payment of common law claims and strictly speaking they are claims which should be dealt with outside the contract directly between the contractor and the employer – hence the name, extra-contractual claims.

One aspect of these claims which is frequently overlooked is that the engineer may not have any, or only limited, power or authority under the contract to deal with them. The powers and authority of the engineer derive solely from the contract and as a general rule they do not extend to extra-contractual matters.

In practice, however, engineers do often become involved in claims made on a common law basis. This is partly because it is not always clear on which basis a claim is being made. It is also partly because the engineer is often the only administrative link between the contractor and the employer. It might be said that when the engineer does become so involved he is in effect acting as the agent of the employer. In this capacity he would not normally be required to be impartial or to make a fair valuation. His task would be to act in the best interests of the employer. This could involve getting the best deal he could for the employer.

Under ICE Conditions, however, the position of the engineer is complicated by the fact that although he is restrained by various clauses to valuing the amount due to the contractor 'under the contract', he is obliged by clause 66 to give decisions on matters arising 'under or in connection with the contract'.

Alternative rights of claim

In most contracts, and ICE Conditions are no exception, there is a measure of overlap between contractual rights of claim and common law rights. Clause 42(3) on failure to give possession of the site is an obvious example of a contractual entitlement covering a potential breach. The question then arises – are the contractual rights the limit of the contractor's entitlement or do common law rights stand alongside as an alternative? The point is relevant in relation to giving, or failing to give, notice of intention to claim; and in relation to the amount recoverable where the contractual amount is less than that recoverable as common law damages.

The position varies from form to form but the answer to the question is sometimes found in the contract itself. Thus in JCT 80 the contract says with reference to claims:

'The provisions of this Condition are without prejudice to any other rights and remedies which the Contractor may possess.'

However, in contrast the IMechE/IEE standard form, MF/1, states:

'The Purchaser and the Contractor intend that their respective rights, obligations and liabilities as provided for in the Conditions shall be exhaustive of the rights, obligations and liabilities of each of them to the other arising out of, under or in connection with the Contract or the Works, whether such rights, obligations and liabilities arise in respect or in consequence of a breach of contract or of statutory duty or a tortious or negligent act or omission which gives rise to a remedy at common law.'

The strict application of this particular exclusion clause was upheld by the Court of Appeal in the case of *Strachan & Henshaw* v. *Stein Industrie (UK) Ltd* (1997) even to the extent of excluding claims under the Misrepresentation Act 1967.

There is nothing in ICE Conditions which expressly states that common law rights are included or excluded. In such circumstances the general rule that legal rights can only be excluded by clear express words would seem to apply – see for example, *Hancock* v. *Brazier* (1966) and *Milburn Services Ltd* v. *United Trading Group Ltd* (1995). However, there is an argument that where the parties have expressly agreed a procedure for dealing with a particular type of breach that is to be taken as the exclusive remedy.

Whatever the correct answer to the question of alternative rights, it is important to note that the point is only relevant to claims where there is both a contractual provision for claim and a corresponding common law right of claim. It has no relevance to claims such as clause 12 for unforeseen physical conditions which give rights additional to breach of contract claims.

21.3 *Common law claims*

Common law claims fall into various categories:

- claims for breach of an express term or an implied term of the contract
- claims for misrepresentation
- claims for *quantum meruit*
- claims for negligence (tort).

Claims for breach

To establish a successful claim for breach of contract it is necessary to prove that damage (in a financial sense) has been suffered as a result of the breach. The fundamentals of such a claim are briefly:

- breach of contract must be proved
- damage must be proved
- the damage must be shown to flow from the breach (causation)
- the damage must not be too remote.

Measure of damages

The principles applied by the courts in measuring damages date back to the case of *Robinson* v. *Harman* (1848) where it was stated:

'The rule of common law is that where a party sustains a loss by reason of a breach of contract, he is, so far as money can do it, to be placed in the same situation, with respect to damages, as if the contract had been performed.'

But this rule is subject to the rules on remoteness of damage.

Remoteness of damage

The law does not allow a claimant to succeed in every case where damage follows breach but draws a practical line by excluding that which is too remote.

The guiding principles of remoteness applying to cases of breach of contract derive from the judgement of Baron Alderson in the very old case of *Hadley* v. *Baxendale* (1854). The rule in *Hadley* v. *Baxendale* is taken as having two branches and is commonly expressed as:

'Such losses as may fairly and reasonably be considered as either arising:
  (1st rule) "naturally", i.e. according to the usual course of things, or
  (2nd rule) "such as may reasonably be supposed to be in the contemplation of both parties at the time they made the contract, as the probable result of breach of it".'

The test for remoteness laid down by Baron Alderson was reformulated in the case of *Victoria Laundry (Windsor) Ltd* v. *Newman Industries* (1949) in terms of foreseeability. However, in *Czarnikow* v. *Koufos* (1969) known as The *Heron II*, the House of Lords moved away from the foreseeability test to one of assumed common knowledge.

The Scottish Power case

Application of the rules on remoteness of damage to a dispute involving civil engineering works was considered by the House of Lords in the case of *Balfour Beatty Construction (Scotland) Ltd* v. *Scottish Power* (1994). Balfour Beatty had a contract with Scottish Power for the supply of electricity to a concrete batching plant. During the construction of a concrete aqueduct the power supply failed preventing continuous concrete pouring. Balfour Beatty demolished the partly constructed aqueduct and claimed damages from Scottish Power for breach of contract. At the first stage of proceedings in the Scottish courts the Lord Ordinary found as a fact that Scottish Power was unaware of the need to preserve a continuous pour for the construction of the aqueduct and that demolition was not a type of loss within its contemplation. The case went to appeal and then on to the House of Lords where the decision of the Lord Ordinary was restored. The House of Lords held:

- Where a party to a contract has broken it, the damages which the other party ought to receive are such as may fairly and reasonably be considered either as arising naturally from such a breach of contract or such as may reasonably be supposed to have been in the contemplation of both parties at the time they made the contract as having a very substantial degree of probability.
- What one party to a contract is presumed to know about the business activities of the other is a question of fact and inference.

In connection with presumed knowledge the Lord Ordinary had said:

'The defenders could certainly contemplate that if the supply failed the plant would not operate and that if it was operating at the time the manufacture of concrete would be interrupted. What they did not know was the necessity of preserving a continuous pour for the purposes of the particular operation. They were not told of the practice of having standby plant for such operations nor were they asked to arrange any specially secure supply of electricity. Furthermore they did not know that a construction joint would not be an acceptable solution if the power was prematurely terminated so that demolition would follow.'

In reviewing that analysis Lord Jauncey said:

'It must always be a question of circumstances what one contracting party is presumed to know about the business activities of the other. No doubt the simpler the activity of the one, the more readily can it be inferred that the other would have reasonable knowledge thereof. However, when the activity of A involves complicated construction or manufacturing techniques, I see no reason why B who supplies a commodity that A intends to use in the course of those techniques should be assumed, merely because of the order for the commodity, to be aware of the details of all the techniques undertaken by A and the effect thereupon of any failure of or deficiency in that commodity. Even if the Lord Ordinary had made a positive finding that continuous pour was a regular part of industrial practice it would not follow that in the absence of any other evidence suppliers of electricity such as the Board should have been aware of that practice.'

Recoverable loss

The rules of remoteness apply to the type of loss not to the amount of the loss. If damage is of a type which is not too remote the actual loss suffered is recoverable whether or not the amount of loss or damage could have been contemplated. See in particular the Court of Appeal ruling in *Parsons (Livestock) Ltd* v. *Uttley Ingham* (1977) where the supplier of a storage hopper with a faulty ventilator was found liable for the financial loss (but not future loss of profit) arising from the death of a herd of expensive pedigree pigs (as opposed to ordinary pigs) caused by mouldy feed nuts.

Damages for defective work

The principles of how 'damages' in the legal sense should be assessed for defective work were reviewed by the House of Lords in the case of *Ruxley Electronics* v. *Forsyth* (1995). Ruxley contracted to build a swimming pool with a maximum depth of 7 ft 6 in for Mr Forsyth. The completed pool had a maximum depth of 6 ft 9 in. A dispute arose on payment for the pool and, by way of counterclaim, entitlement to damages for defective work. The House of Lords ruled:

- The proper application of the general principle that where a party sustains loss by virtue of breach of contract he is so far as money can do it to be placed in the same situation in respect of damages as if the contract had been performed was not the monetary equivalent of specific performance. The court was required to ascertain the loss the plaintiff had in fact suffered by reason of the breach.
- The cost of reinstatement was not the only possible measure of damage for defective performance under a building contract. It is not the appropriate measure of damage where the expenditure would be out of all proportion to the benefit to be obtained even if the alternative measure of value, diminution in value, would lead to only nominal damages because there was no diminution in value.
- While the court was not concerned with what a plaintiff might do with damages if awarded, the plaintiff's intentions were relevant to the question of reasonableness which arose at the stage of considering whether damages should be awarded. The judge's findings of fact to the effect that Mr Forsyth had no intention of rebuilding the pool were relevant because they showed that he had lost nothing except the difference in value (if any).

Mitigation of loss

It is frequently said that a claimant has a duty to mitigate his loss. This is true only to the extent that the claimant seeks to recover his loss as damages. It does not follow that an injured party in a breach of contract should have his conduct determined by the breach.

The following extracts from legal judgments explain this. Viscount Haldane in *British Westinghouse Electric and Manufacturing Co Ltd* v. *Underground Electric Railways of London Ltd* (1912) said:

'A plaintiff is under no duty to mitigate his loss, despite the habitual use by the lawyers of the phrase "duty to mitigate". He is completely free to act as he judges to be in his best interest. On the other hand, a defendant is not liable for all loss suffered by the plaintiff in consequence of his so acting. A defendant is only liable for such part of the plaintiff's loss as is properly to be regarded as caused by the defendant's breach of duty.'

Sir John Donaldson, Master of the Rolls, in *The Solholt* (1983) said:

'The fundamental basis is thus compensation for pecuniary loss naturally flowing from the breach; but this first principle is qualified by a second, which imposes on a plaintiff the duty of taking all reasonable steps to mitigate the loss consequent on the breach, and debars him from claiming any part of the damage which is due to his neglect to take such steps.'

Quantum meruit

Claims for *quantum meruit* – meaning 'what it is worth' – are sometimes called quasi-contractual claims. They are popular with contractors wishing to escape from the rigidity of a lump sum or from contract rates towards payment on a cost-plus basis.

Strictly speaking the phrase *'quantum meruit'* applies to the law of restitution for the value of services rendered where there is no contractual entitlement to payment. But it is also commonly used to describe claims made under a contract for a fair valuation or a reasonable sum. The division between the two is not always clear-cut, as this comment by Mr Justice Goff in *British Steel Corporation* v. *Cleveland Bridge & Engineering* (1981) shows:

'a *quantum meruit* claim straddles the boundaries of what we now call contract and restitution; so the mere framing of a claim as a *quantum meruit* claim, or a claim for a reasonable sum, does not assist in classifying the claim as contractual or quasi-contractual.'

In the broad sense, claims for *quantum meruit* can be made where:

- no contract is ever concluded
- the contract is unenforceable
- the contract is void for mistake
- the contract is discharged by frustration
- the contract is for a lump sum and the employer prevents completion
- no price is fixed in the contract
- the work undertaken falls outside the scope of the contract
- the contract provides for payment of a reasonable sum or a fair valuation.

However, the law of restitution is complex and the cases where claims succeed frequently turn on particular facts. Such cases often prove to be of little lasting authority.

Thus the old case of *Bush* v. *Whitehaven Port and Town Trustees* (1888) sometimes quoted in support of winter working claims was held by the House of Lords in *Davis* v. *Fareham UDC* (1956) not to be authority for any principle of law. Bush, the contractor, was prevented by late possession of the site from commencing work on a pipeline contract until winter when wages were higher and the work was more difficult to construct. The court held that as a summer contract had been contemplated Bush was entitled to recover his extra expenditure as damages or to be paid for all the work on *quantum meruit*.

The proposition that the *Bush* case was said to support was that where the circumstances of a contract are altered, the contract price is no longer binding and can be replaced by *quantum meruit*. But cases mentioned elsewhere in this book show that the courts are not disposed to award *quantum meruit* where there are contractual provisions for compensation.

In *Morrison-Knudsen* v. *British Columbia Hydro and Power Authority* (1985) the contractor could have terminated the contract because of serious breaches by the employer. Instead the contractor elected to continue working and claimed for the value of the work on *quantum meruit*. The court held that such an award could not be given, as an adequate remedy was available under the contract.

In *McAlpine Humberoak* v. *McDermott International* (1992) the contract for the deck structure of an off-shore drilling rig was let on a lump sum basis. The contractor claimed extra cost arising from variations and additional drawings. The judge at first instance relying rather surprisingly on the *Bush* decision, held that the contract had been frustrated by the extent of the variations and awarded the contractor *quantum meruit*. The Court of Appeal overturned this decision and held:

- the variations did not transform the contract or distort its substance or identity
- the contract provided for variations
- the contractual machinery for valuing variations had not been displaced
- an award of *quantum meruit* could not be supported.

See also the case of *ERDC Construction* v. *H. M. Love & Co* (1994) mentioned in Chapter 20, section 20.2.

21.4 *Global claims*

The rule that cause and effect should be linked for every item of claim frequently presents difficulty for the parties to construction contracts – particularly when there has been a succession of claimable events with overlapping and intermingling consequences

Claiming on a global basis obviously alleviates these difficulties; but the question is how acceptable is it for the contractor, or employer, to roll-up his costs and put them forward as the composite loss for a bundle of claims?

Crosby v. Portland

The case of *J. Crosby & Sons Ltd* v. *Portland Urban District Council* (1967) is often quoted as authority for the practice but it needs to be treated with caution. It applies only to quantification and then only when the contractual machinery is no longer effective.

In the *Crosby* case an arbitrator set out his award for the court to consider as a special case. The judge in approving the global award included in his judgment this extract from the arbitrator's findings:

> 'The result, in terms of delay and disorganisation, of each of the matters referred to above was a continuing one. As each matter occurred its consequences were added to the cumulative consequences of the matters which had preceded it. The delay and disorganisation which ultimately resulted was cumulative and attributable to the combined effect of all these matters. It is therefore impracticable, if not impossible, to assess the additional expense caused by delay and disorganisation due to any one of these matters in isolation from the other matters.'

On the strength of these words and far wider application of the supportive judgment than stands examination, a great deal of money has since changed hands, much of it on the dubious proposition that the contractor's total cost less the tender sum is the amount of the employer's liability.

Merton v. Leach

It took until 1985 and the judgment of Mr Justice Vinelott in the case of *London Borough of Merton* v. *Stanley Hugh Leach Ltd* (1985) before a potentially fatal flaw in global claims was openly revealed: that the method of calculation relieves the contractor of any burden of his own additional costs or pricing errors. But the *Merton* case, far from containing the global claims approach opened it up to new horizons, for the case with its wide publicity on many other points of interest confirmed that the principles of the *Crosby* case applied to JCT contracts as well as to ICE contracts. And moreover the judge in describing a global award as a supplement to contractual machinery suggested that in appropriate circumstances the architect had a duty to ascertain global loss.

However, something of a retreat from those cases took place thereafter.

Wharf v. Cumine

Firstly came the Hong Kong case *Wharf Properties Ltd* v. *Eric Cumine Associates* which came before the Privy Council in 1991 and which confirmed that the *Crosby* and the *Merton* cases applied only to quantification and that those cases gave no relief to a claimant from the obligation to plead his case with particularity. Or, as better expressed in the words of Lord Oliver of Aylmerton:

'Those cases establish no more than this, that in cases where the full extent of extra costs incurred through delay depend upon a complex interaction between the consequences of various events, so that it may be difficult to make an accurate apportionment of the total extra costs, it may be proper for an arbitrator to make individual financial awards in respect of claims which can conveniently be dealt with in isolation and a supplementary award in respect of the financial consequences of the remainder as a composite whole. This has, however, no bearing upon the obligation of a plaintiff to plead his case with such particularity as is sufficient to alert the opposite party to the case which is going to be made against him at the trial.'

In the *Wharf* case the employer had paid out 317 million or so Hong Kong dollars to the contractor in claims and he was trying to recover this from his architect whom he claimed had failed to properly manage, control, co-ordinate, supervise and administer the project. However, the best the employer could do with the information available was to make a global approach to his pleadings. Neither the *Crosby* nor the *Merton* cases came to his assistance. Indeed Lord Oliver brought out an aspect of both cases which had been widely overlooked – that those cases say only what it might be proper for an arbitrator or contract administrator to do, not what it might be proper for a claimant to do.

Mid-Glamorgan/ICI cases

Two cases which rapidly followed the *Wharf* case with employers trying to recover monies paid out on cost over-runs were *Mid Glamorgan County Council* v. *Devonald Williams and Partner* (1991) and *Imperial Chemical Industries plc* v. *Bovis and Others* (1992).

Mr Recorder Tackaberry in the *Mid Glamorgan* case gave a useful summary of the legal position following the *Wharf* case:

'1. A proper cause of action has to be pleaded.
2. Where specific events are relied on as giving rise to a claim for monies under the contract then any pre-conditions which are made applicable to such claims by the terms of the relevant contract will have to be satisfied, and satisfied in respect of each of the causative events relied upon.
3. When it comes to *quantum*, whether time based or not, and whether claimed under the contract or by way of damages, then a proper nexus should be pleaded which relates each event relied upon to the money claimed.
4. Where, however, a claim is made for extra costs incurred through

delay as a result of various events whose consequences have a complex inter-reaction which renders specific relation between the event and time/money consequence impossible and impracticable, it is possible to maintain a composite claim.'

The *ICI* case arose out of the refurbishment of ICI's premises in Millbank where the cost rose from an original estimate of approximately £30 m to over £53 m. The contract was let on a management basis and ICI sued the management contractor, the architects and the consulting engineers in respect of £19 m of cost over-runs and associated fees. A global claim was made against each of the defendants. As with the *Wharf* and the *Mid Glamorgan* cases much of the interest of the case is in the approach of the courts to inadequate pleadings but again, as in the *Merton* case, the potential for flaws in global claims was revealed.

The global claim for abortive work in respect of hundreds of items amounted to £840,211. It was said that apportionment was impossible. The defendants asked, if they had a complete defence to all the items save for two minor ones – 'circuits need changing' and 'fire bell repositioned' – what monetary consequences would flow from these two items? The reply was to the effect: 'If any of the events is not proven at trial, the only consequence is that the actual sum paid will fall to be distributed between a lesser number of events, not that the total recoverable will be less'. Judge Fox-Andrews found it 'palpable nonsense that £840,000 could be the cost of repositioning a fire bell'.

Later cases

However, to the extent that *Wharf* and the cases which immediately followed it were heralded as sounding the end of global claims, such belief was premature. A series of cases since 1994 has given new life to global claims.

In *GMTC Tools* v. *Yusa Warwick Machinery* (1994) a manufacturing company sued the firm which had supplied it with an unreliable computer-controlled precision lathe. The claim was global to the extent that it did not deal precisely with the costs arising from each and every breakdown. The judge ordered that more detailed information be provided. The Court of Appeal held that the judge was not entitled to require a party to establish causation and loss by a particular method. Subject to the respondent knowing the case it had to answer, the claimant could formulate its claim for damages as it wished.

In *British Airways Pension Trustees Ltd* v. *Sir Robert McAlpine & Sons Ltd* (1994) disputes arose in connection with diminution in value of a development property due to defects. The amount claimed was £3.1 m. The respondent sought particulars in respect of each and every defect and when not provided the judge ordered that the claim be struck-out. The Court of Appeal reversed the decision with Lord Justice Saville saying:

'The basic purpose of pleadings is to enable the opposing party to know what case is being made in sufficient detail to enable that party properly to prepare to answer it. To my mind it seems that in recent years there has been a tendency to forget this basic purpose and to seek particularisation even when it is not really required. This is not only costly in itself, but is calculated to lead to delay and to interlocutory battles in which the parties and the court pore over endless pages of pleadings to see whether or not some particular point has or has not been raised or answered, when in truth each party knows perfectly well what case is made by the other and is able properly to prepare to deal with it. Pleadings are not a game to be played at the expense of the litigants, nor an end in themselves, but a means to the end, and that end is to give each party a fair hearing. Each case must of course be looked at in the light of its own subject matter and circumstances.'

From *AMEC Building Ltd* v. *Cadmus Investments* Co (1996) it seems that the current approach of the courts is to judge each case on its merits without laying down general principles as to whether global claims are acceptable.

In *Inserco Ltd* v. *Honeywell Control Systems* (1996) the judge made an award based on a global claim; and in *How Engineering* v. *Linder Ceilings* (1999) the court upheld an arbitrator's award based on a global assessment.

For further detailed review of the law and explanation of the character-istics of global claims and total cost claims see the judgments of Judge Byrne in the case of *John Holland Construction & Engineering* v. *Kvaerner R. J. Brown Pty* (1996) and Judge Humphrey Lloyd QC in the case of *Bernhard's Rugby Landscapes* v. *Stockley Park Consortium* (1997).

21.5 Overheads, finance and interest charges

Overheads can be categorised as being on-site or off-site, the latter being another name for head office overheads. A further distinction can be made between overheads incurred in the running of a business and overheads, usually described as prolongation overheads, which are claimed on the basis of loss of opportunity resulting from delay.

On-site overheads

On-site overheads will normally be recoverable without difficulty as part of cost providing adequate records exist to identify the items and amounts in the claim. To the extent that there is difficulty in evaluating on-site over-heads it is usually concern over whether there is duplication between on-site and off-site overheads in respect of particular items.

This arises because there is no standard definition or policy of distinction between the two types of overheads. As a generalisation, large contractors put more items into on-site overheads than do small contractors – aided no doubt by more sophisticated accounting systems. In consequence, large contractors may have a higher level of on-site and correspondingly lower level of off-site overheads than small contractors.

Off-site overheads

Claims for off-site overheads can be far more complex than claims for on-site overheads. There is no objection in principle to such claims whether they be claims under the contract – with overheads included in the definition of cost as in ICE Conditions – or whether they are common law claims. The difficulty is in proving and evaluating the extra cost incurred as a result of the claimable event. This applies whether the head office overheads claimed are of the running cost type or the loss of opportunity type.

Thus in the case of *Tate & Lyle Ltd* v. *Greater London Council* (1982) it was held that in principle Tate & Lyle could recover damages for managerial and supervisory expenses, although in that case it was also held that it was not permissible simply to add a percentage to other items in the claim. It was said that modern office arrangements should permit the recording of time spent by managerial staff on a particular project.

And in *J. F. Finnegan Ltd* v. *Sheffield City Council* (1988) where the contractor claimed additional overheads for delay on a housing improvement contract, the judge said:

> 'It is generally accepted that, on principle, a contractor who is delayed in completing a contract due to the default of his employer, may properly have a claim for head office or off-site overheads during the period of delay, on the basis that the work-force, but for the delay, might have had the opportunity of being employed on another contract which would have had the effect of funding the overheads during the overrun period. This principle was approved in the Canadian case of *Shore & Horwitz Construction Co Ltd* v. *Franki of Canada* (1964), and it was also applied by Mr Recorder Percival QC, in the unreported case of *Whittall Builders Company Limited* v. *Chester-le-Street District Council.*'

Problems of proof/evaluation

With running cost type overheads the problem for the contractor is proving that a particular claimable event has increased or utilised head office overheads. To overcome this some contractors ensure that records are kept by all head office staff, including senior managers and directors, of time spent dealing with problem contracts. For the wisdom in this see the decisions in *Tate and Lyle* and *Babcock Energy* v. *Lodge Sturtevant* (1994).

Under the wording of certain contracts the problem is alleviated by definitions of cost which allow for allocation of head office overheads. All that is required then is evidence from audited accounts of the level of head office overheads. ICE Conditions, including the Seventh edition, are unfortunately somewhat ambiguous on this – clause 1(5) defines cost as expenditure incurred whilst at the same time referring to 'charges properly allocatable'.

It is worth noting here that in *Weldon Plant* v. *Commission For New Towns* (2000) where allocation of overheads (as opposed to proof of expenditure) was allowed under ICE Sixth edition Conditions, that was in connection with a fair valuation and it may have little application to cost-based claims generally.

Formula methods of calculating loss

For prolongation/loss of opportunity type claims there are various formulae for calculating the amount of the loss – Hudson's formula, Emdon's formula, Eichleay formula. All produce a weekly amount of contribution to head office overheads which a particular contract was planned to generate or which was notionally incurred. The calculated rate multiplied by the delay period produces the amount of claim.

A string of legal cases confirms that the courts accept in principle the use of formulae:

- *Ellis-don Ltd* v. *Parking Authority of Toronto* (1978)
- *Whittall Builders* v. *Chester-le-street District Council* (1985)
- *Finnegan* v. *Sheffield City Council* (1988)
- *Alfred McAlpine Homes* v. *Property and Land Contractors* (1995)
- *St Modwen Developments* v. *Bowmer & Kirkland* (1996)
- *AMEC Building* v. *Cadmus Investments Co* (1996)
- *Norwest Holst* v. *Co-operative Wholesale Society* (1998)
- *Beechwood Development Company* v. *Stuart Mitchell* (2001).

However, the point which emerges with regularity from these cases and others is that the contractor's entitlement is subject to proof of cost or lost opportunity. Thus it was said by the judge in the *McAlpine Homes* case, after reviewing the authorities and the application of the various formulae:

'All these observations, like those of Lord Lloyd in *Ruxley*, of Forbes J in *Tate & Lyle,* and of Sir Anthony May in *Keating* suppose, either expressly or implicitly, that there may be some loss as a result of the event complained of so that, in the case of delay to the completion of a construction contract, there will be some "under recovery" towards the cost of fixed overheads as a result of the reduced volume of work occasioned by the delay, but this state of affairs must of course be established as a matter of fact. If the contractor's overall business is not diminished during the period of delay so that whether, for example, as a result of an increase in the volume of work on the contract in question arising from variations etc. or for other reasons, there will be a commensurate contribution towards the overheads which offsets any supposed loss, or if, as a result of other work, there is no reduction in overall turnover so that the cost of the fixed overheads continues to be met from other sources, there will be no loss attributable to the delay. Put another way, this aspect is brought out in the comparable proposition that the contractor has to show that there were no means of reducing the unrecovered cost of the fixed overheads in the circumstances in which he found himself as a result of the events giving rise to the delay. Where a contractor is busy and is taking on work all the time, it will probably not be possible to demonstrate the effect to which I have referred. Furthermore, it has to be borne in mind that as certain overheads are incurred through thick and thin, so a contractor's head office staff may not always be constantly occupied because of, for example, the seasonable or cyclical nature of business in the construction industry.'

In the *Beechwood* case the claimant was a house-builder who was suing a land surveyor for losses arising from a survey error which caused delay to a particular development. Beechwood succeeded under a head of claim explained by the judge as follows:

'The claim under this head is essentially one for reimbursement to Beechwood of contributions to overheads and net profit which would have been made had Beechwood not, as a result of the defender's fault, lost a period of effective working (now assessed by me at 10 weeks). Although Mr Clive presented his primary contention as one for loss of "gross profit" that is, as I understand it, essentially the same as one for loss of contribution to overheads and net profits, gross profit going first to meet overheads and the balance representing net profit.'

The judge was satisfied from the evidence that Beechwood only worked on one development at a time; that accordingly the delay caused by the survey error led to a loss of opportunity to work elsewhere and a corresponding loss of contribution; and that in the circumstances it was appropriate to use a formula method for assessment of Beechwood's losses.

Interest and financing charges

Interest and financing charges may seem on the face of it to be the same but in law they are different.

Interest charges relate to late payment of a sum due. Financing charges relate to costs (or loss and expense) which form the subject of claims for the period from which such costs are incurred to the time of application for payment or certification.

The principle concerned in the payment of financing charges is that the contractor has incurred expense in financing the primary cost involved in his claim and this financing charge, therefore, is not interest on a debt but a constituent part of the debt.

The principle was established by the Court of Appeal in the case of *F. G. Minter Ltd* v. *Welsh Health Technical Services Organisation* (1980). The decision in the *Minter* case was followed by the Court of Appeal in *Rees and Kirby Ltd* v. *Swansea City Council* (1985) where it was confirmed that calculation of financing charges should be on a compound interest basis.

It is now generally accepted following the *Minter* and *Rees and Kirby* cases, that under most standard forms of construction contracts the contractor is entitled to include in his applications for payment the extra cost of financing charges he has incurred up to that time.

ICE Conditions illustrate well the difference between financing charges and interest by including financing charges within the definition of cost in clause 1(5) and by separately dealing with interest on late payments in clause 60(7).

21.6 Additional payment rules of ICE Conditions

Clause 53 of ICE Conditions sets out in six sub-clauses the rules to be followed by the contractor in making claims for additional payments.

Clause 53(1) – variations and rate changes

Clause 53(1) provides that if the contractor intends to claim a higher rate or price than one notified to him by the engineer pursuant to clauses 52(3), 52(4) or 56(2) the contractor shall within 28 days after notification give notice in writing of his intention to the engineer.

This clause is concerned only with disputed valuation of variations and disputed increases or decreases of rates arising from changes of quantities. Furthermore it applies only when the engineer has taken action under clauses 52(3), 52(4), or 56(2) and not generally in respect of disputes under those clauses.

The 28 day notification period is not qualified by saving words such as 'or as soon as may be reasonable'. It would therefore act as a strict barrier on late claims under the clause were it not for the general saving provision later in clause 53 at 53(5) – 'If the Contractor fails to comply with any of the provisions of this Clause [clause 53]'.

Clause 53(2) – notification of claims generally

Clause 53(2) provides that if the contractor intends to claim any additional payment pursuant to any clause of the Conditions other than clauses 52(3) and 52(4) or clause 56(2) he shall give notice in writing of his intention to the engineer as soon as may be reasonable and in any event within 28 days after the happening of the events giving rise to the claim.

It further provides that upon the happening of such events the contractor shall keep such contemporary records as may reasonably be necessary to support any claim he may subsequently wish to make.

This clause is probably intended to cover all claims made under the Conditions, other than those covered by clause 53(1), whether or not the particular clause in the Conditions refers expressly to notification under clause 53 – for instance a claim for recovery of expense under clause 32 (fossils). It is doubtful, however, if the clause has any application to claims for breach of contract.

Again the 28 day notice rule would be a strict condition – at least in respect of entitlement claims – but for the saving provision in clause 53(5).

Clause 53(3) – contractor to keep records

Clause 53(3) provides that without necessarily admitting the employer's liability, the engineer may upon receipt of a notice under clause 53, instruct the contractor to keep contemporary records as reasonable and material to the claim. The clause further provides that the contractor shall keep such records, shall permit the engineer to inspect them, and shall supply copies as instructed.

The importance of records for the substantiation of claims cannot be over-emphasised. Every aspect should be covered – labour and plant records, materials invoices, supervision, programmes, photographs, correspondence and notes of meetings.

Although clause 53(3) is concerned only with the contractor's obligations to keep records, the engineer should ensure that his own record keeping is just as thorough.

Clause 53(4) – interim accounts

Clause 53(4) provides that after giving notice to the engineer under clause 53 the contractor shall, as soon as is reasonable in all the circumstances, send the engineer a first interim account giving full and detailed particulars of the amount claimed to that date and the grounds on which the claim is based.

The clause further provides that thereafter, at such intervals as the engineer may reasonably require, the contractor shall submit further up-to-date accounts giving the accumulated total of the claim and any further grounds upon which it is based.

The obligation on the contractor under clause 53(4) to supply a first interim account but thereafter further accounts only as required by the engineer can create problems and it places a requirement on the engineer which is frequently neglected – not least because the engineer may be reluctant to be seen to be pressing the contractor to increase amounts claimed. However, the result can be unexpectedly large late claims. To counter this a sensible amendment to the clause is to require amounts claimed to be kept up-to-date in monthly applications – as seems to be the intention of clause 60(1).

Clause 53(5) – contractor's failure to comply

Clause 53(5) provides that if the contractor fails to comply with any of the provisions of clause 53 in respect of any claim, then the contractor shall be entitled to payment in respect thereof only to the extent that the engineer has not been prevented from, or subsequently prejudiced by, such failure in investigating the claim.

This is an important clause without which the operation of the Conditions and the practices of contractors in submitting claims might be significantly different. Strict compliance with notice requirements would take effect for many claims – particularly those based on contractual entitlements.

See, for example:

- the decision in the case of *Humber Oils Terminal* v. *Hersent Offshore* (1981) in a claim under ICE Fourth edition Conditions of Contract where it was held that a notice of claim was invalid in failing to meet all three specified notice requirements – i.e. the conditions encountered, the contractor's proposals, and the anticipated delay.
- the decision in the case of *Tersons* v. *Stevenage Development Corporation* (1963) under ICE Second edition Conditions where it was held that compliance with notice requirements was a condition precedent to payment.
- the decision in the case of *Hersent Offshore* v. *Burmah Oil* (1978) under conditions similar to the ICE Fourth edition where it was held that notice given after the varied work had been completed had not been given 'as soon thereafter as is practicable'.

Meaning of notice

The meaning of 'notice' as used in ICE Conditions was one of the questions considered in the *Tersons* v. *Stevenage* case. The Court of Appeal upheld the decision of Mr Justice Roskill who had said:

'I think it is sufficient that the notice should specify that a claim is being made, provided that the notice identifies in general terms the nature of the additional work to which the claim will relate when it is ultimately precisely formulated.'

It was also said in that case that the phrase 'as soon thereafter as is practicable' should be given wide interpretation.

Clause 53(6) – contractor's entitlement to interim payment

Clause 53(6) provides that the contractor shall be entitled to have included in any interim payment certified by the engineer such amount in respect of any claim as the engineer may consider due providing that the contractor shall have supplied sufficient particulars to enable the engineer to determine the amount due.

It further provides that if such particulars are insufficient to substantiate the whole of the claim the contractor shall be entitled to payment in respect of such part of the claim as the particulars may substantiate to the satisfaction of the engineer.

This clause has particular importance when considered in conjunction with the interest provisions in clause 60(7) and the question of failure by the engineer to certify.

Chapter 22
Property and materials

22.1 Introduction

Most construction contracts have clauses which endeavour to protect the employer's interests in the event of the contractor's insolvency. Traditionally this has been done by deeming goods and materials to be the employer's property once brought on to site; by deeming goods and materials stored off-site but paid for to be the employer's property; by deeming the contractor's equipment to be the employer's property once brought on site; and by prohibiting the removal of contractor's equipment from site without approval.

The problem with these clauses is that although often long and complex they are frequently ineffective. It is not simply that the law on title is difficult, it is that when laws on title become entangled with insolvency laws the position is even worse.

The point is illustrated by the case of *Cosslett (Contractors) Ltd* v. *Mid Glamorgan County Council* (1997) which concerned a contract under the ICE Fifth edition where the employer sought to claim ownership of plant being used on the works when the contractor went into receivership. The provisions in the Fifth edition are extensive and apparently thorough but the Court of Appeal held that they were ineffective in passing legal title. It is beyond the scope of this book to comment on the details of the judgment but an indication of the sort of matters which arose in the case can be seen from the summary in *Building Law Reports*, 85 BLR at p. 3:

'Held, dismissing the appeal:

(1) Clause 53(2) did not transfer legal property in the plant to the Council.
(2) The Council's power of sale pursuant to clause 63(1) did not constitute a possessory lien (*Great Eastern Railway Co* v. *Lord's Trustee* (1909) AC 109 considered).
(3) The Council's right to retain possession of the plant pursuant to clause 63(1) for the purpose of completing the works did not constitute an equitable charge and was therefore unaffected by the requirement of registration under section 395 of the Companies Act 1985.
(4) (*obiter*) The Council's power of sale pursuant to clause 63(1) was a floating charge requiring registration under section 395 of the Companies Act 1985.

(5)   (*obiter*) Failure to register the charge made the security created by the power of sale void as against the liquidator but did not affect any other right of the Council including its right to retain possession of the plant and use it to complete the works.'

Policy of ICE Conditions of Contract – Seventh edition

The Seventh edition has moved a long way from the policy of the Fifth edition which tried much but achieved little. The process of simplification started in the Sixth edition and it continues into the Seventh. The clauses, purporting to transfer ownership of contractor's plant and equipment to the employer, are gone and all that remains are the more realistic ambitions of controlling the removal of plant and equipment from site and securing ownership of goods and materials paid for by the employer.

Changes from the Sixth edition

The first point to note on changes from the Sixth edition is that clauses 53 and 54 from the Sixth have been amalgamated into clause 54 in the Seventh edition. Clause 53 in the Seventh is now what was for the most part clause 52(4) in the Sixth – described there as 'notice of claims' but now in the Seventh as 'additional payments'.

The most significant change from the Sixth edition is that clause 54(1) of the Seventh (which corresponds to clause 53(1) of the Sixth) is concerned only with consent for removal of plant and equipment, whereas the Sixth edition clause was concerned also with deemed ownership.

22.2 *Contractor's equipment and the like*

The first three sub-clauses of clause 54 deal with contractor's equipment, temporary works, materials for temporary works and other goods and materials owned by the contractor.

Contractor's equipment is defined in clause 1(1)(w) as 'all appliances or things of whatsoever nature required in or about the construction and completion of the Works but does not include materials or things intended to form or forming part of the Permanent Works'.

Removal of contractor's equipment

Clause 54(1) states that no contractor's equipment, temporary works, materials for temporary works or other goods or materials owned by the contractor brought on to the site for the purposes of the contract shall be removed without the written consent of the engineer, which consent shall not unreasonably be withheld.

The principal purposes of this clause are no doubt to secure the safety of the site and completion of the works. Although there is nothing in the clause to restrict its application to the period up to completion, and there could in some circumstances be some practical reasons for wishing to see it operate in the defects correction period, it is doubtful if this would be consistent with the obligation in clause 33 for the contractor to clear the site on completion.

Although the clause may well be effective against a solvent contractor capable of meeting liability for damages for breach of contract in the event of unauthorised removal, it may not be much of a deterrent to unauthorised removal when it is most needed – i.e. when the contractor is insolvent and various third parties are laying claim to ownership of all or some of the items covered by the clause. And to the extent that the clause is founded on the presumption that the contractor owns that which is defined as 'contractor's equipment' in the contract, or the presumption that the clause is binding on third parties, it may prove less than satisfactory in its operation.

The test for whether the engineer's consent to removal has been reasonably or unreasonably withheld would seem on the wording of clause 54(1) to be a purely practical test – 'the purposes of the Contract'. It is questionable whether withholding consent on commercial grounds would be reasonable.

Liability for loss or damage

Clause 54(2) provides that the employer shall not be liable for loss or damage to the contractor's equipment etc., except as mentioned in:

- clause 20(2) – the excepted risks and
- clause 63 – frustration/war clause.

There is no express mention in clause 20(2) of the contractor's equipment etc.; that clause is concerned with the excepted risks to the contractor's responsibility for care of the works. The only mention in clause 63 of the contractor's equipment etc. is that the contractor has a duty, and the employer has power, under clause 63(3) to remove contractor's equipment from site in the event of war or frustration.

Consequently the meaning of clause 54(2) is less than immediately obvious – particularly as clause 54(2) in the Sixth edition referred to clause 22 and not to clause 20(2) as in the Seventh edition.

What may be intended is that unless the employer damages the contractor's equipment by use or occupation of the works or by removing it from site, then responsibility remains with the contractor for other damage. However it is questionable whether, if the resident engineer's Land Rover drives into the site agent's car, the contractor's insurer would accept the damage claim without seeking redress through the employer's insurances.

Disposal of contractor's equipment

Clause 54(3) states that if the contractor fails to remove any contractor's equipment etc. 'as required by clause 33' within such reasonable time after completion of the works as the engineer may allow, then the employer may:

- sell or otherwise dispose of the items
- retain from the proceeds any costs incurred
- before paying the balance to the contractor.

Clause 33 requires the contractor to remove from site all contractor's equipment etc. on completion.

Clause 54(3) is not an attempt by the employer to claim ownership of the contractor's equipment etc. – even when it remains on site after completion. It is concerned simply with the employer's entitlement to a cleared site and his rights of disposal.

22.3 *Vesting of goods and materials not on site*

Clauses 54(4) to 54(9) detail the arrangements for vesting ownership in the employer of goods and materials not on site so that the contractor can secure payment on interim certificates.

Qualifying goods and materials

Clause 54(4) states that with a view to securing payment under clause 60(1)(c) the contractor may, and shall if the engineer so directs, transfer to the employer before delivery to site the property in goods and materials:

- listed in the appendix to the form of tender
- or as agreed between the contractor and the employer
- and which
  - (a) have been manufactured and are ready for incorporation into the works
  - (b) are the property of the contractor; or the contract of supply makes provision for them to pass unconditionally to the contractor
  - (c) have been marked and set aside as the property of the employer.

Action by contractor

Clause 54(5) requires that as evidence of ownership of off-site goods and materials, the contractor shall take, or cause his supplier to take, action in:

(a) providing the engineer with documentary evidence that property has vested in the contractor

(b) marking and identifying goods and materials to show
   (i)   their destination is the site
   (ii) they are the property of the employer
   (iii) to whose order they are held
(c) setting aside and storing to the satisfaction of the engineer
(d) sending the engineer a list and schedule of values and inviting him to inspect.

Engineers should ensure that these requirements are strictly fulfilled because of the inevitability of conflicting claims to ownership if there is a business failure somewhere along the lines of supply or the contractual chain.

Vesting in the employer

Clause 54(6) states that upon the engineer approving in writing the transfer of ownership of any goods or materials they shall vest in and become the absolute property of the employer provided always that:

(a) approval does not prejudice the power of the engineer to reject any goods or materials not in accordance with the contract
(b) the contractor shall be responsible for loss or damage and shall take out additional insurance.

The provisions in clauses 54(6)(a) and 54(6)(b) confirm that what is intended in clause 54(6) is full transfer of ownership to the employer and not simply deemed ownership.

Lien on goods and materials

Clause 54(7) states that neither the contractor nor a subcontractor nor any person shall have a lien on any goods or materials which have vested in the employer under clause 54(6) and the contractor shall take all necessary steps to ensure that the title of the employer and the exclusion of any such lien are brought to the notice of subcontractors and other persons dealing with such goods or materials.

To the extent that clause 54(7) may be seen as trying to bind or define the rights of third parties it is unlikely to be effective. There are countless legal cases illustrating the difficulty of doing so. A pertinent example is the case of *Dawber Williamson* v. *Humberside County Council* (1979). Here a roofing subcontractor brought slates on to site for which the contractor was paid by the employer. Before the slates were fixed the contractor went into liquidation. The subcontractor attempted to retrieve the slates but the employer prevented this and claimed ownership. The subcontractor then successfully sued the employer for their value. In considering the clause in the main contract which said that when the value of any goods has been included in any certificate under which the contractor has received payment, the goods shall be the employer's property, the judge said:

> 'In my judgement, this presupposes there is privity between the defendants and the subcontractor, which there is not in the present instance, or the main contractor has good title to the material and goods. If the title has passed to the main contractor from the subcontractor, then this clause has force.'

Retention of title clauses

Additionally there is the problem of retention of title clauses. To avoid title passing before payment is made, many contracts for sale of goods often have clauses granting possession and even use to the buyer but retaining ownership with the seller until payment. These are sometimes called Romalpa clauses after the case of *Aluminium Industries* v. *Romalpa Aluminium Ltd* (1976).

It is unlikely that a claim to title by the employer under the provisions of clause 54 would defeat a claim to title under a Romalpa clause.

Delivery of vested goods and materials

Clause 54(8) deals with the situation where goods or materials have been vested in the employer and before completion of the works the employment of the contractor is terminated.

The clause requires the contractor to deliver such goods and materials to the employer, failing which the employer is entitled to enter the premises of the contractor or any subcontractor to remove the goods and materials and to recover the cost of doing so from the contractor. The legal right of the employer to enter the premises of subcontractors and remove goods is questionable and it is not something to be undertaken without legal advice.

The scope of clause 54(8) is clearly intended to be as wide as possible in so far as it is stated to operate upon cessation of the contractor's employment whether under clause 63 (frustration), clause 64 (employer's default), clause 65 (contractor's default) or 'otherwise'.

'Otherwise' may refer to common law determination or termination of the contractor's employment by agreement. Whatever the intention, there is a reminder here that novation agreements need to be carefully drafted in respect of off-site goods and materials since the new contractor will not automatically be bound by or have the benefit of the old contractor's contracts of supply.

Incorporation into subcontracts

Clause 54(9) requires that the contractor shall incorporate provisions equivalent to those provided in clause 54 in every subcontract in which provision is made for the payment of goods or materials before delivery to site. The purpose of clause 54(9) is to achieve contractor's rights against subcontractors which are no less than the employer's against the contractor so that in the event of default in the contractual chain the employer's title is not lost.

The difficulty with such clauses for equivalent provisions is that only rarely will the contractor deal with subcontractors on the same basis as he deals with the employer and there is often no corresponding or similar contractual structure to graft the equivalent provisions into.

Chapter 23
Measurement of the works

23.1 Introduction

Clauses 55 to 57 of the Seventh edition cover, without any significant changes from the Sixth edition (other than in references to particular sets of rules) measurement of the works and dayworks.

The Seventh edition is the first ICE standard form to expressly state in its title, and on its cover, that it is a 'measurement version'. Perhaps it can be gathered from this there may in due course be non-measurement versions or perhaps it is simply putting beyond doubt, not that there should be any from the text, that the standard version of the Seventh edition is a remeasurement contract and not a lump sum contract.

Not all ICE standard forms are remeasurement contracts: ICE Design and Construct Conditions of Contract are lump sum; ICE Minor Works Conditions of Contract are flexible; and the New Engineering Contract provides a range of options from lump sum to cost reimbursable. Until the publication of the ICE Fifth edition in 1973 there was some debate on whether standard ICE Conditions of Contract were lump sum or remeasurement but that was put to rest by drafting changes in the Fifth edition and these have been followed through into the Sixth and Seventh editions. Thus in these later editions there are defined terms making a clear distinction between the 'Tender Total' (the total of the bill of quantities at the date of award) and the 'Contract Price' (the sum to be ascertained in accordance with the provisions of the contract). And clause 56(1) provides – 'The Engineer shall except as otherwise stated ascertain and determine by admeasurement the value in accordance with the Contract of the work done in accordance with the Contract'.

The phrase in clause 56(1) 'except as otherwise stated' is sometimes taken as licence for the employer to state somewhere in the package of contract documents that the works will not be remeasured – which, if effective, would turn the tender total into the basis of the contract price even if not going so far as to turn the contract into a lump sum contract. However it is questionable whether such a simplistic approach is effective. It is possible, and sometimes with good cause on commercial grounds, to turn standard ICE Conditions of Contract into lump sum contracts but a comprehensive set of amendments is necessary.

One of the points to be considered is that under standard ICE Conditions there is no link, other than through the bill of quantities rates, between the tender total and the contract price. The remeasurement process starts from a clean sheet so far as quantities are concerned. However, if the tender total is to form the basis of the contract price then any remeasurement which is necessitated by variations or the like is a process of adjustment to billed quantities. This is standard practice in the building industry but not one commonly used in civil engineering.

A further point for consideration, although more commercial than contractual, is that civil engineering contracts have traditionally been remeasurement rather than lump sum because of perceived difficulties of establishing at tender stage the likely final quantities – particularly of earthworks and the like. One theory is that it is cheaper for the employer to pay for the work actually done than to put the risk of quantities on the contractor and thereby pay a higher price than might otherwise have been necessary. Another theory is simply that where the engineer is responsible for design it is more appropriate for the employer to carry the risk on quantities than the contractor. But whatever the good reasons for ICE Conditions being remeasurement, it does need to be recognised that if conversion to lump sum is undertaken it is essentially a challenge to those reasons and the corresponding allocation of risk in the Conditions.

Standard methods of measurement

Clause 57 of the Seventh edition requires the bill of quantities to be prepared and measurements to be made in accordance with the Civil Engineering Standard Method of Measurement Third Edition 1991 unless otherwise provided in the contract. The Sixth edition Conditions referred to the Second Edition 1985.

Provision is made in part one of the appendix to the form of tender for the method of measurement used to be identified if different from the standard method mentioned in clause 57.

Disputes on measurement

One of the reasons employers and engineers sometimes seek to convert ICE Conditions to lump sum contracts is to avoid the uncertainty of price and the potential for dispute attached to remeasurement.

Such disputes are commonplace. They vary from disputes on measurements, disputes on calculations, and disputes on compilation of the bill of quantities, to disputes on interpretation of the rules of the applicable method of measurement. There is, however, little legal authority and little that can be said by way of guidance on how they should be resolved. They are generally disputes which are particular to the facts of each case and disputes which rarely go beyond arbitration to the courts. Even when there is a legal judgment its effect is likely to be of limited application.

23.2 *Quantities*

Clause 55(1) states that the quantities set out in the bill of quantities are the
estimated quantities of the work but they are not to be taken as the actual
and correct quantities of the works to be carried out by the contractor in
fulfilment of his obligations under the contract.

At first sight clause 55(1) seems designed to protect the employer against
claims by the contractor based on differences between actual quantities and
billed quantities – i.e. it is a statement by the employer to the effect that he
does not warrant that the quantities are correct and/or a device to prevent
any such term being implied. However, although this may in part be the
intention it is doubted if this is the full intention of the clause. Greater
importance can probably be attached to the way in which the clause
endeavours to prevent the contractor from linking his obligations under the
contract to merely carrying out the quantities of work set out in the bills.

Billing errors

The propositions, if they should be advanced, that the contractor has no
remedy or is intended to have no remedy other than remeasurement for
inaccurately billed quantities are defeated by various clauses of the Conditions, in particular:

- clause 55(2) which provides for correction of billing errors, and
- clause 56(2) which provides, in certain circumstances, for increase or
  decrease of rates where the actual quantities differ from those billed
- clause 57 which requires the work to be measured in accordance with the
  standard method of measurement.

Given the benefit of these clauses it is unlikely that the contractor under
Seventh edition Conditions would find any better remedy in law, but there
are a few cases of general interest which show that the courts have been
inclined in some instances under less explicit conditions of contract to allow
claims arising from inaccurate or imprecisely billed work. For example:

- *Bryant & Son Ltd* v. *Birmingham Hospital Saturday Fund* (1938)
- *A. E. Farr Ltd* v. *Ministry of Transport* (1960)
- *Mitsui Construction Co* v. *A-G for Hong Kong* (1986).

Estimated quantities

It should not be assumed from clause 55(1) that because the billed quantities
are not warranted as correct and accurate they are not required to be prepared with care. Comment is made in Chapter 3, Section 3.1 on the Canadian
case of *Edgeworth Construction* v. *Lea & Associates* (1993) where it was held

that a consulting engineer owed a duty of care to the contractor in preparing the bill of quantities.

English law seems to be some way behind Canadian law on this at the present time and action by the contractor against the engineer on an inaccurate bill of quantities might be optimistic. On the wording of clause 55(1), action against the employer would seem to be ruled out except that although the quantities are not required to be actual and correct, they are required to be the 'estimated quantities'. If it can be proved that the quantities in the bills are not the true estimated quantities, but are perhaps quantities entered in error or are estimated quantities adjusted for some reason or other, then any protection given to the employer by clause 55(1) may be seriously undermined.

Contractor's obligations

The contractor's obligation as stated in clause 8(1) and elsewhere is to construct and complete the works in accordance with the contract. The contract by definition in clause 1(1)(e) includes, amongst other things, the bill of quantities, specification and drawings. There is no stated order of precedence and clause 5 states that all documents are to be taken as mutually explanatory of one another. It is clear, however, from the reference in clause 55(1) to the contractor's obligations and the provisions which follow in clause 55(2) that when it comes to determining the contractor's obligations the specification and the drawings take precedence over the bill of quantities.

One effect of this, by way of example, is to cast doubt on how clause 44(1)(b) (extension of time for increased quantities) should be operated when additional work is shown on drawings but is not included in the billed quantities.

23.3 Correction of errors

Clause 55(2) is complementary to clause 55(1) in so far as whilst clause 55(1) deals with the numerical aspects of quantities, clause 55(2) deals with item descriptions.

Clause 55(2) provides that:

- no error in description in the bill of quantities
- nor any omission
- shall vitiate the contract
- nor release the contractor from his obligations

and that

- any error or omission shall be corrected by the engineer and

- the value of work actually carried out shall be ascertained in accordance with clause 52(2) or 52(3)

with the proviso that

- there shall be no rectification of errors, omissions or wrong estimates in the descriptions, rates and prices inserted by the contractor in the bill of quantities.

Purpose of clause 55(2)

Clause 55(2) is not concerned with the correction of errors in any academic sense; its importance lies in the phrase 'the value of the work actually carried out shall be ascertained'. That is to say if work is wrongly billed or omitted from the bills it has to be measured and valued as though it had been properly billed. And that, by reference to clause 57, includes measurement in accordance with the stipulated standard method of measurement.

Without this clause there might be some faint argument that the contractor was not entitled to have work which was not billed measured and valued. It might be said that under clause 5 the engineer is to explain discrepancies on the basis that the contract documents are mutually explanatory of one another and under clause 11 the contractor's rates and prices are to cover all his obligations under the contract. Clause 55(2), however, dispels such contentions by giving the drawings and specification precedence over the bill of quantities in determining what shall be measured.

Conflict aspects of clause 55(2)

The correction of errors under clause 55(2) can be a source of concern to engineers/employers. It is sometimes argued in disputes where the omissions or errors in the bills are said to be patently obvious that the contractor should either have alerted the engineer/employer at tender stage or should have allowed for the relevant work in his tender. The counter-argument that the contractor prices only what he is required to price for in the bills and that the conditions of contract deal with the rest can seem insensitive but it is difficult to refute.

Many engineers/employers try to avoid the problem by including statements in instructions to tenderers to the effect that they shall bring to notice any errors or omissions in the bill of quantities relating to work shown in the drawings and specifications. However, it is difficult to give this pre-contractual obligation any legal/contractual effect such that the employer has rights against the contractor for pre-contract breach.

Similarly it is difficult to impose on the contractor a duty of care in respect of pricing the tender when the real issue is the engineer's duty of care in compiling the bill of quantities.

Interpretation of clause 55(2)

Clause 55(2) creates a few problems of interpretation in its reference to errors or omissions being 'corrected' by the engineer and the value of work 'actually carried out' being ascertained in accordance with clause 52(2) or 52(3). This appears to place the status of variations on the correction of errors and omissions and adds weight to the argument that clause 51(4) intends changes in quantities to be regarded as variations.

However, it is contrary to clause 55(2) to regard as a variation something already deemed by the clause to be the contractor's obligation. The correction of an error in the bill of quantities can never properly be described as a variation.

Another problem is that whereas in the Sixth edition the valuation of variations was based primarily on bill rates, the reference in clause 55(2) to clause 52 simply linked the correction of errors and omissions to bill rates. In the Seventh edition, where quotations from the contractor are at the forefront of valuation of variations, the link to bill rates is not as clearly maintained.

No rectification of errors by contractor

The proviso in clause 55(2) that there is to be no rectification of errors made by the contractor emphasises that the purpose of the clause is to deal with errors and omissions in the bill of quantities as issued – not as returned when priced.

The contractor's rates and prices are therefore excluded. The 'descriptions' referred to are presumably preliminaries and the like where the method of measurement allows the contractor flexibility to enter his own items. The contractor cannot obtain correction of these after acceptance of his tender.

If there is an error in computation between the rates and grossed-up totals there is little doubt that the rates stand as effective notwithstanding the fact that the employer may be misled by a wrong tender total into acceptance of the tender. There is an argument that to maintain the equity of the situation the rates should be adjusted backwards from the totals but there is nothing in the contract to support this, and if the employer has suffered a loss his remedy is against the engineer for negligence in checking the bills.

23.4 *Measurements and valuations*

Clause 56 is in four parts:

- 56(1) which states the important rule that the engineer shall value the works by admeasurement
- 56(2) which provides in somewhat obscure terms that the contractor's rates and prices can be increased or decreased in certain circumstances
- 56(3) which requires the contractor to assist in the measurement process
- 56(4) dayworks.

Valuation by admeasurement

Clause 56(1) states that the engineer shall, except as stated otherwise, ascertain and determine by admeasurement the value, in accordance with the contract, of the work done in accordance with the contract.

Admeasurement is an old fashioned word meaning the process of applying a measure to ascertain dimensions.

The burden is clearly on the engineer to undertake the admeasurement but that does not absolve the contractor from a similar responsibility. Under clause 60(1)(a) the contractor is to submit monthly estimates of contract value and under clause 60(4) he is to submit a final account with supporting documentation showing in detail the amounts he considers due.

The scheme is that the engineer measures on behalf of the employer, and the contractor measures for himself. If the contractor neglects to measure he is not expressly in breach of contract or disqualified from payment. However, he could jeopardise his right to interim payments and he would be reliant on the engineer's figures for the final valuation. If the engineer neglects to measure that is failure to carry out his duties under the contract. That might amount to a technical breach of contract although it is unlikely that the contractor could prove his loss; more likely it would amount to negligence as between the engineer and the employer.

'except as otherwise stated'.

The inclusion of this phrase in clause 55(2) suggests two possibilities: one that within the Conditions there are provisions excluding some work from being valued by admeasurement; the other that outside the Conditions but within the contract there may be special clauses for valuation other than by admeasurement. The first could apply to provisional sums, prime cost items or dayworks; the second to rules in standard methods of measurement or to contracts converted to lump sums.

Work done 'in accordance with the contract'

The engineer is only required to measure and value work done 'in accordance with the contract'.

Taken together with clause 39 (removal of improper work and materials) it confirms that under the Conditions there is no power for the engineer to accept sub-standard work on a reduced value basis. Some standard forms such as the New Engineering Contract have express terms to deal with this. In practice under ICE Conditions engineers and contractors do sometimes agree to artificial variations in order to downgrade both specification and prices but a formal arrangement accepting sub-standard work is more satisfactory.

23.5 *Increase or decrease of rate*

Clause 56(2) provides that should the actual quantities be greater or less than those in the bill of quantities and if, in the opinion of the engineer, such increase or decrease of itself shall so warrant, the engineer shall, after consultation with the contractor, determine an appropriate increase or decrease of any rates or prices rendered unreasonable or inapplicable.

'increase or decrease of itself'

The precise meaning of 'increase or decrease of itself' is elusive but it is generally understood to apply to a change of quantity which itself changes the method of working or the economics of a working method. It is not thought to apply to a quantity change which does no more than alter the overall profitability of the contract by yielding more or less volume of a particular work item.

'after consultation with the Contractor'

The wording of clause 56(2) suggests that it is the engineer who takes the lead in any re-rating exercise, but more often than not in practice it is the contractor. Nevertheless the power of the engineer to change rates on his own initiative, particularly to reduce them, must be recognised.

The engineer is constrained by a requirement to consult with the contractor and this would seem to oblige the contractor to co-operate in making available the breakdown of his rates and prices.

'appropriate increase or decrease'

It is not always the case that an increase of quantity will reduce the cost of production and justify a rate reduction. Consider the costs of disposal of off-site material which can increase dramatically once the contractor's planned tipping facility has been exhausted. In every case it is necessary to examine the economics of working.

It is suggested that what the clause contemplates is the discovery of the change of rate necessary to restore the same level of unit profitability to an item as in the tender rate.

There is usually no justification for re-rating resulting from increases or decreases in quantities to be undertaken by building up new rates from scratch. The starting point should normally be the bill rates.

'any rates or prices'

The generality of this phrase suggests that it is not only the rates and prices of items which have increased or decreased which may be reviewed. It can include other rates. For example, if the contractor intended to use lorries carting topsoil off the site for return loads of imported fill, with the cost of transport spread over both rates, then if the volume of topsoil decreases that could directly increase the unit cost of imported fill.

The re-rating exercise can go further so as to include changes in prices for preliminary items and, in extreme cases of changes in quantities, it is not unknown for the contractor to seek total re-rating.

'rendered unreasonable or inapplicable'

It is easy enough to see what is meant by rates rendered 'inapplicable' – the assumptions made in pricing at tender stage no longer apply. But rates rendered 'unreasonable' is a vaguer concept.

Contractors may say that to be asked to do more of a loss making item is unreasonable; but in truth it is not the rate which is rendered unreasonable but the amount of the loss. Correcting that is outside the scope of clause 56(2).

More difficult to refute is the proposition sometimes put forward by engineers that, where an increase in quantities provides the contractor with an opportunity to change his method of working and reduce his unit rates, it is reasonable that there should be a reduction in the rates, whether or not the contractor actually changes his methods. In theory the argument may have some substance but in practice the emergence of increased quantities is often a gradual process and it is questionable if it is appropriate for the engineer to use hindsight where neither he nor the contractor had the benefit of foresight.

An interesting aspect of clause 56(2) resulting from the wording that it is the increase or decrease (in work items) which itself renders rates unreasonable or inapplicable, is that the clause operates in a sequenced manner. Thus rates for work items undertaken before the increase or decrease cannot be rendered unreasonable or inapplicable by the change.

Attending for measurement

Clause 56(3) provides:

- the engineer shall give reasonable notice to the contractor when he requires any part of the works to be measured
- the contractor shall send a qualified agent to assist
- the contractor shall furnish all particulars required
- should the contractor fail to attend, the measurements made by the engineer or approved by him, shall be taken as the correct measurement of the work.

The clause should be read in conjunction with clause 38(1) which requires the contractor to give notice to the engineer before covering up permanent work and requires the engineer to attend for the purposes of examining and measuring.

The express statement in clause 56(3) that should the contractor fail to attend then the measurements made by the engineer are to be taken to be the correct measurements, casts doubt on whether an adjudicator or arbitrator is normally entitled to decide otherwise. However, if the engineer's measurements are patently incorrect the position may be different.

23.6 Daywork

Clause 56(4) sets out the procedures, the requirements for records and the basis of payment for work which is to be valued on dayworks.

Clause 56(4) says nothing about the engineer's power to order dayworks. That is to be found in clause 52(5) which states that the engineer may, if in his opinion it is necessary or desirable, order in writing that additional or substituted work be carried out on a daywork basis in accordance with clause 56(4).

Contractor's entitlement to be paid on a daywork basis

Clause 56(4) says little about the contractor's entitlement to be paid on a daywork basis other than to say that 'Where any work is carried out on a daywork basis the Contractor shall be paid for such work under the conditions ... etc'. In the absence of an order in writing from the engineer under clause 52(5) it is difficult for the contractor to find valid grounds for claiming payment on a daywork basis. However, the absence of a written order may not be fatal to a claim if the contractor can establish an oral instruction or clear intention that work should be done on a daywork basis or if it can be shown that, in the circumstances, daywork is the proper way of establishing a fair valuation. See for example the *Laserbore* case mentioned in Chapter 20, Section 20.5.

Daywork schedules

The first part of clause 56(4) states that where any work is carried out on a daywork basis the contractor shall be paid for such work under the conditions and at the rates and prices set out in the daywork schedule included in the contract, or failing the inclusion of a daywork schedule he shall be paid at the rates and prices and under the conditions contained in the 'Schedules of Dayworks carried out incidental to Contract Work' issued by The Civil Engineering Contractors Association (formerly issued by The Federation of Civil Engineering Contractors) current at the date of carrying out the daywork.

There is no space provided in the appendix to the form of tender to identify the applicable schedules but normally it will be apparent from the bill of quantities or elsewhere if the standard schedules mentioned in clause 56(4) are displaced by a contract particular schedule.

Contract particular schedules are frequently the source of dispute – not least when they comprise a mix of special schedules and the standard schedules. Engineers are advised to follow strictly the guidelines in the standard civil engineering method of measurement in preparing special schedules.

Records and returns

The second part of clause 56(4) requires that the contractor shall supply the engineer with such records and receipts and other documentation as necessary to prove amounts paid and/or costs incurred and that such returns shall be in the form and shall be delivered at such times as the engineer shall direct. This part of the clause concludes with the requirement that records and returns shall be agreed within a reasonable time.

Long time users of ICE Conditions may remember when under earlier editions (the Fifth and previous) the requirement was for all daywork records to be delivered daily. That was changed in the Sixth edition by putting the burden on the engineer to direct the form and times of delivery of daywork returns. There is much to be said for the practice of such direction either being stated in the contract documents or given at the onset of the contract.

The requirement that returns be agreed within a reasonable time puts a burden on both the engineer and the contractor. The greater burden is perhaps on the engineer because the contractor is entitled to include daywork values in his interim applications for payment whether or not they are agreed and failure by the engineer to certify leaves the employer liable for interest charges.

Ordering materials

The third and final part of clause 56(4) states simply that before ordering materials the contractor shall, if so required, submit to the engineer quotations for approval.

23.7 *Method of measurement*

Clause 57 provides that:

● unless otherwise provided in the contract, or
● unless general or detailed description in the bill of quantities, or
● any other statement clearly shows to the contrary

the bill of quantities shall be deemed to have been prepared and measurements made, according to the procedure set out in the Civil Engineering Standard Method of Measurement Third Edition 1991 or such later or amended edition as may be stated in the appendix to the form of tender.

The effect of clause 57 when taken in conjunction with clause 55(2) is to limit the application of the contractor's tendered rates and prices to the work as billed unless there is some proviso, description or statement to the contrary.

The importance of the clause lies in its phrase 'measurements shall be made according to the procedure'. In short, the general rule is that irrespective of how the work has been billed the contractor is entitled to have it measured in accordance with the standard method of measurement.

Chapter 24
Provisional sums and prime cost items

24.1 Introduction

The use of provisional sums and prime cost items in contracts under ICE Seventh edition Conditions is subject to three control mechanisms:

- the definitions given in clauses 1(1)(k) and 1(1)(l)
- the provisions in clauses 58(1), 58(2) and 58(3) of the Conditions
- the rules for measurement stated in the Civil Engineering Standard Method of Measurement Third Edition 1991.

Together these state the meaning of the terms in the Conditions; the powers of the engineer to order work under provisional sums and prime cost items; design responsibility under the Conditions in respect of such sums and items; how the bills of quantities should be prepared; and how the work undertaken should be measured. With the exception of minor changes, such as 'carried out' in place of 'execution', the Seventh edition deals with provisional sums and prime cost items in like manner to the Sixth edition.

General meanings

There is general acceptance in the construction industry that the meaning of prime cost is cost which is reimbursable on the basis of invoices plus a percentage allowance for the contractor's overheads and profit and sometimes attendances. Prime cost items are normally used for specialist works and mostly where work is to be undertaken by nominated subcontractors.

Provisional sums are not as easily defined. In some contracts they are no more than contingency allowances for work which may or may not be required – either specific or general. In other contracts they are sums for works, the extent of which is not fully known at the outset or the details of which are not fully developed, or they are allowances for dayworks.

Provisional items differ from provisional sums in that they are usually items included in the bills of quantities for work which may or may not be encountered but which will be required if it is encountered – for instance, rock in excavations. Work which is the subject of provisional items does not normally fall under the contractual rules for provisional sums and prime cost items.

24.2 *Provisional sums*

Clause 1(1)(l) of the Conditions states that a 'provisional sum' means a sum included and designated in the contract as a specific contingency for the carrying out of work or the supply of goods, materials or services which may be used in whole, in part, or not at all, at the direction and discretion of the engineer.

Clause 58(1) on the use of provisional sums provides that in respect of every such sum the engineer may order either or both of the following:

(a) work to be carried out, or goods, materials or services to be supplied by the contractor, the value being determined in accordance with clause 52
(b) work to be carried out, or goods, materials or services to be supplied by a nominated subcontractor in accordance with clause 59.

Rule 5.17 of CESMM3 (Civil Engineering Standard Method of Measurement) reads as follows:

'Provision for contingencies shall be made by giving Provisional Sums in the Bill of Quantities and not by increasing the quantities beyond those of the work expected to be required. Provisional Sums for specific contingencies shall be given in the general items of the Bill of Quantities. A Provisional Sum for a general contingency allowance, if required, shall be given in the Grand Summary in accordance with paragraph 5.25.'

Additionally rule 5.7 of CESMM3 states that provisional sums for work executed on a daywork basis may be given comprising separate items for labour, plant, materials and supplementary charges; and rule 5.25 states that a provisional sum for a general contingency shall be given, if required, in the grand summary.

The position, therefore, under the Seventh edition is that the engineer can:

- include items in the bills of quantities for specific contingencies
- use all or part or nothing of such items
- direct the contractor to carry out any work ordered under the items, or
- direct that the work be done by a nominated subcontractor.

Terminology

The references in CESMM3 to provisional sums for dayworks and for general contingencies should not be taken as references to provisional sums as defined in the Conditions. The terminology of CESMM3 is not wholly consistent with the terminology of the Conditions.

Valuation

Neither the Conditions nor CESMM3 give detailed guidance on the valuation of work carried out under provisional sums (unlike the position for prime cost items) and questions can arise on how such matters as overheads, profit and allowances should be dealt with. Given the possibility of the work being done either by the contractor or a nominated subcontractor is it not surprising that there is no single set of rules. The intention is clearly that overheads, profit and allowances for the contractor's own work should follow the valuation rules of clause 52 and, for nominated subcontractor's work, should follow the rules of clause 59 – which are effectively the same as for prime cost items.

Compilers of bills of quantities should be careful not to add complication to the rules of valuation – see for example the case of *St Modwen Developments* v. *Bowmer & Kirkland* (1996) under a building contract where disputes arose as to the application of bill entries stating 'included' when the rules for provisional sums and prime cost items had been mixed in relation to works to be undertaken by domestic subcontractors.

Programming of provisional sums

The question of whether the contractor is obliged to include in his programme for work in a provisional sum has not been satisfactory resolved. Nor has the follow-on question – is the contractor entitled to an extension of time for carrying out work in a provisional sum?

It is sometimes argued that the contractor should not be expected to programme work which may never be carried out and the ordering of any work under a provisional sum should be regarded as extra.

It is suggested that this argument is flawed. The work in a provisional sum is part of the contract work and it is not a variation when it is ordered albeit that it may be valued as one. Moreover there are no provisions in clause 44 for extending time for the work in provisional sums. Probably the best the contractor can do is to argue that an instruction to carry out work under a provisional sum is also an instruction under clause 13. Then to the extent that the timing of the instruction or its contents delays or disrupts his arrangements he is entitled to recover any extra cost which could not have been foreseen at the time of tender. He is also entitled to an extension of time.

The contractor is, of course, on stronger ground if the value of work exceeds the provisional sum. It can then be argued that the provisional sum is the limit of what the contractor has undertaken to perform and any excess must be in the nature of a variation.

24.3 *Prime cost items*

Clause 1(1)(k) of the Conditions states that a prime cost item means an item which contains, wholly or in part, a sum referred to as prime cost which will be used for the carrying out of work or the supply of goods, materials or services for the works.

Clause 58(2) on the use of prime cost items provides that in respect of every such item the engineer may order either or both of the following:

(a) the contractor to employ a nominated subcontractor in accordance with clause 59, or
(b) the contractor himself to carry out the work or provide the goods and services himself – but only if the contractor consents to do so.

Where the contractor himself is to carry out the work etc. he is to be paid in accordance with any quotation accepted by the engineer or in the absence thereof, in accordance with a valuation under clause 52.

Rule 5.15 of CESMM3 reads as follows:

'The estimated price of work to be carried out by a Nominated Sub-contractor shall be given in the Bill of Quantities as a Prime Cost Item. Each Prime Cost Item shall be followed by

(a) an item for a sum for labours in connection therewith which, in the absence of any express provision in the Contract to the contrary, shall include *only*
    (i) in any case in which the Nominated Sub-contractor is to carry out work on the Site for allowing him to use temporary roads, scaffolding, hoists, messrooms, sanitary accommodation and welfare facilities which are provided by the Contractor for his own use and for providing space for office accommodation and storage of plant and materials, for disposing of rubbish and for providing light and water for the work of the Nominated Sub-contractor and,
    (ii) in any case in which the Nominated Sub-contractor is not to carry out work on the Site for unloading, storing and hoisting materials supplied by him and returning packing materials, and
(b) an item expressed as a percentage of the price of the Prime Cost Item in respect of all other charges and profit.'

Rule 5.16 of CESMM3 reads:

'Where any goods, materials or services supplied by a Nominated Sub-contractor are to be used by the Contractor in connection with any item, reference shall be made in the description of that item, or in the appropriate heading or sub-heading, to the Prime Cost Item under which the goods or materials or services are to be supplied.'

Use of prime cost items

Note that by its definition a prime cost item 'will be used'. The engineer must therefore attempt in the first instance to follow either the nominated subcontractor route or the contractor-by-consent route. Only if the contractor declines to take on the work himself or objects under the provisions of clause 59 to nomination can the engineer omit the work by variation under clause 51.

The use in clause 58(2)(b) of the phrase 'the contractor himself' is not thought to exclude subcontracting under clause 4 as the means of carrying out the work.

Payment for prime cost items

Where the work is a prime cost item undertaken by a nominated sub-contractor the rules of valuation as stated in the Conditions are those in clause 59(5) which in summary are:

- actual price (prime cost) net of discounts other than prompt payment discounts
- any sum provided in the bill of quantities for labour (attendances)
- overheads and profit at the percentage rate quoted in the bill of quantities, or on the absence of any such rate at the rate inserted by the contractor in the appendix to the form of tender.

These are much the same rules as those stated in CESMM3.

Where the contractor undertakes the work himself neither the valuation rules of clause 59(5) of the Conditions nor rule 5.15 of CESMM3 applies. Instead the contractor is entitled to be paid on an accepted quotation or failing that in accordance with the valuation rules of clause 52(3) of the Conditions.

In effect under the Seventh edition, because clause 52 itself provides for quotations for variations, the payment rules for provisional sums and prime cost items merge so that:

- prime cost plus uplifts as clause 59(5) and CESMM3 5.15 applies to nominated subcontractor work
- quotations or clause 52(3) rules apply to the contractor's own work.

The position was slightly different under the Sixth edition since there was no provision for a quotation in clause 52.

A point to note is that if work is billed as a provisional sum and valued under the nominated subcontract rules it is more likely that the percentage for overheads and profit stated by the contractor in part two of the appendix to the form of tender will apply than if the work had been billed as a prime cost item and the percentage included in the bill of quantities. And, since the percentage in the appendix does not obviously attach to any sum of money, contractors may be disposed to insert a higher rate in the appendix than would be inserted in the bill.

Omission of a prime cost item

To the extent that the engineer has power to omit prime cost work from the contract such power comes from clause 51 (variations) rather than from definitions/rules relating to prime cost sums.

In the event of an omission variation it is suggested that the contractor remains entitled to his percentage for overheads and profit. That would be consistent with clause 59(2)(c) which provides that where nominated sub-contract work is omitted because the contractor declines to accept the proposed subcontractor (or there is nominated subcontractor default) then 'there shall nevertheless be included in the Contract Price such sum (if any) in respect of the Contractor's charges and profit being a percentage of the estimated value of such omission as would have been payable had there been no such omission and the value thereof had been that estimated in the Bill of Quantities or inserted in the Appendix to the Form of Tender as the case may be'.

Increase in a prime cost item

Where the amount payable under a prime cost item exceeds the stated amount of the item it is unlikely that a variation order will be required unless there has been a significant change in the work content of the item. Where there is such a change the contractor may be able to press for a variation if only to preserve his rights to entitlements of time under clause 44.

Usually the contractor will be compensated for any increase in a prime cost item through the payment mechanism of the Conditions and, in particular, through the percentage uplift for overheads and profit. However, in the Canadian case of *Cana Construction Ltd* v. *The Queen* (1973) the contractor's tender included a sum for overheads and profit based on an estimated prime cost given by the employer – which turned out to be half the final amount. The court decided that on the facts the change amounted to additional work under the variation clause and the contractor was entitled to an uplift on the larger amount as reasonable remuneration.

Programming of prime cost works

In principle, since prime cost works form part of the contract works, the contractor should allow for such works in his programme. In practice, particularly where nominated subcontractors are involved, the contractor may have inadequate information, or inadequate control, to perform the task with any degree of certainty or confidence. That applies most obviously when the identities and the timing requirements of the nominated subcontractors are not known until after commencement of the works.

The provisions of clause 59 (nominated subcontractors) give some relief to the contractor even to the extent that problems accommodating the time requirements of a particular nominated subcontractor may be a valid ground for objection to the nomination.

24.4 Design requirements

Clause 58(3) states that if:

- in connection with any provisional sum or prime cost item
- the services to be provided include any matter of design or specification
- of any part of the permanent works or equipment or plant to be incorporated
- such requirement shall be expressly stated in the contract, and
- shall be included in any nominated subcontract.

The clause concludes with the statement that 'the obligation of the Contractor in respect thereof shall be only that which has been expressly stated in accordance with this sub-clause'.

The intentions of clause 58(3) are not easy to assess. The clause carries the side note 'design requirements to be expressly stated' and at first sight its wording is consistent with the provision in clause 8(2) that the contractor shall not be responsible for the design or specification of any part of the permanent works except as expressly provided for in the contract.

However, it must be presumed that clause 58(3) does something more than repeat the provision in clause 8(2) and closer inspection of the wording of clause 58(3) suggests that it is concerned not so much with responsibility for design but with obligation for design. For comment on the distinction see Chapter 6, Section 6.1.

Thus clause 58(3) does not use the word 'responsibility' – it refers to 'requirement' and to 'obligation'. On one interpretation therefore clause 58(3) has nothing to say about responsibility for design (that being covered by clause 8(2)) but it is stating that no obligation to provide a design service can be imposed through a provisional sum or a prime cost item unless expressly stated. That may explain why in clause 58(3) only 'services' are mentioned whereas in the preceding clauses 58(1) and 58(2) the word 'services' is consistently put in the phrase 'work to be carried out or goods, materials or services to be supplied'.

There is an alternative interpretation of clause 58(3) to the effect that the clause is concerned with both obligation and responsibility for the design of specialist equipment and as such it is permissible to state fitness for purpose requirements in the contract without creating conflict with clause 8(2). From a practical point of view this has much to recommend it.

Whatever the proper interpretation of clause 58(3) it does seem unlikely that the engineer can by order under clause 58(1) or 58(2) impose design obligations and/or design responsibilities which are not expressly stated in the contract.

Chapter 25
Nominated subcontractors

25.1 Introduction

The Seventh edition retains with only one significant change the lengthy provisions on nominated subcontractors which first appeared in 1973 in the Fifth edition. Given the pitfalls for the employer and the engineer in using nominated subcontractors and the general trend in the construction industry away from their use, there is a case for saying that the Seventh edition would be a better and more modern contract without such provisions.

Problems with nominated subcontracting

The basic question with nominated subcontracting is how much responsibility the employer should take for the defaults or failures of the subcontractor he or the engineer has selected and imposed on the contractor. That question was very much in the mind of the drafting committee of the Fifth edition because of the decision of the House of Lords in 1970 in the case of *North West Metropolitan Hospital Board* v. *T. A. Bickerton & Sons* (1970). The case illustrated the problem that if the employer indemnifies the contractor against loss caused by the nominated subcontractor, the contractual chain of responsibility is broken since the contractor suffers no loss. The defaulting subcontractor then escapes scot-free and the employer cannot recover his loss.

Additionally there are potentially serious practical and financial problems for the employer and the engineer if the main contractor validly objects to a particular nomination or the nominated subcontractor actually appointed goes out of business or falls into default to the extent that his appointment is terminated.

Policy of the ICE Conditions

Standard forms of construction contracts issued since the *Bickerton* decision have approached the problem of indemnities with varying degrees of caution but this is how Sir William Harris, Chairman of the Joint Contracts Committee, described the policy of the Fifth edition:

'The Committee took the view that, as far as possible, the chain of liability should not be broken and that the responsibilities and liabilities should be spelt out clearly: also that employers and engineers should face up to this situation and consider carefully in each case, whether "nomination" is in fact necessary.'

In explaining why the clauses incorporating this policy were so long and complicated Sir William Harris said:

'These were undoubtedly the most difficult clauses to deal with, because the whole concept of nominated subcontractors raises many complex difficulties, as recent decisions in the courts show. One of the most serious of these difficulties is the problem of the defaulting or insolvent subcontractor who is a source of great potential loss and increased expense. The Committee had to decide upon whom that risk should rest: upon the employer who chose the subcontractor, and ordered the contractor to employ him or the contractor who was obliged to employ him. The Committee decided that, despite the power of objection given to the contractor, he who called the tune should pay the piper and the clauses are designed so that if loss should ultimately be suffered as a result of his chosen specialist's fault or bankruptcy, that loss should be upon the employer. The contractor, however, is under an obligation to do all he can to avoid such loss and to recover it from the subcontractor. Where one has, as a basic premise, an artificial situation, it must follow that clauses designed to give it legal enforceability will themselves appear to be somewhat tortuous and complicated.'

The only policy change of note from the Fifth through the Sixth to the Seventh edition has been on the contractor's entitlement to extension of time for delay consequent upon a nominated subcontractor's default. Such entitlement was provided for in the Fifth, omitted from the Sixth and then reinstated in the Seventh by the additional clause 59(4)(f). An indication perhaps, of the complexities described by Sir William Harris.

Liability of consulting engineers

When things go wrong with nominated subcontractors and the employer finds he has liabilities which would not have arisen with domestic subcontractors it is not unusual for the consulting engineer responsible for the adoption of the nomination process to find himself under pressure to explain and justify his actions. If he cannot provide good reasons he may stand accused of breach of his duty of care to the employer.

It is unlikely that the consulting engineer can gain protection from suit by the argument that as the Conditions provide for nominating subcontracting, use of nominated subcontractors cannot, of itself, be negligent. It is more likely that the consulting engineer, whether he be the named engineer to the contract or not, is deemed as a professional to be aware of the special risks of nominating and to have a duty to avoid them if possible.

Structure of clause 59

Clause 59 is no longer the lengthiest clause in the ICE Conditions, that honour having passed, in the Seventh edition, to clause 60 (certificates and payment). It remains, however, a clause of daunting dimensions and challenging construction. The marginal notes, which cannot by virtue of clause 1(3) be used in construction, do however assist in breaking down the clause to manageable proportions and, in short, the seven main sub-divisions of the clause are as follows:

- 59(1) – objections to nomination
- 59(2) – engineer's action on contractor's objection to nomination or determination of a nominated subcontract
- 59(3) – contractor's responsibility for nominated subcontractors
- 59(4) – nominated subcontractor's default and termination of nominated subcontracts.
- 59(5) – provisions for payment
- 59(6) – production of vouchers etc.
- 59(7) – payments to nominated subcontractors.

Changes in clause 59

The significant change from the Sixth to the Seventh edition is the inclusion in the Seventh edition of a new clause 59(4)(f) which provides for extension of time for delay 'consequent upon the Nominated Sub-Contractor's default'. For comment on this see Section 25.5 below.

There is a minor change to clause 59(4)(a) with the additional requirement at the end of the clause that the contractor shall give reasons when notifying the engineer of his opinion that a nominated subcontractor is in default.

Other changes are small points of terminology such as substitution of the words 'will indemnify' in place of 'will save harmless and indemnify' in clause 59(1).

25.2 *Objection to nomination*

Clause 59(1) states that the contractor shall not be obliged to enter into a subcontract with any nominated subcontractor:

- against whom the contractor may raise reasonable objection, or
- who declines to enter into a subcontract containing terms:

(a) that the subcontractor will undertake such obligations and liabilities as will enable the contractor to discharge his own liabilities to the employer

(b)   that the subcontractor will indemnify the contractor against all claims arising out of failure in performing the obligations of the subcontract
(c)   that the subcontractor will indemnify the contractor against claims for negligence
(d)   that the subcontractor will provide the contractor with security for performance of the subcontract
(e)   equivalent to those in clause 63.

The reference to clause 63 in clause 59(1)(e) is thought to be a mistake. The intended reference is probably to clause 65, which was clause 63 in the Sixth edition.

Reasonable objection

The test for reasonable objection is fraught with difficulty – not least because what is reasonable from the contractor's viewpoint may be unreasonable from the employer's or engineer's viewpoint. However, since it is for the contractor to raise the objection it is probably the contractor's viewpoint which prevails.

Consequences of an unreasonable objection

The Conditions do not deal expressly with either how and by whom a reasonable objection is to be assessed or what consequences follow from an unreasonable objection.

There are various possibilities. One is that for the operation of clause 59 and completion of the works any objection is to be taken as an effective objection subject to the employer's right to raise a dispute on the reasonableness point. Another is that the engineer, using his powers of instruction under clause 13, can effectively rule whether an objection is reasonable or unreasonable – in which case it is for the contractor to implement the dispute procedures of the contract.

Matters which might form the basis of reasonable objection could include:

- commercial conflict
- previous difficulties
- lack of confidence
- financial instability
- poor safety record
- inadequate insurances.

Whether knowledge that a particular subcontractor is claims-oriented can be the basis for reasonable objection is a moot point. However, it is difficult to see how the employer can complain about such an objection if he has used 'a hard attitude to claims' as a disqualifying factor in his own selection of the list of potential main contractors. And questions on the main contractor's attitudes to claims do often feature on forms for references.

Like obligations and liabilities – clause 59(1)(a)

The commonest problems with the subcontractor declining to take on like obligations and liabilities as the contractor, come in relation to time and liquidated damages. The contractor probably cannot object to a nominated subcontractor who will not operate within the limits of a shortened programme but he has firm grounds for objection if the subcontractor's work cannot be programmed to fit within the contract time. The employer's dilemma in such a situation was highlighted in the case of *Trollope & Colls* v. *North West Metropolitan Regional Hospital Board* (1973).

In that case the time remaining for phase III of a hospital building contract, after extensions granted on phases I and II, was 16 months instead of the 30 months originally intended. The employers, finding themselves unable to nominate for phase III subcontractors who could complete in the shorter time, argued for an implied term in the contract that an extension should be granted to phase III to accommodate the delays in phases I and II. The contractors opposed the granting of any such extension. They were, in the words of Lord Pearson:

> 'turning the situation to their own advantage, because if the contract could not be carried out, a new arrangement would have to be made for the work to be done at the prices prevailing in or about 1971, which were considerably higher than the contract prices. The difference between the contract prices and the prices prevailing in or about 1971 is said to be in the region of one million pounds.'

In a later building case *Fairclough Building Ltd* v. *Ruddlan Borough Council* (1985) it was held that an architect's instruction in nominating a subcontractor who could not complete within the time remaining after an earlier nominated subcontractor had defaulted was invalid and the contractor was entitled to refuse the nomination.

It is not unusual when such circumstances arise under ICE Conditions for the engineer to resolve the problem by extending the time for completion under the main contract. Under the Sixth edition the stated grounds for extension used were often 'other special circumstances' but under the Seventh edition the new ground in clause 44(1)(e) is probably more appropriate – delay, impediment, prevention by the employer.

Indemnity against claims – clause 59(1)(b)

The indemnity against claims that the contractor is entitled to under clause 59(1)(b) is not restricted to claims from the employer. It can and should include the contractor's own costs and claims from other subcontractors.

Indemnity against negligence – clause 59(1)(c)

The contractor would do well to check the nominated subcontractor's insurances before accepting the nomination since an indemnity against negligence could be a worthless document if lacking financial support.

Security for performance – clause 59(1)(d)

The provision in clause 59(1)(d) for the subcontractor to provide the contractor with security says nothing about the level of security or the form it should take.

Security under the main contract is regulated by clause 10 and is limited to 10% of the tender total. It is questionable whether the contractor can require a higher level than this from the nominated subcontractor.

Similarly it is questionable whether the contractor can require an on-demand bond rather than a default bond of the type in the main contract.

Provisions equivalent to clause 63 – clause 59(1)(e)

As noted above it is thought that the proper reference in clause 59(1)(e) is to clause 65.

Clause 65 provides a detailed procedure for determination of the contractor's employment upon certain defaults. For the non-financial matters the engineer plays a crucial role in its operation.

The question is should the 'equivalent' provisions include a role for the engineer, with the contractor substituting for just the employer; or should the contractor substitute for both the engineer and the employer? Given the wording of clause 59(4)(a) which says 'which in the opinion of the contractor justifies the exercise of his right', it would seem that the engineer is excluded from the procedure. That still leaves some uncertainty as to whether the 'opinion of the engineer' remains a constituent of the grounds for determination.

25.3 *Engineer's action on objection or determination*

Clause 59(2) combines the actions the engineer is to take when the contractor declines to enter into a subcontract with a subcontractor nominated by the engineer or when there is valid termination of the employment of a nominated subcontractor.

The wording of the opening part of clause 59(2) deserves scrutiny. The clause commences 'If pursuant to sub-clause (1) of this Clause the Contractor declines to enter into a sub-contract with a sub-contractor nominated by the Engineer'.

The first point to note is the use of the phrase 'declines to enter'. In clause

59(1) the contractor can either raise reasonable objection or decline to enter into a subcontract. The two are distinct alternatives. Clause 59(2) on the face of it is concerned only with the second alternative 'declines to enter'. Although it is arguable that the clause has no application to the situation which applies when there is reasonable objection, the gap which would then be left in the contract points to its application – if only on an implied basis.

The second point to note is the use of the phrase 'nominated by the Engineer'. By clause 1(1)(m) a nominated subcontractor means a firm (etc.) 'nominated in accordance with the Contract'. Nominated by the engineer can perhaps be taken, therefore, as a particular method of nomination distinct, for instance, from nomination in the contract documents. But once again if this distinction is carried through into the operation of clause 59(2) it is difficult to assess the practical and contractual effect – unless it is intended that the contractor cannot decline to enter into a subcontract with a subcontractor nominated otherwise than 'by the Engineer'.

Courses of action

Putting aside the above textual difficulties, clause 59(2) gives the engineer five possible courses of action when there is need to re-nominate. These are:

(a)   to nominate an alternative subcontractor
(b)   to vary the works under clause 51
(c)   to omit the nominated subcontract work by order under clause 51
(d)   to instruct the contractor to secure a subcontractor of his choice
(e)   to invite the contractor to carry out the work himself.

Re-nomination

The express power of the engineer to re-nominate covers the problem in the *Bickerton* case mentioned in Section 25.1 above where the employer maintained, after the first nominated subcontractor had defaulted, that he was not bound to re-nominate or pay more than the original price.

The provisions for payment in clause 59(5) make clear that under ICE Conditions the actual price incurred must be paid.

Variation

If the engineer can overcome the problem of objection or termination by varying the works so as to avoid re-nominating he can do so under clause 59(2)(b). Such a variation would be valued under the rules of clause 52.

However, the contractor does not remain entitled to his percentage for charges and profit where there is a variation as he does for certain omissions (see below).

Omission

The express power in clause 59(2)(c) relates only to an omission where the work is to be carried out by the employer either concurrently or at some later date. The power is given to avoid the argument that any omission of work which is work necessary for completion is a breach of contract. If such an omission is a breach, as it probably is, then it cannot be ordered under clause 51 without consideration of the payment of damages to the contractor. In recognition of this, clause 59(2)(c) provides that the contractor is entitled to his charges and profit on the estimated value of the work omitted.

If the engineer can devise a way of omitting the work permanently, such an omission can be made under clause 59(2)(b). The contractor then has no entitlement to charges and profit.

Contractor's own choice of subcontractor

The intention of clause 59(2)(d) is presumably that the engineer can instruct the contractor to secure a 'nominated' subcontractor of his own choice; that is, the employment of the subcontractor within the provisions of clause 59 and not under clause 4 (domestic subcontractors).

The point is not fully clear but the employment of domestic subcontractors seems to come under clause 59(2)(e).

Contractor to carry out the work himself

The provision in clause 59(2)(e) for the engineer to invite the contractor to undertake the work himself is simply a return to provisions in clause 58 for the use of provisional sums and prime cost items.

It might be said that if this is a solution to a problem of objection to nomination or termination of the subcontract, it is a solution which could, and should, have been used to avoid nomination in the first place.

But against this, there is the point that the work to be completed after termination of a subcontract may be within the capability of the contractor and in such circumstances re-nomination would be unnecessary.

25.4 *Contractor's responsibility for nominated subcontractors*

Clause 59(3) states that except as otherwise provided in clause 58(3) the contractor shall be as responsible for the work carried out by a nominated subcontractor as if he had carried out the work himself.

The proviso referring to clause 58(3) relates only to design and, this apart, the clear purpose of clause 58(3) is to maintain the chain of contractual responsibility so that the contractor and not the employer is responsible for the work actually performed by a nominated subcontractor.

That responsibility extends to delays to completion caused by a nominated subcontractor in the performance of his works. The main contractor has no right to an extension of time for such delays and his remedy for loss suffered, including liquidated damages paid under the main contract, is against the nominated subcontractor.

However, there is a distinction to be made between the contractor's responsibility for the performance of a nominated subcontractor under clause 59(3) and the contractor's responsibility for the consequences of delay caused by termination of the employment of a nominated subcontractor. Clause 59(4) deals with the latter.

In short, under clause 59(3) the contractor is responsible for a nominated subcontractor's poor performance where under clause 59(4) the employer accepts some responsibility for a nominated subcontractor's non-performance (or default).

25.5 *Nominated subcontractor's default*

The provisions of clause 59(4) can be summarised as follows:

- 59(4)(a) – the contractor shall notify the engineer in writing with reasons if any event arises which in the opinion of the contractor justifies termination of the subcontract.
- 59(4)(b) – with the consent in writing of the engineer the contractor may expel the nominated subcontractor from the works; if consent is withheld the contractor is entitled to instructions under clause 13.
- 59(4)(c) – if the nominated subcontractor is expelled the engineer shall take action under clause 59(2) – i.e. re-nominate or take some other course of action.
- 59(4)(d) – where the nominated subcontract is terminated the contractor shall take all necessary steps and proceedings to recover from the nominated subcontractor all expenses incurred including those of the employer.
- 59(4)(e) – if the contractor fails to recover from the nominated subcontractor all his reasonable expenses the employer shall reimburse the balance.
- 59(4)(f) – the engineer shall take any delay to completion of the works consequent upon a nominated subcontractor's default into account when determining the contractor's entitlement to extension of time under clause 44.

Notice of default

Under clause 59(4)(a) if, in the opinion of the contractor, there are grounds for termination, the contractor shall 'at once' notify the engineer in writing.

This is stated as a duty, so the contractor seems bound to give notice whether or not he intends to proceed further.

Clearly notice is required if the contractor does wish to terminate. The purpose of notice in lesser circumstances is probably to alert the engineer to impending problems and to give the engineer the opportunity to intervene or apply such pressure as appropriate on the nominated subcontractor.

Consent required for termination

Clause 59(4)(b) effectively precludes the contractor from terminating the nominated subcontractor's employment without the engineer's consent.

It is difficult to envisage circumstances where the contractor might think it appropriate to terminate without consent; to do so would not only be breach of contract rendering the contractor liable for damages but it would also deprive the contractor of his rights under clauses 59(4)(e) and 59(4)(f).

Additionally the contractor would lose the benefit of the concluding part of clause 59(4)(b) whereby the contractor is entitled to instructions under clause 13 if consent to termination is withheld.

A point of some uncertainty on clause 59(4)(b) generally, and the clause 13 reference in particular, is whether a distinction is to be made between consent withheld because the engineer considers the contractor's request for consent unjustified and consent withheld because the engineer sees no alternative to persevering with the existing nominated subcontractor.

Engineer's action on termination

Clause 59(4)(c) requires the engineer to take action 'at once' under clause 59(2) in the event that a nominated subcontractor is expelled from the works.

There is no express reference in the clause to expulsion/termination with the engineer's consent, as there is in clause 59(4)(d), but it can probably be implied.

Failure by the engineer to act 'at once' under clause 59(2) could leave the employer liable to damages claims from the contractor additional to those provided for in clause 59(4)(e).

In practical terms the engineer's scope for action under clause 59(2) may be less in the event of termination than in the event of objection to nomination.

Recovery of additional expense

Clause 59(4)(d) is further maintenance of the chain of contractual responsibility. It places an obligation on the contractor to take 'all necessary steps and proceedings' available to recover the additional expenses of both the contractor and the employer. 'All necessary steps and proceedings' could include actions in adjudication, arbitration or litigation as appropriate.

The reference in clause 59(4)(d) to recovering 'security' would require the contractor to call on any bond he had obtained from the nominated sub-contractor.

The additional expenses of the contractor could include his own costs of delay and disruption and claims from other subcontractors plus legal costs. The additional expenses of the employer would most likely be payments to the contractor and payments to professional advisers.

Reimbursement of contractor's loss

The indemnity in clause 59(4)(e) that the employer will reimburse the contractor's unrecovered expenses following termination of a nominated sub-contract goes a long way towards relieving the contractor of the apparent burden of full responsibility imposed by clause 59(3).

However, the timescale for settling the amounts due is likely to be pro-tracted. The contractor may have to wait to see what he can recover as an unsecured creditor when there has been a liquidation or he may have to sue in arbitration or in litigation. The contractor is unlikely to know what his position is at the date the submission of the final account becomes due and how much he can claim from the employer.

Fortunately the wording of clause 60(4) on the submission of the final account leaves an opening in referring to amounts due 'up to the date' of the defects correction certificate.

Consequent delay

Clause 59(4)(f) entitles the contractor to an extension of time if completion of the works is delayed by a nominated subcontractor's 'default'. However, it is reasonably clear from the overall structure of clause 59 and the context in which clause 59(4)(f) appears that 'default' within the clause has the restricted meaning of default which has led to termination of the sub-contract. It may possibly extend to 'default' where the engineer has refused consent to terminate and has given instructions under clause 13 but that clause has its own direct link to extension of time under clause 44 and the establishment of a link through clause 59(4)(f) would only be relevant in the event of a dispute on entitlement under clause 13.

Provisions similar to those now found in clause 59(4)(f) of the Seventh edition were originally included in the Fifth edition but then omitted from the Sixth edition. The problem facing draftsmen in knowing what to do for the best can be seen from the building case, *Mellowes PPG Ltd* v. *Snelling Construction Ltd* (1989). The contract provided that an extension of time should be granted to the main contractor for delay caused by the sub-contractor. The contractor was therefore unable to recover damages from the subcontractor because, by his entitlement to an extension of time, he had suffered no loss himself.

The difficulty is in striking a balance between fair entitlements of the parties and in strictly maintaining an unbroken contractual chain.

The position now under the Seventh edition is that the contractor is entitled to an extension of time and corresponding relief from liquidated damages but the contractor has an obligation to endeavour to recover his own costs and the employer's from the defaulting subcontractor.

25.6 *Provisions for payment*

Clause 59(5) provides that for all work carried out or goods, materials or services supplied by a nominated subcontractor, the contractor shall be entitled to have included in the contract price:

(a)   the actual price paid, or due to be paid in accordance with the terms of the subcontract net of all trade discounts
(b)   the sum provided in the bill of quantities for labours
(c)   a sum for charges and profit based on the rate in the bill of quantities or, in its absence, the rate in the appendix to the form of tender for adjustment of prime cost items.

Default of the contractor

The phrase in brackets in clause 59(5)(a) 'unless and to the extent that any such payment is the result of a default of the Contractor' is to ensure that the contractor cannot pass on to the employer any extra payments he is obliged to make on his own account to the nominated subcontractor arising from his (the contractor's) defaults.

'Defaults' is possibly too narrow a word to be fully effective since the contractor may have other obligations to the nominated subcontractor which are neither defaults nor the responsibility of the employer.

Discounts

Clause 59(5)(a) reduces the price payable by the employer for trade discounts etc. but allows the contractor to retain the benefit of any discount he can obtain for prompt payment.

This has been criticised as open to abuse and prompt payment discount would certainly be a commercial bargaining factor if the contractor was free to make his own choice of subcontractor.

Percentage for other charges and profit

Under clause 59(5)(c) the contractor is entitled to a percentage of the actual price at the rate in the bill of quantities or 'where no such provision has been made' at the rate in the appendix.

Engineers should note that while the percentage stated in the bill of quantities will be computed in the tender total the percentage in the appendix will not normally influence the tender total.

Production of vouchers

Clause 59(6) provides, sensibly in the light of employer's payment obligations, that the contractor shall produce to the engineer, when required, all quotations, invoices etc.

25.7 *Payment to nominated subcontractors*

Clause 59(7) fulfils two functions. Firstly, it empowers the engineer to check that payments are being properly made to a nominated subcontractor, and secondly, it allows the employer to make direct payments if the contractor defaults in his payments.

The clause does not place any obligation on the employer to pay a nominated subcontractor, even if there is default; nor does it create a right of action against the employer. It does, however, provide comfort to nominated subcontractors and it is certainly of benefit if the main contractor becomes insolvent because the employer can pay directly for work which has been certified.

In summary the provisions of clause 59(7) are:

- before issuing a payment certificate to the contractor under clause 60 the engineer is entitled to proof that all nominated subcontractors have been paid amounts certified in previous certificates
- if any nominated subcontractor has not been paid the contractor is required to give reasons and proof that he has informed the nominated subcontractor in writing of his reasons for non-payment or set-off
- if the contractor fails to give such reasons and proof, the employer is entitled to pay the nominated subcontractor the amount certified by the engineer which the contractor has failed to pay
- the employer can deduct any direct payments so made from amounts due to the contractor by way of set-off
- amounts so set off by the employer shall be deducted from the amounts due on future certificates but the issue of such certificates shall not be delayed.

Note that under clause 60 amounts payable in respect of nominated subcontractors are to be shown separately in the contractor's monthly statements and the certificates of the engineer.

Legal position on direct payments

It is usual for receivers or liquidators of an insolvent contractor to claim from the employer the full value of unpaid certificates and uncertified work in progress. That is because, as legal successors of the contractor, they have acquired his rights.

If the employer pays any subcontractor, nominated or otherwise, for work, goods or services undertaken or provided prior to receivership he will, unless special circumstances apply, remain liable to the receiver or liquidator for the same sum. He cannot discharge his legal debt to the receiver or liquidator by paying someone else (e.g. the nominated subcontractor).

The provisions in clause 59(7), if operated properly, protect an employer who has made direct payments, since they are made under a contract which is itself still legally effective. However, the issues involved are unusually complex and justify taking legal advice before any direct payments are made.

Chapter 26
Certificates and payments

26.1 Introduction

The certification and payment provisions of the Seventh edition are for the most part the same as those in the Sixth edition with the 1998 amendments. Those were the amendments introduced to make the Sixth edition compliant with the requirements of the Housing Grants, Construction and Regeneration Act 1996.

The only change comes in clause 60(6)(c) which relates to payment of the final tranche of retention money. The clumsy wording of the 1998 amendment clause is replaced by a readily understandable arrangement for certification and payment.

Housing Grants, Construction and Regeneration Act 1996

For explanation of the background to the Act and the scope of its application see Chapter 28, Section 28.2. The parts of the Act relating to payments are:

- Section 109 – entitlement to stage payments
- Section 110 – dates for payment
- Section 111 – notice of intention to withhold payment
- Section 112 – right to suspend performance for non-payment
- Section 113 – prohibition of conditional payment provisions.

The 1998 amendments to the Sixth edition dealt only with the statutory requirements imposed by Sections 109–111 of the Act. The Section 112 right to suspend performance for non-payment was not addressed nor was the Section 113 prohibition on conditional payment provisions. Omission of the latter creates no problems since no edition of ICE Conditions contains conditional payment provisions of the type prohibited by the Act. Omission of the Section 112 right to suspend is something of a problem since it now overhangs the contractual provisions on suspension and payments rather than being incorporated into them.

Compliance with the Act

It seems to be generally accepted that the amendments to the Sixth edition (now part of the Seventh edition) in respect of Sections 109–111 are compliant with the Act and that so far as payment provisions are concerned there is no case for operation of the Scheme for Construction Contracts Regulations.

But in any event it is worth noting that if there are any non-compliant payment provisions in the Conditions, the provisions of the Scheme apply only to the non-compliant parts. This is significantly different from the position in respect of adjudication. Any non-compliant contractual provisions on adjudication lead to adoption of the full set of the Scheme provisions for adjudication.

Entitlement to stage payments

The relevant parts of Section 109 of the Act read as follows:

'(1)   A party to a construction contract is entitled to payment by instalments, stage payments or other periodic payments for any work under the contract unless –

(a)   it is specified in the contract that the duration of the work is to be less than 45 days, or

(b)   it is agreed between the parties that the duration of the work is estimated to be less than 45 days.

(2)   The parties are free to agree the amounts of the payments and the intervals at which, or circumstances in which, they become due.

(3)   In the absence of such agreement, the relevant provisions of the Scheme for Construction Contracts apply.'

Dates for payment

Section 110 of the Act reads:

'(1)   Every construction contract shall –

(a)   provide an adequate mechanism for determining what payments become due under the contract, and when, and

(b)   provide for a final date for payment in relation to any sum which becomes due.

The parties are free to agree how long the period is to be between the date on which a sum becomes due and the final date for payment.

(2)   Every construction contract shall provide for the giving of notice by a party not later than five days after the date on which a payment becomes due from him under the contract, or would have become due if –

(a)   the other party had carried out his obligation under the contract, and

(b)   no set-off or abatement was permitted by reference to any sum claimed to be due under one or more other contracts, specifying the amount (if any) of the payment made or proposed to be made, and the basis on which the amount was calculated.

(3) If or to the extent that a contract does not contain such provision as is mentioned in subsection (1) or (2), the relevant provisions of the Scheme for Construction Contracts apply.'

Notice of intention to withhold payment

Section 111 of the Act reads:

'(1) A party to a construction contract may not withhold payment after the final date for payment of a sum due under the contract unless he has given an effective notice of intention to withhold payment.

The notice mentioned in Section 110(2) may suffice as a notice of intention to withhold payment if it complies with the requirements of this section.

(2) To be effective such a notice must specify –
  (a) the amount proposed to be withheld and the ground for withholding payment, or
  (b) if there is more than one ground, each ground and the amount attributable to it,

and must be given not later than the prescribed period before the final date for payment.

(3) The parties are free to agree what that prescribed period is to be.

In the absence of such agreement, the period shall be that provided by the Scheme for Construction Contracts.

(4) Where an effective notice of intention to withhold payment is given, but on the matter being referred to adjudication it is decided that the whole or part of the amount should be paid, the decision shall be construed as requiring payment not later than –
  (a) seven days from the date of the decision, or
  (b) the date which apart from the notice would have been the final date for payment,

whichever is the later.'

Right to suspend for non-payment

For details and comment on Section 112 of the Act, see Chapter 15 Section 15.2, on suspension of work.

26.2 Monthly statements

Clause 60(1) requires the contractor to submit, unless otherwise agreed, statements at monthly intervals showing:

(a)   the estimated contract value of the permanent works carried out up to the end of the month

(b)   a list of goods or materials, and their value, delivered to site but not incorporated into the permanent works

(c)   a list of goods or materials, and their value, identified in the appendix to the form of tender, not yet delivered to site but property in which has vested in the employer

(d)   the estimated amounts the contractor considers himself entitled to in respect of other matters under the contract including amounts in the bill of quantities for temporary works or contractor's equipment.

Amounts payable in respect of nominated subcontractors are to be listed separately.

A monthly statement need not be submitted if, in the opinion of the contractor, the amount will not justify the issue of an interim certificate. This is dependent on the amount stated as the minimum in the appendix to the form of tender.

Due dates for payment

The opening part of clause 60(1) is important in establishing by the phrase 'monthly intervals commencing one month after the Works Commencement Date' the 'due dates' for payment under the contract for the purpose of compliance with Section 110 of the Act.

Value of permanent works

The essence of the scheme for payments is that each statement should show cumulative value and not incremental adjustment of the previous statement. Where the engineer has prescribed a form in the specification, as contemplated by clause 60(1), the statement will invariably be submitted in cumulative form. Where the contractor is left with freedom to devise his own form he may slip into habits used elsewhere in the construction industry of incremental statements.

It is suggested that the phrase 'estimated contract value' does not relieve the contractor of presenting a measured account each month. The Seventh edition, like the Sixth, is a measure and value contract and whilst stage payments, activity schedules and percentage completions may be appropriate for lump sum contracts, that is only because the final sum payable is broadly known at the outset.

Clause 60(1)(a) retains the phrase 'up to the end of that month' although the requirement for statements is for statements 'at monthly intervals'. Perhaps the wording could have been clearer. It is not thought that the value should be to the end of a calendar month.

Goods on site

The only certainty about ownership of goods on site is that when they are fixed on the employer's land he has good title. For the rest, whatever contracts of sale or construction contracts may say, possession remains the strongest point from which to argue. Since average payment times on invoices are in the UK in excess of two months, it follows that more often than not the contractor is paid for materials on site before the supplier is paid. Not surprisingly, suppliers often turn up on site and take their goods away when a contractor goes out of business.

Security of the site should therefore be an important consideration in the payment for unfixed goods and materials. The engineer should be aware of this in extending the site under clause 1(1)(v).

Goods off-site

Payment for goods off-site is even more risky than payment for goods on-site. Receivers and liquidators are not readily persuaded that things within the curtilage they control are not within their power to dispose of. Clause 54 lays down an elaborate procedure for vesting goods and materials not on site in the employer but it often proves to be of no consequence.

If the employer is prepared to take the risk of paying for off-site goods, the key point for the engineer is to establish ownership. Payment to the contractor for goods he does not own can only give the employer good claim to title in exceptional circumstances.

Value of goods and materials

Unfixed goods and materials may have two widely differing values – that as invoiced, and that as marked-up in the bill of quantities. The basis of payment until they are fixed should be as invoiced and net of discounts.

'other matters'

The provision in clause 60(1)(d) for inclusion in the contractor's statement of the estimated amounts in connection with 'other matters' covers most obviously the claim clauses of the contract. Thus clauses 13(3), 14(8) and others refer expressly to payment in accordance with clause 60. Clause 53(6) on claims generally confirms that the contractor is entitled to have included in any interim payment certified under clause 60 such amount as the engineer considers due.

Other clauses which can give the contractor an entitlement to payment but make no express reference to clause 60, such as clauses 17 (setting-out) and 38 (uncovering), will also be within the scope of 'other matters' in clause 60.

Inclusion of claims

The provision that the contractor 'shall' submit at monthly intervals a statement showing the estimated amounts to which he considers himself entitled is sometimes used as the basis for rejecting a contractor's claim which is lodged late or post-completion.

The provision needs to be read in conjunction with clause 53 and it is difficult to see how in the ordinary run of things failure by the contractor to include a claim in a monthly statement does any more than deprive him of his right to interim payment.

Extra-contractual claims

It is doubtful if claims for breach of contract, not expressly covered in the Conditions, should be included as 'other matters' within clause 60. Such claims could include failures by the engineer to perform his duties or by the employer to fulfil his obligations.

The wording in clause 60(1)(d) – 'for which provision is made under the Contract' – certainly indicates that only contractual claims are covered by the clause and suggests that submissions and payments for breach are to be dealt with elsewhere. There is, however, a complication.

Although a contract administrator would not normally have power to adjudge and certify on extra-contractual claims, the engineer in ICE Conditions is under a duty by clause 66 to decide on all matters of dissatisfaction referred to him 'in connection with or arising out of the contract'. This clearly includes extra-contractual claims. If the engineer finds in favour of the contract, it is then arguable that payment is due because provision has been 'made under the Contract'.

Retention on claims

A small financial point against including extra-contractual claims in the clause 60 procedure is that retention is deducted under 60(5) from the amount due under 60(1)(d).

In principle, retention should not be deducted from any payment which is payment for breach. But in so far as deduction is allowed under clause 60(5) in relation to amounts included under clause 60(1)(d) in applications for payments, then the contractor can be deemed to have accepted deduction. However, settlement of an extra-contractual claim, whether under clause 66 or by agreement, should be free of retention.

Failure to submit a monthly statement

The contractor is obliged by clause 60(1) to submit a monthly statement unless 'otherwise agreed' or unless in the opinion of the contractor the amount will not justify the issue of an interim certificate.

In normal circumstances it is unlikely that the contractor will fall into breach by not submitting a monthly statement when some payment is due. And, even if he does, the employer has no obvious remedy except that in exceptional circumstances the employer might be able to prove loss arising from the breach – where, for example the employer's funding is itself related to expenditure criteria.

Interest on late payment

By clause 60(7) the contractor is entitled to interest on late payments. The wording of the clause can be taken as an obligation on the employer to pay interest whether or not there is an application from the contractor.

In practice contractors will usually as a matter of prudence include any interest they consider due under clause 60(7) in their clause 60(1) statements.

Value added tax

Finally, there is a complication on value added tax as to whether it applies equally to the provision of goods and services and damages for breach of contract when payments are made together under the contract. Court awards for goods and services attract VAT; court awards for damages do not. Resolution of the position in respect of contract payments is best left to accountants and the Customs and Excise.

26.3 *Monthly payments*

Clause 60(2) provides a mechanism as required by section 110(1)(a) of the Act, for determining what payments become due under the contract.

Clause 60(2) provides:

- within 25 days of the date of delivery of the contractor's monthly statement
- the engineer or the engineer's representative shall certify
- and within 28 days of the same date (delivery of the contractor's statement) the employer shall pay:
    - the amount due in the opinion of the engineer (less retention) in respect of the value of the permanent works, and
    - amounts the engineer considers proper in respect of goods and materials.

The clause concludes by stating that payments 'become due' on certification with the final date for payment being 28 days after the date of delivery of the contractor's monthly statement and that amounts certified in respect of nominated subcontractors shall be shown separately in the certificate.

Amounts to be paid

Clause 60(2)(a) and 60(2)(b) use different terminology for the amounts to be paid to the contractor. Clause 60(2)(a) uses the phrase, for work and claim items, 'the amount which in the opinion of the Engineer on the basis of the monthly statement is due to the Contractor'. Clause 60(2)(b) uses the phrase, for unfixed goods and materials 'such amounts (if any) as the Engineer may consider proper'.

Nothing much may turn on this since clause 60(2) goes on to say 'The payments become due on certification...'. Nevertheless given the impact of the Act and its use of the phrase 'a sum due' and the reference in clause 60(10) of the Conditions to 'a sum due', it is questionable whether the Conditions should make the distinction in clauses 60(2)(a) and 60(2)(b) between amounts 'due' and amounts considered 'proper'.

Notice of sum due

Section 110(2) of the Act requires that every construction contract shall provide for the giving of notice not later than 5 days after the date on which a payment becomes due, of the payment proposed to be made and the basis on which it is calculated.

Under clause 60(2) of the Conditions payments become due on certification.

Under clause 60(9) every certificate issued by the engineer under clause 60 is to be sent to the employer and to the contractor on the employer's behalf. By the certificate the employer gives notice to the contractor specifying the amount proposed to be made and the basis on which it is calculated.

By this means the Conditions provide for giving notice as required by the Act.

Notice of withholding

Section 111(1) of the Act stipulates that payment of a sum due may not be withheld after the final date for payment unless effective notice is given of intention to withhold payment.

Clause 60(10) of the Conditions addresses this statutory requirement. It states:

'where a payment under Clause 60(2) (4) or (6) is to differ from that certified or the Employer is to withhold payment after the final date for payment of a sum due under the Contract the Employer shall notify the Contractor in writing not less than one day before the final date for payment specifying the amount proposed to be withheld and the ground for withholding payment or if there is more than one ground each ground and the amount attributable to it.'

In clause 60(2) it is stated that the final date for payment is 28 days after the date of delivery of the contractor's monthly statement. Thus the employer has only the period between the date of certification and the final date for payment to serve any notice of intention to withhold payment of any part of the amount due. This period can be as short as 3 days if the engineer takes the full 25 days for certification permitted by clause 60(2).

26.4 Late payments and under-payments

Suspension

Section 112 of the Act gives a statutory right to suspend performance where a sum due under 'a construction contract' is not paid in full by the final date for payment and no effective notice to withhold payment has been given.

No attempt is made to incorporate in the Conditions a contractual right equivalent to the statutory right. The effect is that for contracts which come within the definition of a 'construction contract' the contractor is left with his statutory right. Where the contract is not a 'construction contract' within the meaning of the Act the contractor has neither a statutory right nor a contractual right.

Adjudication

Clause 66 of the Conditions provides a mechanism whereby disputes on under-payments can be referred to the engineer and thereafter to conciliation, adjudication and arbitration.

For contracts subject to the Act a speedier approach to resolution of any dispute is exercise of the statutory right to adjudication 'at any time' given by section 108 of the Act. For comment on whether the provisions of clause 66 are compliant with the Act and thereby replace exercise of the statutory right to adjudication under the Scheme for Construction contracts see Chapter 28.

Termination

The Conditions do not include payment failures as employer's defaults in clause 64 so the contractor has no contractual right to terminate his employment under the contract for non-payment, late payment or under-payment.

Repeated failure to pay on certificates or evidence of the employer's intention not to pay could provide the contractor with common law rights of determination but that is a matter for legal advice.

Employer's legal rights to set-off

The general right of the employer to set off against amounts certified was confirmed in *Gilbert-Ash* v. *Modern Engineering* (1973). And in *Enco Civil Engineering* v. *Zeus International* (1991) it was held, in relation to a dispute under an ICE Fifth edition contract, that summary judgment could not be given for a certified sum where there were disputes between the parties and the employer had set in motion the procedure under clause 66 whereby the engineer would have to reconsider his certificate and possibly replace it.

For contracts which are 'construction contracts' within the meaning of the Act the employer's position is now modified to the extent that withholding 'a sum due under the contract' is subject to the notice of intention to with-hold provisions in the Act or corresponding compliant terms in conditions of contract. For all contracts under the Seventh edition the notice of intention to withhold provisions in clause 60(10) is applicable.

26.5 *Interest on overdue payments*

The provisions in clause 60(7) of the Seventh edition for interest on overdue payments are the same as in the original Sixth edition clause except that a decision of an adjudicator is included alongside the finding of an arbitrator in clause 60(7)(b).

Clause 60(7) provides that in the event of:

(a)   failure by the engineer to certify or the employer to make payment in accordance with clauses 60(2), (4) or (6)

or

(b)   any decision of an adjudicator or any finding of an arbitrator to such effect

the employer shall pay the contractor interest compounded monthly for each day on which any payment is overdue or which should have been certified and paid at a rate equivalent to 2% per annum above the base lending rate of the bank specified in the appendix to the form of tender.

Findings of an arbitrator

Clause 60(7) goes on to provide that if an arbitrator finds that a sum should have been certified by a particular date but was not so certified, it shall be regarded as overdue for payment from either:

- 28 days after the date the arbitrator holds the engineer should have certified, or
- from the date of the certificate of substantial completion of the whole of the works where no certifying date is identified by the arbitrator.

This provision is consistent with, and follows, the judgment in the case of *Morgan Grenfell (Local Authority Finance) Ltd* v. *Seven Seas Dredging Ltd* (1990).

In that case under the ICE Fifth edition the arbitrator found in favour of the contractor for unforeseen work in dredging for the Port of Sunderland. The contractor was awarded £1,954,811 plus interest. The arbitrator awarded interest of £967,604 calculated at 2% above bank rate under clause 60(6). If the arbitrator had calculated interest under section 19(A) of the Arbitration Act 1950 the award of interest would have been £187,449 less.

The employer appealed and the court had to decide whether the contractor was entitled to interest under clause 60(6) in respect of amounts included in a statement under clause 60(1) but not certified by the engineer, when an arbitrator later decided that they should have been certified. The appeal was dismissed with the court finding that interest under clause 60(6) was allowable.

Meaning of failure to certify

At the heart of the *Morgan Grenfell* case was the long argued point of whether an engineer, in rejecting all or part of the contractor's statement, leaves the employer open to liability for interest if he (the engineer) or an arbitrator subsequently revises upwards the evaluation of the contractor's claim. A common view was that providing the engineer acted in good faith in forming 'his opinion' under clause 60(2), there was no 'failure to certify'. Judge Newey QC in the *Morgan Grenfell* case seemed to go against this in finding in favour of the contractor when saying:

> 'I have reached the same conclusions as the learned arbitrator. In my judgement clause 60(2) requires the engineer, who has been provided with a statement in accordance with clause 60(1), to "certify" the amount which the contractor should be paid. Obviously, that amount should be the amount, which in his "opinion" the contractor should be paid. If the word "opinion" had not been used in the sub-clause some similar word or words would have had to have been used. For the sub-clause to work the engineer has to decide how much the contractor should receive and that is a matter of "opinion", "judgement", "conclusion" or the like. If the engineer certifies an amount which is less than it should have been, the contractor is deprived of money on which he could have earned money.'

It had been widely expected that the *Morgan Grenfell* case would go on to the Court of Appeal but this never happened and it was left to the case of *The Secretary of State for Transport* v. *Birse-Farr Joint Venture* (1993) to re-examine how the provisions for interest in the ICE Fifth edition should apply.

The court in that case held that an engineer will only have failed to certify if:

- he fails to certify altogether
- he under-certifies on an erroneous view of the contract.

The judge said:

> 'The contractor must demonstrate not merely that a sum which has later
> been allowed by the engineer or the arbitrator was not included in an
> earlier certificate but must show that there was a failure by the engineer
> to perform his duties under the contract in deciding what sum to
> certify.'

The court also held that:

- compound interest is not due under the ICE Fifth edition; but only interest
  on arrears of interest
- until completion the contractor can claim interest on arrears of interest,
  but thereafter only simple interest accrues.

The *Birse-Farr* decision has some impact on the Seventh edition Conditions in
regard to the meaning of failure to certify but does not affect the express
entitlement in the Conditions to compound interest.

Extra-contractual claims

It is worth noting that the interest provisions of clause 60(7) apply only to
amounts due under the contract. The contractor therefore has no contractual
entitlement to interest for late payment of an extra-contractual claim.

Interest not financing charges

The provision for interest in clause 60(7) should not be confused with the
provision for finance charges in 'cost' as defined in clause 1(5). Interest
applies to late payment on a contractor's application; finance charges are
part of the cost included in such an application.

Payment of interest

Clause 60(7) makes no provision for the engineer to certify interest on
overdue payments. It is the responsibility of the employer to pay interest
directly it becomes due.

However, the contractor would be wise to include any interest due in his
monthly application under clause 60(1)(d) as being an amount for which
provision is made under the contract. The engineer would then be obliged to
certify such interest under clause 60(2)(a).

26.6 Minimum amount of certificate

Clause 60(3) provides that until the whole of the works have been certified as substantially complete in accordance with clause 48 the engineer shall not be bound to issue an interim certificate for a sum less than that stated in the appendix to the form of tender. Thereafter the engineer shall be bound to do so and the certification and payment of amounts due to the contractor shall be in accordance with the time limits contained in clause 60.

26.7 Final account

Clause 60(4) endeavours to regulate the timely completion of accounts by requiring the contractor to submit a statement of final account and supporting documentation not later than three months after the date of the defects correction certificate. The statement is to show the value of the works executed together with all further sums which the contractor considers due.

By clause 61(1) there is stated to be only one defects correction certificate and that follows expiration of the last defects correction period.

Failure to include sums due

Failure by the contractor to include sums due under the contract in his final account may deprive him of entitlement to payment since it is no longer a matter of securing interim payment. The contractor would, of course, retain his right to sue for damages for breach where that was an alternative.

Engineer's final certificate

Within three months after receipt of the final account and verifying information the engineer is required to issue a certificate stating the amount which in his opinion is finally due to either the contractor or the employer as the case may be.

Clause 60(4) does not use the phrase 'final certificate' and there is no status of finality attached to the certificate as there often is in building contracts where the final certificate is binding unless challenged within a specified time. Moreover since the 'final' certificate covers only sums considered due up to the date of the defects correction certificate, there may well be later certificates as disputes are settled or insurance related matters are resolved.

Engineer not *functus officio*

There is nothing in clause 60(4) to suggest that the engineer is rendered *functus officio* after the issue of his 'final' certificate and indeed he may still have duties to perform under the contract such as giving decisions under clause 66 which might lead him to revising his final certificate.

Credits to the employer

The wording of clause 60(4) leads to some interesting questions on how the engineer should handle credits due to the employer.

The clause says that the engineer is to certify after giving credit to the employer 'for all sums to which the Employer is entitled under the contract'. It then says that the amount certified shall 'subject to Clause 47 be paid'. That suggests the amount to be paid is the amount certified after deduction of any liquidated damages for late completion.

This confirms that the engineer shall not deduct in his final certificate, and certainly not in earlier certificates, for liquidated damages due to the employer for late completion.

26.8 Retention

Clauses 60(5) and 60(6) deal respectively with amounts of retention and release of retention.

Amount of retention

Clause 60(5) fixes the retention held under clause 60(2)(a) as the difference between:

(a)   an amount calculated at the rate and up to the limit stated in the appendix to the form of tender on the amounts due to the contractor under clauses 60(1)(a) (measured work) and 60(1)(d) (other matters) and
(b)   any payment due to the contractor under clause 60(6) (payment of retention).

Note that retention is not held on amounts considered 'proper' under clauses 60(1)(b) and 60(1)(c) for unfixed goods and materials.

Payment of retention

The release of retention is due, at its simplest, in two stages:

- one half upon the issue of any certificate of substantial completion
- the balance upon the expiry of the final defects correction period.

Clauses 60(6)(a) and 60(6)(b) deal with the first part and clause 60(6)(c) deals with release at the end of the defects correction period.

Clause 60(6)(a) provides that the total of payments of retention consequent upon the issue of certificates of substantial completion for sections or parts of the works shall not exceed one half of the limit of retention set out in the appendix to the form of tender.

Clause 60(6)(b) provides that on the issue of the certificate of substantial completion for the whole of the works one half of the stipulated limit becomes payable less any amounts already paid in respect of sections or parts of the works.

Under clause 60(6)(b) the engineer is to certify the amount of retention to be released on the issue of the certificate of substantial completion for the whole of the works within 10 days of the date of issue of the completion certificate. And payment becomes due on certification with the final date for payment 14 days after the issue of the completion certificate.

Clause 60(6)(c) provides that at the end of the defects correction period, or the end of the last such period if there is more than one, the balance of the retention money becomes due to the contractor.

The engineer is to certify the amount due within 10 days of the end of the period and payment becomes due on certification of the amount due with the final date for payment being 14 days after the end of the period notwithstanding that at that time there may be outstanding claims by the contractor against the employer. This is subject to the proviso that if at any time there remains to be carried out by the contractor any outstanding work under clause 48 or any work ordered pursuant to clauses 49 or 50, the engineer may withhold certification until completion of such work of so much of the remainder as shall in the opinion of the engineer represent the cost of the work remaining to be done.

Retention in trust

There is no express provision in ICE Conditions for the employer to hold retention monies in a trust fund. Nor is there such a provision in other standard forms of civil engineering contract.

However, in the case of *Wates Construction (London) Ltd* v. *Franthom Property Ltd* (1991) the Court of Appeal decided on a building contract that even in the absence of an express obligation to place retention money into a separate account, the employer was obliged to do so.

It would no doubt be unusual for a contractor in civil engineering to press the employer to establish a separate retention account but it is clear from the case of *MacJordan Construction Ltd* v. *Brookmount Erostin Ltd* (1991) that if the employer's insolvency pre-dates the establishment of such a fund, the contractor is in no better position than an ordinary creditor in respect of any monies held as retention.

26.9 *Correction and withholding of certificates*

Clause 60(8) on the correction and withholding of certificates remains the same as in the unamended Sixth edition.

Clause 60(8) deals only with the engineer's powers and the employer's obligations in respect of payments to nominated subcontractors and to that extent it is not in conflict with the new clause 60(10) which deals with the employer's intention to withhold monies from certified amounts.

Engineer's power to omit

Clause 60(8) commences by providing that:

- the engineer shall have power to omit from any certificate
- the value of any
  - work done
  - goods or materials supplied
  - services rendered
- with which he may for the time being be dissatisfied
- and for that purpose
- or for any other reason which may to him seem proper
- by any certificate
  - delete
  - correct
  - modify
- any sums previously certified.

Engineer's dissatisfaction

Subject to application of the dispute resolution procedures of the contract and the Act it is clearly the engineer's view of whether work is satisfactory such that its value should be included in any certificate which is determinative. This is consistent with clause 60(2) in confirming that the engineer not the contractor determines 'the amount due'.

The provision allowing the engineer to apply adjustments to values previously certified recognises that matters may come to light which diminish or eliminate the value of work already paid for in that reconstruction or demolition may be necessary at the contractor's expense.

Other reasons

The engineer's power to act in respect of 'any other reason' appears to be limited by the structure of clause 60(8) to modifying amounts previously certified. It is doubted if it extends to omission of value in the first instance. Such omission seems to be limited to the engineer's dissatisfaction.

As to what 'any other reason' might properly be, the most likely explanation is correction of computation errors in earlier statements and certificates. It is not thought to be a general power for the engineer to coerce or punish the contractor for contractual failings and it is most unlikely that it could apply to matters outside the contract – such as knowledge of the contractor's financial difficulties.

Possibly it relates to the withholding, particularly from interim certificates, of amounts due under the contract to the employer from the contractor; for example, in the case of clause 39(2) where the employer has carried out corrective work at the contractor's expense. However, that clause, and others similar, say the employer may deduct 'from any monies due' which suggests that the deduction should be by way of set-off in accordance with clause 60(10) (notice of intention to withhold payment) and not by reduced certification.

Application to nominated subcontractors

In respect of payments to nominated subcontractors, the engineer's powers under clause 60(8) are restricted.

By clause 60(8)(a) the engineer may not reduce in an interim certificate the sum previously certified for a nominated subcontractor if the contractor has already paid over that sum.

By clause 60(8)(b) if the engineer, in the final certificate, reduces the sum previously certified to a nominated subcontractor and that sum has already been paid by the contractor, then the employer is obliged to reimburse the contractor for the amount overpaid to the extent that the contractor is unable to recover the same from the subcontractor.

Numerous complex scenarios can be advanced whereby, under these provisions, the employer comes out the unfortunate and innocent loser. Such is the price of nominated subcontracting.

26.10 Certification and payment notices

Clause 60(9) requires that every certificate issued by the engineer under clause 60 shall be sent to the employer and, on the employer's behalf, to the contractor.

The clause goes on to state, emphasising its compliance with the Act, that by the certificate the employer shall give notice to the contractor specifying the amount, if any, of the payment proposed to be made and the basis on which it is calculated.

26.11 *Notice of intention to withhold payment*

Clause 60(10) requires that where a payment is to differ from the amount certified or the employer intends to withhold payment after the final date for payment, the employer shall notify the contractor in writing not less than one day before the final date for payment specifying:

- the amount proposed to be withheld
- the grounds for withholding and the amount attributable to each ground.

By this clause the Conditions are made compliant with Section 111 of the Act.

Chapter 27
Defaults and determination

27.1 Introduction

The provisions in the Seventh edition for defaults and determination are expanded from those in the Sixth edition to include defaults of the employer. This has produced a re-arrangement of clauses 63 to 65 of the Conditions as follows:

	ICE Sixth edition	ICE Seventh edition
• clause 63	determination of contractor's employment	frustration and war clause
• clause 64	frustration	default of the employer
• clause 65	war clause	default of the contractor

Frustration and war clause

Clause 63 of the Seventh edition which now combines frustration and the war clause is considered in Chapter 29 under miscellaneous clauses.

Employer's defaults

It has long been a point of criticism of earlier editions of ICE Conditions that they provided the employer with remedies for the contractor's default but provided no corresponding remedies to the contractor for employer's default. This imbalance was rectified in the ICE Design and Construct Conditions of Contract published in 1992 where not only was the contractor given express rights to determine for certain employer's defaults but additionally he was given the express right to suspend work or reduce the rate of work for failure by the employer to pay on certificates.

The Seventh edition only includes provisions for the contractor to terminate for employer's defaults – and those defaults are limited to assignment without consent and insolvency. The contractor's rights in the event of failure by the employer to pay on certificates are not covered, except by way of interest under clause 60(7). However, the contractor does now have the statutory right to suspend performance for non-payment under Section 112 of the Housing Grants, Construction and Regeneration Act 1996 – providing, of course, that the particular contract is a construction contract within the meaning of the Act.

One change in detail between the Sixth and Seventh editions worth noting is that the new provisions for payment after termination in clause 65(5) expressly provide for interim payments, albeit only to the employer from the contractor, whereas the old provisions in clause 63(4) of the Sixth edition effectively froze all payments until after the expiry of the defects correction period.

27.2 *Termination generally*

There are three sets of circumstances by which a contract can come to be prematurely terminated:

- termination by agreement of the parties
- termination arising from events beyond the control of the parties
- termination resulting from the default of one of the parties.

Termination by agreement may be by an ad hoc arrangement or under a clause of the contract which gives the employer the right to terminate at will – sometimes described as a 'convenience' clause. The Seventh edition does not contain such a clause but they are not uncommon in other construction contracts.

Clauses dealing with termination for events beyond the control of the parties are often called *force majeure* clauses. The events they deal with are various and given the generality of the term *force majeure* the events they cover do need to be defined for each contract. The Seventh edition avoids the phrase *force majeure* and considers only two events – frustration and war.

Termination arising from default can either be termination under the contract or termination under common law rights. The terminology used for termination for default attracts a variety of phrases but only rarely will the terminology itself be decisive. However, note the case of *Dyer Ltd* v. *Simon Build/Peter Lind Partnership* (1982) where it was held that expulsion from the site under clause 63 of the ICE Fifth edition did not amount to determination or termination of the contract.

Terminology

The ICE Fifth edition used the phrases 'forfeiture' and 'expulsion'; the Sixth edition used 'determination' and again 'expulsion'; the Seventh edition uses 'termination' and 'termination of the Contractor's employment'.

Phrases such as 'termination of the Contractor's employment' and 'determination of the Contractor's employment under the Contract' perhaps best describe what happens when there is termination or determination on default grounds. Rarely will the contract itself be terminated. Usually it lives on to regulate the consequences arising from the termination and the parties' ongoing rights and liabilities.

Another phrase sometimes used, particularly in legal actions, is repudiation.

Repudiation

Repudiation is an act or omission by one party which indicates that he does not intend to fulfil his obligations under the contract. In civil engineering works, a contractor who abandons the site, or an employer who refuses to give possession of the site, are examples of a party in repudiation.

Determination generally

The ordinary remedy for breach of contract is damages but there are circumstances in which the breach not only gives a right to damages but also entitles the innocent party to consider himself discharged from further performance.

Determination at common law

When there has been repudiation or a serious breach which goes to the heart of the contract so that it is sometimes called 'fundamental' breach, common law allows the innocent party to accept the repudiation or the fundamental breach as grounds for determination of the contract. The innocent party would then normally sue for damages on the contract which had been determined.

The problems with common law determination are that it is valid only in extreme circumstances and it can readily be challenged.

Determination under contractual provisions

To extend and clarify the circumstances under which determination can validly be made and to regulate the procedures to be adopted, most standard forms of construction contract include provisions for determination. Building forms have traditionally given express rights of determination to both employer and contractor, but ICE Conditions prior to the Seventh edition have given express rights only to the employer.

Many of the grounds for determination in standard forms are not effective for determination at common law. Thus failure by the contractor to proceed with due diligence and failure to remove defective work are often to be found in contracts as grounds for determination by the employer. But at common law neither of these will ordinarily be a breach of contract at all since the contractor's obligation is only to finish on time and to have the finished work in satisfactory condition by that time.

The commonest and the most widely used express provisions for determination relate to financial failures. Again at common law many of these are ineffective and even as express provisions they are often challenged as ineffective by legal successors of failed companies.

The very fact that grounds for determination under contractual provisions are wider than at common law leads to its own difficulties. A party is more likely to embark on a course of action when he sees his rights expressly stated than when he has to rely on common law rights. This itself can be an encouragement to error. Some of the best known legal cases on determination concern determinations made under express provisions but found on the facts to be lacking in validity.

In *Lubenham Fidelities* v. *South Pembrokeshire District Council* (1986) the contractor determined for alleged non-payment whilst the employer concurrently determined for failure to proceed regularly and diligently. On the facts, the contractor's determination was held to be invalid. But in *Hill & Sons Ltd* v. *London Borough of Camden* (1980), with a similar scenario, it was held on the facts that the contractor had validly determined.

Parallel rights of determination

Some construction contracts expressly state that their provisions, including those of determination, are without prejudice to any other rights the parties may possess. That is, the parties have parallel rights – those under the contract and those at common law – and they may elect to use either.

ICE Conditions do not have such a stated alternative but the omission is not significant. The general rule is that common law rights can only be excluded by express terms. Contractual provisions, even though comprehensively drafted, do not imply exclusion of common law rights.

The point came up in the case of *Architectural Installation Services Ltd* v. *James Gibbons Windows Ltd* (1989) where it was held that while a notice of determination did not validly meet the timing requirements of the contractual provisions, nevertheless there had been a valid determination at common law.

Legal alternatives

To take advantage of the parallel rights of determination and to avoid as far as possible the danger that one course of action might be found to be defective, determination notices are sometimes served under both the contract and at common law as legal alternatives.

But on that it must be said that contractual rights which are not also common law rights attract only contractual remedies. So a defective notice under a contractual provision which has no common law equivalent cannot be salvaged by a common law action.

Determination – a legal minefield

Two aspects of determination operate together to make it a legal minefield. One is that determination is top of the scale in terms of the significance/ seriousness of its consequences and it is therefore likely to lead to formal dispute. The other is the uncertainty of outcome of legal actions on determination. It is not just that the law is complex; it is also that allegations have to be supported by evidence given in a cooler atmosphere than sometimes prevails when determination clauses are activated.

Determination is a matter for lawyers and it should never be commenced without legal advice.

27.3 *Default of the employer*

Clause 64 of the Seventh edition provides for termination by the contractor of his own employment under the contract in the event of certain specified defects on the part of the employer. As noted above there was no equivalent clause in the Sixth edition.

Clause 64 is in three parts:

- 64(1) – specified defaults
- 64(2) – removal of contractor's equipment
- 64(3) – payment on termination.

Specified defaults

The defaults specified in clause 64(1) are:

- assignment or attempted assignment by the employer of the contract, or any part thereof, or any interest or benefit thereunder without the prior written approval of the contractor
- financial failures of the kind:
  - becoming bankrupt
  - being subject to a receiving order or administration order
  - making an arrangement or assignment in favour of creditors
  - agreeing to a committee of inspection
  - going into liquidation, receivership or administration
  - having an execution levied on goods which is not stayed or discharged within 28 days.

Assignment by the employer

Employers need to be particularly alert to the possibility of inadvertently falling foul of clause 64(1) by entering into commercial arrangements for the future of the works which could be construed as assignment of the contract.

Action by the contractor

In the event of a specified default the contractor may under clause 64(1):

- give notice in writing to the employer specifying the default
- after seven days' notice terminate his own employment under the contract
- extend the period of notice to give the employer an opportunity to remedy his default.

Purpose of the notice period

It is a matter of some legal importance as to whether the notice given by the contractor is simply a notice of his intention to vacate the site or whether it is a warning for the employer to remedy his default. Similar questions arise in respect of notices given to contractors by employers in respect of contractor's default.

The wording of the clause leans to the probability that the notice period should be seen as a warning period. That however leads to difficult questions as to who is to judge whether the default has been remedied and what procedure is to be followed if there is a dispute on whether it has or has not been remedied.

To say that these are difficult issues is an understatement. In the case of *Attorney General of Hong Kong* v. *Ko Hon Mau* (1988), on conditions similar to the ICE Fifth edition, both parties served notice of determination. The court held that the notices were provisional in the sense that their final validity would not be determined until arbitration, but pending the arbitration, providing they were given in good faith, they were both effective.

Removal of contractor's equipment

Clause 64(2) states that upon the expiry of the seven day notice period the contractor shall with all reasonable despatch remove from the site all contractor's equipment.

This applies notwithstanding the provision of clause 54 which debars the removal of any of the contractor's equipment without the consent of the engineer.

The wording of clause 64(2) is slightly odd in that it says 'shall ... remove'. If taken in their usual mandatory sense these words would put an obligation on the contractor to remove his equipment after seven days whether or not he intended to extend the period of notice to give the employer more time to remedy his defect.

Payment upon termination

Clause 64(3) provides that upon termination of the contractor's employment pursuant to clause 64(1) (employer's default) the employer shall be under the same obligations as to payment as if the works had been abandoned under clause 63 (frustration) and shall pay the contractor the amounts due under clause 63(4) plus the amount of any loss or damage arising from the termination.

In total, therefore, the contractor is entitled to:

- the value of all work carried out
- proportionate amounts in respect of preliminaries
- costs of goods and materials the contractor is legally obliged to accept
- any expenditure reasonably incurred in expectation of completing the works
- the reasonable cost of removing contractor's equipment
- loss or damage arising from the termination.

The final heading 'loss or damage' would include loss of overheads and profit on the uncompleted work in accordance with the usual principles. So in theory, on termination due to the employer's default, the contractor should receive not only the value of the work done plus costs incurred but also lost overheads and profit.

In practice the contractor may receive little or nothing if the default is financial failure of the employer and the contractor is only an unsecured creditor.

27.4 *Default of the contractor*

Clause 65 of the Seventh edition corresponds with clause 63 of the Sixth edition. Mostly it is much the same although the wording is rearranged in parts. However, it does contain an additional contractor's default – sub-contracting the whole of the works without consent.

Clause 65 is in five parts:

- 65(1) – specified defaults
- 65(2) – completing the works
- 65(3) – assignment to the employer
- 65(4) – valuation at the date of termination
- 65(5) – payment after termination.

Specified defaults

The defaults specified in clause 65(1) fall into four categories:

- assignment without consent
- subcontracting the whole of the works without consent
- financial failures
- performance failures.

Financial failures

The financial failures are the same as those specified for the employer under clause 64(1)(b):

- becoming bankrupt
- being subject to a receiving order or an administration order
- making an arrangement or assignment in favour of creditors
- agreeing to a committee of inspection
- having an execution levied which is not stayed or discharged within 28 days.

To some extent the provisions are perverse in that a contractor who has applied for administration or a creditors voluntary arrangement may be showing by his conduct that, far from repudiating his obligations under the contract, he is taking legal steps to enable him to complete them.

Failures of performance

Under clause 65(1) the specified defaults are:

- abandoning the contract without due cause
- failing to commence without reasonable excuse
- suspending progress without due cause for 14 days after notice to proceed
- failing to remove or replace defective work, goods or materials after notice to do so
- failing to proceed with due diligence despite previous warnings
- being persistently or fundamentally in breach of obligations under the contract.

'has abandoned the Contract'

Abandonment may be patently obvious where the contractor has left the site or has otherwise given notice of his intention not to fulfil his obligations. Such abandonment is repudiation and grounds for common law determination.

However, there can also be situations where the contractor's conduct in ceasing work is neither abandonment nor repudiation. In *Hill* v. *Camden* (1980) the contractor, unable to obtain prompt payment, cut his staff and labour. The employer took this action as repudiatory and served notice of determination. Lord Justice Lawton had this to say of the contractor's action:

'The plaintiffs did not abandon the site at all; they maintained on it their supervisory staff and they did nothing to encourage the nominated sub-contractors to leave. They also maintained the arrangements which they had previously made for the provision of canteen facilities and proper insurance cover for those working on the site.'

A point worth noting is that the Seventh edition adds the phrase 'without due cause' to clause 65(1)(e) which reads 'has abandoned the Contract without due cause'. This is not in the Sixth edition. The explanation may lie in clause 63 of the Seventh edition which uses the word 'abandoned' in connection with frustration (i.e. with due cause).

'has failed to commence'

By clause 41(2) the contractor shall start the works on or as soon as is reasonably practicable after the works commencement date. The default under clause 65 is failing to commence in accordance with clause 41 without reasonable excuse.

Failure to commence within a reasonable time is not of itself a repudiatory breach. The contractor might say that he can complete in a fraction of the time allowed and has every intention of finishing on time. And there is plenty of scope for argument on how much time the contractor needs off-site to prepare himself for starting construction work on site and how much time, having started to set-up on site, is reasonable before starting other site activities.

Suspension of progress

It is not clear whether the default of suspending progress relates only to inactivity after a clause 40 suspension or whether it has some wider application. Nor is it clear if the words 'without reasonable excuse' apply to this default as well as to failure to commence.

The words 'after receiving from the Engineer written notice to proceed' suggest that the suspension is related to some procedure in the contract, most obviously, an ordered suspension under clause 40. If the provision has wider application, the notice to proceed is an administrative requirement of clause 65 which is additional to other notices.

Failure to remove defective work, goods or materials

At common law, failure by the contractor to remove or replace defective work, goods or materials is not, of itself, a breach of contract. It is only a

breach of contract if the contractor has failed to do so by the completion date.
In *Kaye Ltd* v. *Hosier & Dickinson Ltd* (1972) Lord Diplock said:

> 'During the construction period it may, and generally will, occur that from time to time some part of the works done by the contractor does not initially conform with the terms of the contract either because it is not in accordance with the contract drawings or the contract bills or because the quality of the workmanship or materials is below the standard required by condition 6(1)... Upon a legalistic analysis it might be argued that the temporary disconformity of any part of the works with the requirements of the contract, even though remedied before the end of the agreed construction period, constituted a breach of contract for which nominal damages would be recoverable. I do not think that makes business sense. Provided that the contractor puts it right timeously I do not think that the parties intended that any temporary disconformity should of itself amount to a breach of contract by the contractor.'

Under ICE Conditions (clause 39 of the Seventh edition) the engineer is empowered to order the removal of unsatisfactory work and materials and failure to comply with such an instruction is a breach of contract. The remedy of determination in clause 65 is extreme and in many cases it might be expected that the mere serving of a notice under clause 65 will be effective in getting action from the contractor under clause 39.

Prior to the case of *Tara Civil Engineering* v. *Moorfield Developments Ltd* (1989) it was generally thought that under the ICE Fifth edition the engineer could not serve notice under clause 63 (the default clause) until he had acted under clause 39. However, the judge in the case did not accept that proposition. He said:

> 'I am in no doubt at all that clause 63 can and should be construed without any suggestion that it is limited by clause 39 or that it should be preceded by a notice which is in some way identifiably referable to clause 39. The engineer and the employers have various options open to them under the contract and those options should not be restricted by the sort of argument that has been put in this case.
>
> I therefore find that the engineer has issued documents which on their face appear to put in motion the machinery of clause 63.'

Failure to proceed with due diligence

This is a matter which should be approached with the greatest of caution. The courts have been most reluctant to impose on contractors any greater obligation than to finish on time.

In *Greater London Council* v. *Cleveland Bridge* (1986) the Court of Appeal refused to imply a term into a building contract that the contractor should proceed with due diligence notwithstanding the inclusion of that phrase in the determination clause of that contract. The point was repeatedly made in the case that the contractor should be free to programme his work as he thought fit. For other cases showing the difficulty of defining due diligence and failure to proceed with it, see *Hill* v. *Camden* (1980) and *Hounslow* v. *Twickenham Garden Developments* (1971).

Failure by the contractor to proceed in accordance with this approved clause 14 programme might provide some evidence to support a charge of failing to proceed with due diligence although failure to proceed to the programme is not itself normally a breach of contract.

The reference to 'previous warnings' in clause 65(1,) should, perhaps, be read in conjunction with clause 46. That clause places a duty on the engineer to notify the contractor if he considers progress too slow to achieve completion by the due date.

One of the particular problems of ICE Conditions, Seventh edition included, is that nowhere in the Conditions is there an express obligation to proceed with 'due diligence'. The obligation on progress as stated in clause 41 is to proceed 'with due expedition and without delay'.

Persistently or fundamentally in breach

The defaults of 'persistently or fundamentally in breach of his obligations under the Contract' may be mutually exclusive – that is, persistent breaches of minor obligations or a single breach which goes to the heart of the contract. It is not clear whether the 'previous warnings' apply to these defaults as well as to failing to proceed with due diligence.

Failure by the contractor to provide a bond as required by clause 10 could be an example of a fundamental breach; failure by the contractor to supply a programme under clause 14 would not be fundamental although persistent refusal to supply a programme might satisfy clause 65.

Action by the employer

Clause 65(1) entitles the employer to expel the contractor from the site after giving seven days' notice in writing specifying the contractor's default.

As in clause 64 the period of notice may be extended. But this time the extension is allowed by the employer to give the contractor the opportunity to remedy his default.

Clause 65(1) further provides that where a notice of termination is given pursuant to a certificate issued by the engineer, such notice shall be given 'as soon as is reasonably possible after receipt of the certificate'.

This should avoid the situation in the case of *Mvita Construction* v. *Tanzania Harbours Authority* (1988), under FIDIC Conditions which matched those of the ICE Fifth edition. There, the employer took three months before acting on an engineer's certificate. It was held that the employer was bound to give his notice within a reasonable time after the engineer's certificate to avoid a change in circumstances.

Engineer to certify in writing

For the performance defaults in clauses 65(1)(e) to 65(1)(j), the engineer must certify in writing to the employer, with a copy to the contractor, that in his opinion the contractor is in default before the employer can give notice of termination.

The question arises, can this be challenged and what happens if it is? Can the employer enter the site and expel the contractor or must he await the outcome of adjudication or arbitration proceedings?

These were amongst the questions considered by the court in the *Tara Civil Engineering* case under an ICE Fifth edition contract. The engineer certified that Tara was in default and the employer gave notice of its intention to expel Tara from the site. Tara obtained an ex parte injunction restraining the employer from expelling them to which the employer responded by applying to the court for the order to be discharged. Granting the order requested by the employer, the court held that it should not go behind the engineer's certificate under clause 63 or any other documents relied on unless there was proof of bad faith or unreasonableness. The judge said:

> 'The most important of the three documents setting clause 63 in motion is the certificate of the engineer as to his opinion. It is important that the certificate is of his opinion only and not of fact. I take the view that I should only go behind that certificate, or behind any of the other documents relied on as setting in motion the clause 63 procedure, if there is either a lack of documents which on their face appear to set the procedure in motion or there is proof of bad faith or proof of unreasonableness.'

And later in the judgment, commenting on the policy of the courts, he said:

> 'At this stage there is no intention by the court to take sides in the determination of the ultimate disputes between the parties. The concern of the court is far from seeking to assist either party to break the contract. It is impossible to decide at this stage what is the conduct which would be in breach of the substantive terms of the contract. The court's present concern is to enforce the terms of the contract with regard to the only matters presently under consideration, namely, the regulation of the conduct of the parties pending the resolution of the substantive dispute by arbitration.'

The employer can, therefore, act on the certificate of the engineer even if it is challenged. But, of course, the employer will be liable if an adjudicator or arbitrator later overturns the certificate.

Completing the works

Clause 65(2) provides that where the employer has entered the works after expelling the contractor, the employer may:

- complete the works himself
- employ another contractor to complete the works.

The employer can use for completion any of the contractor's equipment and temporary works goods and materials on any part of the site. And he can sell any of the contractor's equipment, temporary works, unused goods and materials and apply the proceeds towards the satisfaction of sums due to him from the contractor.

These are greater powers in theory than in practice. They only work to the full if the contractor has title to all the equipment, temporary works and unused goods and materials. He may, in fact, have little title, with most equipment and temporary works on hire and most unused goods and materials not paid for.

Assignment to the employer

Clause 65(3) empowers the engineer to require the contractor to assign to the employer the benefit of any agreement for the supply of goods, service or execution of work. But this must be done within 7 days of the employer's entry upon the works.

It is not wholly clear what is meant by assigning the benefit of any agreement. It cannot be thought that any supplier or subcontractor will provide goods or services to the employer with the burden of payment left with the contractor.

Almost certainly if any assignment is going to work in practice the employer will have to take on both the benefits and the burdens. And since the burdens will include payment of sums outstanding, the employer will usually be better off simply by reaching new agreements with suppliers and subcontractors.

Valuation at date of termination

In the Sixth edition clause 63(4) covered payment after determination and clause 63(5) covered valuation at the date of determination. The order of these clauses is reversed in the Seventh edition with clause 65(4) covering valuation and clause 65(5) covering payment.

The engineer is required by clause 65(4) to value the work at the date of termination by fixing and certifying:

- the amount earned by the contractor in respect of work done, and
- the value of unused goods and materials and any contractor's equipment and temporary works under the control of the employer.

The valuation is to be done as soon as practicable after the employer's entry and can be carried out *ex parte* or however the engineer thinks fit.

The purpose of the task is not obvious. The amount which may eventually become due to the contractor is arrived at by the method of calculation given in clause 65(5) and the valuation at the date of termination is not obviously included in the calculation.

Payment after termination

Clause 65(5) of the Seventh edition corresponds with clause 63(4) of the Sixth edition but it is redrafted. The old clause stated that the employer was not liable to pay the contractor any money on account until after expiration of the defects correction period. The new clause does not expressly state this restriction and possibly it is no longer intended to apply. However, the provision in the new clause for interim payments is stated only in terms of payment due from the contractor to the employer and it may be that in practice engineers will be reluctant to certify monies due to the contractor until all other accounts are finally settled.

Clause 65(5) provides:

(a) if the employer enters the site and expels the contractor for default under clause 65 he shall not be liable to pay the contractor any money under the contract (whether previously certified or not) unless and until the engineer certifies that an amount is due under clause 65(5)(b).

(b) the engineer shall certify the difference between:
  (i)   such sum as would have been due to the contractor if he had completed the works – plus any proceeds of sales by the employer under clause 65(2)
  (ii) the costs of completing the works (whether or not the works are completed under a separate contract) plus damages for delay and all other expenses properly incurred by the employer

(c) the difference as certified by the engineer under clause 65(5)(b) shall be a debt due to the employer or the contractor as the case may be

(d) if the engineer is satisfied that prior to the completion of the works the sum as calculated under 65(5)(b)(ii) exceeds the sum as calculated under 65(5)(b)(i) he may issue an interim certificate showing a debt due from the contractor to the employer

(e) every certificate issued by the engineer under clause 65(5) shall be sent to the employer and copied to the contractor with such detailed explanation as necessary.

Sums already certified

Note that the employer can refuse to pay on interim certificates coming due for payment after termination. Even if there are certificates which are overdue, as frequently happens when termination is anticipated and payments are held back, the employer should still be able to mount a successful defence against any writ for payment.

Damages for delay

There is no certainty on the wording of clause 65(5) whether the damages for delay in completion are liquidated damages accrued to the date of termination plus general damages thereafter, or liquidated damages for the full period to completion. It is suggested that the latter, which is more consistent with the provisions of clause 47 (liquidated damages), is probably correct.

Value of bonds

It is not clear from clause 65(5) how any sums the employer obtains from security bonds are to be taken into account in settling the final sums due. The sums due are stated only in terms of amounts due on the contract, the costs of completion and expenses.

The engineer probably has no power to consider payments made under bonds. In any event such payments are often long delayed as bondsmen not infrequently mount a thorough examination of their liability under the exact terms of the bond.

Chapter 28
Settlement of disputes

28.1 Introduction

Traditionally the settlement of disputes under ICE Conditions was founded on a well-understood two stage procedure. The first stage was the engineer's clause 66 decision on any matter in dispute; the second stage was arbitration if either party was dissatisfied with the engineer's decision. The procedure had some clearly defined rules, most notably that the engineer's decision became final and binding if the dispute was not referred to arbitration within a specified time (1 month or 3 months depending on the circumstances) and that the scope of any arbitration was fixed by the matters referred to the engineer for decision (subject to the inclusion of issues connected with and necessary to determination of such matters).

The procedure had its problems, for instance in the Fifth edition the employer was not required to give effect to the engineer's decisions and in the Sixth edition the drafting seemed to assume that a dispute between one of the parties and the engineer was automatically a dispute between the parties. Generally, however, it was a workable even if not a wholly straightforward procedure. To the extent that it could be criticised as incorporating, by the need for the engineer's decision, delay in dealing with disputes there was always some merit in the opposing view that any delay was a breathing space and that the majority of disputes went no further than the engineer's decision.

In the Sixth edition a novelty was added in that either party could require the dispute to be put to a conciliation process – but only if the request was made after receipt of the engineer's decision and before reference to arbitration. The conciliation process in place at the time (the ICE Conciliation Procedure 1988) was a merit-based process similar to what is now called adjudication although later versions of the Procedure (1994 and 1999) changed the process to commercially-based mediation.

The introduction of conciliation into clause 66, forward looking and welcome as it was, illustrated the potential for over-complicating what was already quite a complex dispute resolution procedure. The drafting, unintentionally it was thought, failed to state any time limit on a request for conciliation and thereby put at risk the time limits which had previously applied in making the engineer's clause 66 decision final and binding. That was corrected in the August 1993 corrigenda.

Clause 66 as it now stands in the Seventh edition is illustration on a far greater scale of the dangers of over-complication. The clause encompasses matters of dissatisfaction, engineer's decisions, definitions of dispute, and provisions for conciliation, adjudication and arbitration. A great deal has been added by way of variety and detail but the cost of the complexities introduced by the changes may prove to be that the clause has progressed (or regressed) from being the clause in the contract which regulated dispute resolution to being a clause concerned with dispute resolution but not in full control of it.

Changes to clause 66

In so far as the text of clause 66 in the Seventh edition is identical (or nearly so) to the text of the March 1998 amendments to the Sixth edition, it can be said that there is no change between the Sixth and the Seventh editions. However, comparing the Seventh edition with the standard (as first published) version of the Sixth edition the changes can be summarised as follows:

Sixth edition	*Seventh edition*
Dispute deemed to arise when one party serves a notice of dispute on the engineer.	Parties entitled to refer a matter of dissatisfaction to the engineer for decision.
Engineer to state his decision on the dispute within 1 month (or 3 months if after substantial completion).	Engineer to state his decision within 1 month
Either party may after receipt of the engineer's decision (or after the time for it to be given has elapsed) require the dispute to be put to conciliation – providing no notice to refer to arbitration has been given.	Parties agree no matter shall constitute a dispute until the engineer's decision is given (or the time for it to be given has elapsed) and one party has served on the other a notice of dispute.
Either party may within 3 months after receipt of the engineer's decision (or after the time for it to be given has elapsed) or, within 1 month of receipt of a conciliator's recommendation, refer the dispute to arbitration.	Either party may seek the agreement of the other for the dispute to be put to conciliation – providing the request is made before notice to refer to arbitration is given.
	Either party has the right to refer a dispute on 'a matter under the Contract' to adjudication. The adjudicator's decision shall be binding until finally determined by legal proceedings or arbitration.

Sixth edition	*Seventh edition*
	All disputes (other than a failure to give effect to an adjudicator's decision) to be finally determined by reference to arbitration.
	Notice to refer to arbitration to be given with 1 month of a conciliator's recommendation, or within 3 months of an adjudicator's decision.

Effect of changes

At first sight it might seem that clause 66 of the Seventh edition does little more in practical terms than provide the additional option of adjudication. However, on closer examination it can be seen that because there are no restrictions on when a dispute can be referred to adjudication, this has the effect that the engineer's decision does not become final and binding (as the old rule) if not referred to arbitration within 3 months. In short, an engineer's decision under the Seventh edition, once given, lies open for possible reference to conciliation, adjudication or arbitration for as long as allowed by legal limitation periods. This is a major policy change but one which was perhaps unavoidable having regard to the statutory right to adjudication created by the Housing Grants, Construction and Regeneration Act 1996.

A further significant change, unintentional it is thought rather than policy, may have occurred in the possible deletion of the old rule that the scope of any arbitration is fixed by matters put to the engineer for a clause 66 decision. The wording of the Sixth edition was quite clear: every dispute was to be settled by the engineer – whose decision was to be final and binding unless revised in arbitration. The wording of the Seventh edition is nothing like as clear. The engineer does not make decisions on disputes; he makes decisions on matters of dissatisfaction. Disputes do not arise until after the engineer's decision has been given.

Background to the changes

To be fair to the draftsmen of the ICE Conditions it must be said that they faced a difficult task in deciding how to cope with the statutory right to adjudication imposed by the Housing Grants, Construction and Regeneration Act 1996 in relation to UK contracts falling within the Act's definition of construction contracts. They had a choice: either to let the dispute resolution procedures of the Conditions stand alongside the parallel statutory right to adjudication under the Scheme for Construction Contracts Regulations 1998 (the Scheme); or to include rights of adjudication within the Conditions which would satisfy the requirements of the Act. They chose the latter whilst at the same time trying to retain the engineer's decision as the first stage of dispute resolution.

The problem the draftsmen then faced was reconciling the requirement in the Act for adjudication 'at any time' with the imposition of a condition precedent in first obtaining the engineer's decision. The draftsmen sought to resolve that problem by defining a dispute as something which followed the engineer's decision on a newly introduced concept – a matter of dissatisfaction. By that device the draftsmen felt able to include in clause 66 provisions for both the engineer's decision and for adjudication under a scheme compliant with the Act.

Further effect of changes

It was always questionable whether it would be effective to endeavour to give in the Conditions a particular meaning to the term 'dispute' as mentioned in the Act. Some statutes do permit the parties to exclude or modify their application but there is nothing to this effect in the Housing Grants, Construction and Regeneration Act 1996. Moreover the courts have generally taken a robust view as to what constitutes 'a dispute' – see for example, *Enco* v. *Zeus* (1991) and *Halki* v. *Sopex* (1997).

It was no surprise, therefore, when the courts intimated in *Mowlem* v. *Hydra-Tight Ltd* (2000) and in *Carter* v. *Nuttall* (2000) that contractual attempts to impose conditions precedent to adjudication render contractual provisions non-compliant with the Act and that in such circumstances a referring party can exercise its statutory right to adjudication under the Scheme. The *Mowlem* decision comes particularly close to the Seventh edition Conditions in that the point at issue was the effect of a matter of dissatisfaction procedure in the New Engineering Contract – a procedure similar to that in the Seventh edition.

On one view the position now under the Seventh edition is that a referring party can either stay within the bounds of the contractual procedures such that the notice of dispute follows the engineer's decision on a matter of dissatisfaction and the adjudication is conducted under the stipulated Institution of Civil Engineers' Adjudication Procedure, or the referring party can step outside the contractual procedure and seek adjudication under the Scheme at any time. Benefits of adjudication under the Scheme include speedier references to adjudication by by-passing the matter of dissatisfaction procedure and the right of either party to ask for reasons for the decision under Rule 22 of the Scheme. (Rule 6.1 of the ICE Procedure states that the adjudicator shall not be required to give reasons.)

On another view, which relies largely on the proposition that the courts will be supportive of procedures agreed by the parties, the contractual scheme for dispute resolution is the only scheme available to the parties.

Scope of adjudication

Whether adjudication be under the Scheme or under the provisions of clause 66 of the Conditions it is important to note that it does not encompass the full range of disputes which might arise 'under or in connection with the contract'.

The Scheme (following Section 108 of the Act) refers only to disputes arising 'under the contract'. Clause 66(6) of the Conditions refers only to 'a dispute as to a matter under the Contract'. This is noticeably different from the wider phrase used in Clause 66(2)(b) in respect of matters of dissatisfaction, i.e. 'arising under or in connection with the Contract or the carrying out of the Works'.

In *Ashville Investments* v. *Elmer Contractors Ltd* (1987) the Court of Appeal ruled that claims for rectification and claims based on alleged innocent or negligent misstatements were claims arising 'in connection with' the contract but it is reasonably clear from the judgment that they cannot be taken as claims 'under' the contract.

Claims 'under' the contract will generally relate to disputes on contractual entitlements and liabilities and claims for breach of contract. However, disputes as to the agreed terms of a contract were also held in *Carter* v. *Nuttall* (2000) to be disputes 'under' the contract.

Timing of adjudication

An essential requirement for compliance with the Act is the right to adjudication 'at any time'. One odd, if not disturbing, effect of this is that adjudication can be commenced notwithstanding the commencement of other dispute resolution procedures. Thus in *Herschell Engineering* v. *Breen Property Ltd* (2000) it was held that a party who starts court proceedings does not waive his right to adjudication and a dispute may be referred to adjudication whilst proceedings are pending.

28.2 *Housing Grants, Construction and Regeneration Act 1996*

In 1993 the Government established a joint review of procurement and contractual arrangements in the UK construction industry. The outcome of that review was the publication in 1994 of Sir Michael Latham's report *Constructing the Team*. The report included recommendations that adjudication should be the normal method of dispute resolution and there should be a Construction Contracts Bill to give statutory backing to proposed changes in contractual arrangements.

For various reasons the Government decided to put into law only a few of the report's recommendations and instead of a Construction Contracts Bill those matters were passed, with other measures, in a composite bill – the Housing Grants, Construction and Regeneration Act 1996.

The construction related matters in the Act include:

- a statutory right to adjudication
- entitlement to stage payments
- requirements for dates for payments
- notices on intention to withhold payments
- rights to suspend performance for non-payment
- prohibition of conditional payment provisions.

Application of the Act

The Act applies to UK construction contracts, made in writing, and entered into after 1 May 1998.

The Act defines the construction contracts to which it applies by reference to the term 'construction operations'. These are described in some detail in the Act and broadly they include ordinary works of civil engineering but with notable exceptions for operations relating to mineral extraction, power generation, water treatment and process plants.

The borderline between construction operations and excluded operations is to some extent a matter of interpretation as evidenced by the flow of cases to the courts. Decisions already given on the subject include:

- *Homer Burgess Ltd v. Chirex* (1999)
  - pipework was part of a pharmaceutical plant and not a construction operation within the meaning of the Act
- *Nottingham Community Housing Association v. Powerminster Ltd* (2000)
  - work of servicing heating systems in domestic premises constituted construction operations within the meaning of the Act
- *Palmers Limited v. ABB Power Construction* (1999)
  - possible for a contractor's operations to fall outside the definition of a construction operation yet for a subcontractor's work providing building or painting services to come within the definition
  - scaffolding operations constituted construction operations within the meaning of the Act
- *ABB Power Construction v. Norwest Holst Engineering* (2000)
  - ordinary construction operations on a power generation site not within the scope of the Act
- *ABB Zantingh v. Zedal Building Services* (2000)
  - necessary to look at the nature of a site as a whole to determine its primary purpose
  - works of constructing a generating station on a printing complex were not excluded works and came within the definition of construction operations under the Act
- *Staveley Industries v. Odebrecht Oil and Gas Services* (2001)
  - a subcontractor involved in the manufacture of modules for oil and gas production platforms not able to pursue adjudication under the Act.

Right to refer disputes to adjudication

Section 108 of the Act provides a statutory right to adjudication to parties to construction contracts within the meaning of the Act. The statutory right does not make adjudication mandatory but it does make it unavoidable if one party exercises its right.

Section 108 reads as follows:

'(1)   A party to a construction contract has the right to refer a dispute arising under the contract for adjudication under a procedure complying with this section.

For this purpose "dispute" includes any difference.

(2)   The contract shall –

    (a)   enable a party to give notice at any time of his intention to refer a dispute to adjudication;

    (b)   provide a timetable with the object of securing the appointment of the adjudicator and referral of the dispute to him within 7 days of such notice;

    (c)   require the adjudicator to reach a decision within 28 days of referral or such longer period as is agreed by the parties after the dispute has been referred;

    (d)   allow the adjudicator to extend the period of 28 days by up to 14 days, with the consent of the party by whom the dispute was referred;

    (e)   impose a duty on the adjudicator to act impartially; and

    (f)   enable the adjudicator to take the initiative in ascertaining the facts and the law.

(3)   The contract shall provide that the decision of the adjudicator is binding until the dispute is finally determined by legal proceedings, by arbitration (if the contract provides for arbitration or the parties otherwise agree to arbitration) or by agreement.

The parties may agree to accept the decision of the adjudicator as finally determining the dispute.

(4)   The contract shall also provide that the adjudicator is not liable for anything done or omitted in the discharge or purported discharge of his functions as adjudicator unless the act or omission is in bad faith, and that any employee or agent of the adjudicator is similarly protected from liability.

(5)   If the contract does not comply with the requirements of subsections (1) to (4), the adjudication provisions of the Scheme for Construction Contracts apply.

(6)   For England and Wales, the Scheme may apply the provisions of the Arbitration Act 1996 with such adaptations and modifications as appear to the Minister making the scheme to be appropriate.

For Scotland, the Scheme may include provision conferring powers on courts in relation to adjudication and provision relating to the enforcement of the adjudicator's decision.'

The wording of Section 108(5) is interesting when compared with the wording of Section 110(3) on payments. Section 110(3) reads 'If or to the extent that a contract does not contain such provisions ... the relevant provisions of the Scheme for Construction Contracts apply'. In short if contractual provisions for adjudication are non-compliant with the Act then for any adjudication under the Act the statutory Scheme applies in its entirety – whereas non-compliant payment provisions are displaced only on a partial basis.

The Scheme for Construction Contracts

The Schemes referred to in Sections 108(5) and 108(6) of the Act were published as Regulations in March 1998 and came into force on 1 May 1998. The official title of the English law version is the Scheme for Construction Contracts (England and Wales) Regulations 1998.

The Scheme provides a detailed set of rules for adjudication under the Act including:

- notices of adjudication
- appointment of the adjudicator
- service of referral notice
- powers of the adjudicator
- time allowed for adjudication
- adjudicator's decision
- effects of the decision.

It is not intended to review the details of the Scheme in this book; numerous other works are devoted entirely to that task. However, it is worth noting that the courts have shown a firm determination to make adjudication an effective dispute resolution process and they positively enforce adjudicator's decisions, notwithstanding evidence of errors in the decision providing they are satisfied that the adjudicator has jurisdiction.

28.3 *Avoidance of disputes and matters of dissatisfaction*

Clause 66 of the Seventh edition commences at 66(1) with an explanatory provision which carries the marginal note 'avoidance of disputes'. Clause 66(1) states:

> 'In order to overcome where possible the causes of disputes and in those cases where disputes are likely still to arise to facilitate their clear definition and early resolution (whether by agreement or otherwise) the following procedures shall apply for the avoidance and settlement of disputes.'

The purpose of this is presumably to emphasise, with a view to showing that the adjudication provisions are compliant with the Act, that clause 66 as a whole is not solely concerned with the settlement of disputes and parts of it precede the formation of disputes.

Well intentioned as the clause may be it is potentially either heavy-handed or not to be taken literally in its mandatory tone. The reality is that the majority of disputes are avoided by give and take, negotiations or common-sense, or withdrawal from untenable positions. It cannot be the intention of the Conditions to put all attempts at avoidance of disputes on the formal footing laid out in clause 66.

Matters of dissatisfaction

Clause 66(2) provides that if at any time:

(a)   the contractor is dissatisfied with any act or instruction of the engineer's representative or any other person responsible to the engineer or
(b)   the employer or the contractor is dissatisfied with any decision, opinion, instruction, direction, certificate or valuation of the engineer or with any other matter arising under or in connection with the contract or the carrying out of the works

the matter of dissatisfaction shall be referred to the engineer who shall notify his written decision to the employer and the contractor within one month of the reference to him.

As explained earlier in this chapter the reason for the inclusion of these provisions referring matters of dissatisfaction to the engineer for his decision rather than referring disputes for decision, is to make reference to the engineer a condition precedent to the formation of a dispute. This, it was hoped by the draftsmen, would preserve the traditional first stage procedure of the ICE dispute resolution process without rendering the newly intro-duced adjudication procedures non-compliant with the Act. In the light of the judgment in *Mowlem* v. *Hydra-Tight* (2000), that hope may not have been realised but nevertheless the provisions on dissatisfaction remain applicable (subject to certain doubts expressed below) as the starting point for the contractual dispute resolution procedures which follow in clause 66 (as distinct from necessarily being applicable for statutory adjudication).

Act or instruction of the engineer's representative

In the unamended version of the Sixth edition (i.e. pre the 1998 amendments) the provisions now in clause 66(2)(a) requiring the contractor to refer dis-satisfaction with acts or instructions of the engineer's representative to the engineer appeared in clause 2(7). Under that clause however the decision of the engineer did not carry the status of a clause 66 decision. Clause 2(7) served as a clause of practical rather than contractual effect – albeit that it would have been improved by the imposition of a time limit on the contractor in registering his dissatisfaction so as to prevent him acting on an instruction first and complaining about it later.

As the clause now stands in the Seventh edition it puts the contractor's dissatisfaction with acts or instructions of the engineer's representative (or other persons responsible to the engineer) on a par with matters of dissatisfaction between the parties or with the engineer. It is questionable whether this is sensible or necessary.

Dissatisfaction on other matters

Clause 66(2)(b) includes much of the wording of the old clause 66(1) in its reference to decisions, opinions, certificates etc. of the engineer but it deliberately avoids reference to disputes between the parties. Instead it refers to dissatisfaction of either of the parties with decisions etc. of the engineer and 'any other matter arising under or in connection with Contract or the carrying out of the Works'. This somewhat artificial and tangential approach to determination of whether the parties are in dispute may not be sufficient to make the reference to the engineer a condition precedent to arbitration in the same way that reference of a dispute to the engineer was a condition precedent to arbitration under earlier editions of the Conditions. In those editions the link between the engineer's decision on disputes and arbitration of disputes was firm and clear. In the Seventh edition a link is obviously intended between the engineer's decision on a matter of dissatisfaction and the arbitration of any corresponding dispute but it will probably take a court ruling to establish whether a decision on something which the contract says is not a dispute can be a condition precedent to the arbitration of the dispute.

'at any time'

The opening words of clause 66(2) 'If at any time' are expressed without restriction as to whether they mean at any time before completion, before issue of the defects correction certificate, or before limitation periods expire. They might even be taken to run time beyond normal limitation periods.

At the very best the words convey an unfortunate, and no doubt unintended, lack of urgency. Good contract/project management usually includes for the prompt resolution of all matters in dispute and restriction on the late generation of disputes. It may be arguable that there is an implied term in clause 66(2) that if 'at any time' a matter of dissatisfaction arises it shall be referred to the engineer for decision without delay or within a reasonable time. This is an attractive interpretation but there are sufficient arguments against the implied term to make it unlikely to be correct. One is that if the operation of clause 66(2) is intended to affect the parties' legal rights, particularly by way of exclusion, the words need to be expressed and not implied.

A minor practical point on 'any time' is how long the engineer remains in post after the issue of the defects correction certificate so as to be able to decide on matters of dissatisfaction referred to him.

'any decision'

The inclusion in clause 66(2)(b) of the words 'any decision' cannot, it is suggested, be intended to apply to a decision given by the engineer under clause 66 itself. If it did so apply either party could frustrate the operation of the subsequent provisions of the clause by repeated notices of dissatisfaction.

28.4 The engineer's decision

Clause 66(2) requires the engineer to notify the employer and the contractor in writing of his decision on any matter of dissatisfaction within one month of its referral to him. The three month limit stated in earlier editions of the Conditions for decisions given after substantial completion of the whole of the Works is not retained in the Seventh edition.

Failure by the engineer to give his decision within the one month time limit or to give it at all is clearly breach of his duty under clause 2(1)(a) – unless of course, as sometimes happens, the matter for decision is put to the engineer in such vague terms that it cannot be decided without explanation or clarification.

Late decisions

The question of whether a late decision is a valid decision came up in the case of *ECC Quarries Ltd* v. *Merriman Ltd* (1988) under an ICE Fifth edition contract. The engineer failed to give his clause 66 decision within three months of being requested to do so by the contractor. The contractor then wrote to the engineer giving him a reasonable time to deal with the matter. The engineer's decision, which should have been given in May, was eventually given in July. Nine months later the contractor sought to refer the matter to arbitration. The employer successfully obtained from the court a declaration that the contractor was precluded from pursuing his claim as he had failed to give notice to refer within three months of the decision.

Some interesting points came out of the case. The judge held that, if the engineer had purported to give his decision after either party had referred the matter to arbitration, the decision would be null and void because the engineer would have been rendered *functus officio* at the time of the referral. Late delivery of the engineer's decision did not itself however render the decision null and void. The time for giving a decision could be effectively extended by one party acting unilaterally – but note that in this case there was no opposition on that particular point.

Engineers would be well advised to seek the agreement of both parties if they need an extension of time for their decision and the parties would also do well to clarify their respective positions.

In the *ECC* case the employer wisely had the position clarified by the court rather than risk a costly arbitration which would be vulnerable to appeal.

Effect of engineer's decisions

Clause 66(4) of the Sixth edition stated in clear terms that the parties were to give effect to the engineer's clause 66 decisions and that they were final and binding on the contractor and the employer unless and until the recommendation of a conciliator was accepted or the decision was revised by an arbitrator. The Seventh edition states only, again in clause 66(4), that the employer and the contractor shall give effect to every decision of the engineer unless and until that decision is revised by agreement of the parties or pursuant to clause 66 (i.e. by conciliation, adjudication or arbitration).

It is questionable, given the change in wording, if the Seventh edition confers the same status on the engineer's decisions as did the Sixth. It was clear, and established in law, that under the Sixth edition if the parties failed to commence conciliation or arbitration proceedings within the contract stipulated time limits, they were left with decisions to which they had to give effect and which were final and binding. It may be intended that under the Seventh edition decisions become similarly final and binding by the passage of time but it cannot be said with certainty that that is the proper construction of the written terms and it is doubted if a term can be implied on the point. The problem is that an obligation to give effect to a decision does not necessarily carry with it the implication that in giving effect to a decision the parties agree with it or accept it as final and binding.

Status of clause 66 decisions

Whether or not clause 66 decisions have the same effect and status under the Seventh edition as under the Sixth edition, they are still decisions of greater significance than routine decisions made by the engineer under other clauses of the Conditions. It is essential therefore that they are clearly recognisable as clause 66 decisions and that referrals to the engineer for such decisions are clear in their purpose.

In the case *Monmouth County Council* v. *Costelloe & Kemple Ltd* (1965) the Court of Appeal had to consider whether a letter written by an engineer was a clause 66 decision under the ICE Fourth edition. If it was, the contractor was time barred from commencing arbitration. In ruling that the letter was not a clause 66 decision because, amongst other things, the contractor had never asked for one, Lord Justice Harman said this:

> 'The other consideration which moves me is this. This is a process by which the defendants can be deprived of their general rights at law and therefore one must construe it with some strictness as having a forfeiting effect. It is not a penal clause, but it must be construed against the person putting it forward who is, after all, trying to shut out the ordinary citizen's right to go to the courts to have his grievances ventilated. Therefore, I think it would require very clear words and a very clear decision by the appointed person, namely the engineer, to shut the defendants out of their rights.'

The decision-making process

The fact that the engineer's clause 66 decision is different in status from his other decisions under the contract suggests that the decision-making process should be different.

There is nothing fixed on this. Some engineers do take representations from both parties on the matters put to them; others do no more than review their files. The engineer is certainly not acting as an arbitrator, nor as a quasi-arbitrator – a misused phrase but explained by Lord Morris of Borth-y-Gest in *Sutcliffe* v. *Thackrah* (1974) as follows:

> 'There may be circumstances in which what is in effect an arbitration is not one which is in the provisions of the Arbitration Act. The expression quasi-arbitrator should only be used in that connection.'

The prime requirement in the decision-making process is that the engineer must act fairly and impartially. It is essential therefore that if the parties do wish to make representations they are permitted to do so.

The prudent engineer will invite representations, not merely to cover his own position, but because by doing so he improves the prospects of the parties accepting his decision.

Directly employed engineers

Directly employed engineers may sometimes seem to be in a difficult position in reaching a balanced decision. This is how Mr Justice Macfarlan in *Perini* v. *Commonwealth of Australia* (1969) saw the matter (albeit under a different clause). He said:

> 'The second matter which I must mention is the entitlement of the director to consider departmental policy. This point must be judged against a background that the director is the senior officer of the department in New

South Wales, that he is obliged to carry out the orders of his superiors and that he has many duties under this very agreement which he performs as the servant of the Commonwealth, and in the performance of which he is obliged to execute and give effect to departmental policy. I am of the opinion that in discharging the duties imposed upon him by clause 35 he is entitled to consider departmental policy but I am also of the opinion that he would be acting wrongly if he were to consider himself as controlled by it. His overriding duty in performing the function imposed by clause 35 is to give his own decision having regard to the rights and interests of the parties as I have described them. He is thus obliged to consider each application having regard to those rights and interests; he may also consider it from the point of view of departmental policy but the rights and interests must be the only matters involved in the decision. It is irrelevant if departmental policy coincides with the rights and interests of the parties under the agreement, but it would be quite wrong in my opinion, for department policy to govern a particular decision, unless the personal decision of the director having regard to the rights and interests of the parties under the agreement was that rights and interests required it to be applied.'

The decision maker

The engineer is not permitted to delegate the making of his clause 66 decision, and clause 2(4)(c) expressly prohibits this.

There can, however, be a fine line between taking advice from subordinates or colleagues and reaching one's own decision. The engineer who takes a hands-off role in the administration of the contract can be particularly vulnerable to the charge that he has delegated – and if such a charge can be upheld, the decision would, of course, be invalid.

For years many engineers and arbitrators have thought the practice of allowing the phrase 'This matter is being dealt with by Mr . . .' to be used in letters on matters which could not be delegated was stepping too close to the line. However, in *Anglian Water Authority* v. *RDL Contracting* (1988), a case relating to a clause 66 decision under the ICE Fifth edition, Judge Fox-Andrews QC said this:

'In the commercial world many decisions are made by people such as Mr Rouse, who append their signatures to letters drafted by others. It would require compelling evidence to establish in such circumstances that the decision was not that of the signatory. The facts that Mr Baxter was the project engineer and had taken an active part previously in the contract had no probative value.'

The decision might well have been different had the engineer allowed a subordinate to sign his name – a not unusual occurrence but one which should be avoided.

Taking advice

On the matter of the engineer taking advice, this further extract from the judgement in the *Perini* case is worth noting. Mr Justice Macfarlan said:

> 'I cannot accept all the arguments submitted by learned counsel for the plaintiff that the director is bound to investigate every dependent fact himself; this conclusion would, I think, be to ignore the realities of the situation. I am of the opinion, though, that by this agreement and by his mandate he may act upon the findings and opinions of other persons, be they subordinates or independent persons such as architects or meteorological observers; he may also consider and pay attention to the recommendations of subordinates with respect to the very application he is considering. I do agree though that the actual decision must be one which flows from the volition of his own mind and I am of the opinion that it is quite irrelevant that that decision is expressed by the placing of his initials upon the recommendation of a subordinate officer.'

28.5 Disputes

Clause 66(3) states that the employer and the contractor agree that no matter shall constitute a dispute, nor be said to give rise to a dispute, unless and until in respect of that matter:

(a)   the time for the giving of a decision by the engineer on a matter of dissatisfaction under clause 66(2) has expired, or the decision given is unacceptable, or has not been implemented, and, in consequence, the employer or the contractor has served on the other and on the engineer a notice in writing (the notice of dispute)

(b)   an adjudicator has given a decision on dispute under clause 66(6) and either the employer or the contractor is not giving effect to the decision, and in consequence one has served on the other and the engineer a notice of dispute.

The clause concludes by stating that the dispute shall be that stated in the notice of dispute and that for the purposes of all matters arising under or in connection with the contract or the carrying out of the works the word 'dispute' shall be construed accordingly and shall include any difference.

Printing error

The first point to note is that there is almost certainly a printing error in the Conditions in that between clauses 66(3)(a) and 66(3)(b) there should be the word 'or'. That can be seen by comparison with the 1998 amendment to the

Sixth edition which includes the word 'or'. Without the word 'or' clause 66(3) is arguably of very limited effect.

Meaning of dispute

The second point to note is that 'dispute' is not a defined term and that the clause makes a distinction between 'a dispute' and 'the dispute'.

'A dispute' is something which the parties agree does not arise until the engineer has given a decision on a matter of dissatisfaction or the time for giving it has expired. 'The dispute' is that which is stated in the notice of dispute.

Although 'Notice of Dispute' is a capitalised term it is not included in clause 1 of the Conditions as a defined term as might be expected from the convention on capitals employed in the Conditions.

Purpose of clause 66(3)

Clause 66(3)(a) links in with clause 66(2) with the intention of retaining the engineer's decision as a condition precedent to more formalised dispute resolution procedures – including adjudication. As discussed earlier in this chapter it probably does no more than fix the meaning of dispute for the contractual procedures – leaving the meaning of dispute for adjudication under the statutory Scheme to be fixed at law.

Clause 66(3)(b) is difficult to follow from its wording. It appears to suggest that a dispute is not a dispute until an adjudicator has given a decision on the dispute, but what it probably means is that where the parties are in dispute as to whether one or the other has complied with an adjudicator's decision there is no need to go back to the engineer under clause 66(2) with a matter of dissatisfaction before serving a further notice of dispute.

28.6 *Effect on contractor and employer*

Clause 66(4)(a) states that notwithstanding the service of a notice of dispute the parties shall continue to perform their obligations unless the contract has already been determined or abandoned.

Clause 66(4)(b) states:

'(b) The Employer and the Contractor shall give effect forthwith to every decision of
(i) the Engineer on a matter of dissatisfaction given under Clause 66(2) and
(ii) the adjudicator on a dispute given under Clause 66(6)
unless and until that decision is revised by agreement of the Employer and Contractor or pursuant to Clause 66.'

The broad intention of these clauses is reasonably clear although the wording of clause 66(4)(b) might be improved if the word 'or' separated 66(4)(b)(i) from 66(4)(b)(ii) rather than the word 'and'. However, despite the obvious merit of requirements that the parties should honour their obligations under the contract even when in dispute and should give effect to decisions made unless and until revised, there can be difficulties in their application.

Enforcement

For example, the contractor might have difficulty in enforcing payment of an amount considered to be due even with the support of the engineer's certificate and the engineer's decision. Thus in *Enco Civil Engineering Ltd* v. *Zeus International Development Ltd* (1991) under the ICE Fifth edition the employer challenged an engineer's payment certificate and declined to pay, and asked for a decision under clause 66. Before the decision was given the contractor issued a writ for summary judgment on the amount certified. The employer issued a summons to stay proceedings. It was held that the procedure in clause 66 did not prevent the court from ordering a stay; that summary judgment could not be given when the certificate was under review; and that the employer had arguable defences and there was nothing in the ICE Conditions to prevent him setting off cross-claims against sums certified.

Legal rights

Additionally the parties may be able to rely on legal rights outside the contract, such as the right to suspend performance for non-payment conferred by the Housing Grants, Construction and Regeneration Act 1996, to avoid continuance of contractual obligations or giving effect to decisions.

28.7 Conciliation

Clause 66(5)(a) provides that either party may at any time before service of a notice to refer to arbitration under clause 66(9) seek in writing the agreement of the other for the dispute to be considered under The Institution of Civil Engineers' Conciliation Procedure 1999 or any amendment or modification thereof being in force at the date of the notice.

Clause 66(5)(b) provides that if the parties agree to conciliation, any recommendation of the conciliator shall be deemed to have been accepted as finally determining the dispute by agreement unless a notice of adjudication under clause 66(6) or a notice to refer to arbitration under clause 66(9) has been served in respect of that dispute not later than one month after receipt of the recommendation.

Conciliation by agreement only

Under the Sixth edition a party in receipt of an engineer's clause 66 decision could enforce conciliation proceedings providing he served notice of conciliation before the other party served notice of arbitration. In practice that usually meant that the party dissatisfied with the engineer's decision could select whether or not to go to conciliation.

Clause 66(5)(a) makes it clear that under the Seventh edition conciliation is by agreement only.

Conciliation procedures

Clause 66(5)(a) refers to the ICE Conciliation Procedure 1999 or 'any amendment or modification' in force at the date of notice.

Parties should view this, and similar statements which occur in 66(b) in respect of adjudication procedure and clause 66(11) in respect of arbitration procedure, with care. If a new or revised procedure is published whilst the contract is in progress the parties may find themselves bound by rules which they had no knowledge of at the time of making the contract.

The change from the ICE Conciliation Procedure 1988 to the ICE Conciliation Procedure 1994 is one example. The 1988 procedure was merit-based; the 1994 procedure was not. Another example is the change from the ICE Arbitration Procedure 1983 to the ICE Arbitration Procedure 1997. The 1983 procedure was based on the Arbitration Acts 1950 and 1979; the 1997 procedure was based on the Arbitration Act 1996.

When the change is considered significant it is of course open to the parties to argue, as is sometimes done, that the later procedure is not an amendment or modification of the earlier procedure but is a new procedure.

Binding recommendation

The provision in clause 66(5)(b) that a conciliator's recommendation becomes final and binding unless the dispute is referred to adjudication or arbitration within one month, also needs to be watched with caution. Under clause 66(9)(b) an adjudicator's decision does not become final and binding until after 3 months has elapsed, and there is no time limit on an engineer's decision becoming final and binding.

'any recommendation'

Also to be watched with care is the application of the phrase in clause 66(5)(b) 'any recommendation'.

Conciliation is frequently a less structural process than other forms of dispute resolution and the phrase 'any recommendation' could provide scope and status to more than the parties expect or intend.

Conciliation generally

The procedure whereby a neutral person assists the parties to reach a settlement to their dispute can be known as either conciliation or mediation. The terms are not fixed and there is a good argument that flexibility should be maintained in the use of the terms so that the various procedures they encompass can converge and adapt to suit particular circumstances. Nevertheless there is a move, at least in some countries, to ascribe the terms conciliation and mediation to distinct procedures.

This is how the Master of the Rolls, Lord Donaldson, in an address in 1991 to the London Common Law and Commercial Bar Association, explained the distinction:

> 'Conciliation: This is a process whereby a neutral third party listens to the complaints of the disputants and seeks to narrow the field of controversy . . . He moves backwards and forward between the parties explaining the point of view of each to the other. He indulges in an onion peeling operation. He peels off each individual aspect of complaint, inquiring whether that aspect really matters. In the end he and the parties are left with a core dispute which, so much having been discarded under the guidance of the conciliator, at once seems more capable of settlement on common-sense lines.
>
> Mediation: This is what a PR man would describe as "conciliation plus". The mediator performs the functions of a conciliator, but also expresses his view on what would constitute a sensible settlement. In putting forward his suggestion, which the parties will be free to reject, the mediator will in most cases be guided by what he believes would be the likely outcome if no settlement was reached and the matter went to a judicial or arbitral hearing.'

In short, the neutral person takes a more interventionist or positive role in mediation than in conciliation. Sometimes the terms evaluation mediation (for mediation) and facilitative mediation (for conciliation) are used.

Generally the courts have no role to play in conciliation/mediation processes and procedures, although signed agreements reached in such processes are normally legally enforceable. However, the courts have indicated, for example, in the case of *Channel Tunnel Group Ltd* v. *Balfour Beatty Construction Ltd* (1993) that they will stay proceedings to support alternative dispute resolution procedures. Eurotunnel argued that the court could not stay the matter to arbitration because it was to a panel of experts that the matter must first be sent. Lord Mustill granting the stay said:

> '. . . I believe that it is in accordance, not only with the presumption exemplified in the English cases . . . that those who make agreements for

the resolution of disputes must show good reason for departing from them, but also with the interests of the orderly regulation of international commerce, that having promised to take their complaints to the experts and if necessary to the arbitrators, that is where the appellants should go.'

ICE Conciliation Procedure 1999

The ICE Conciliation Procedure 1999 is what many would now describe and recognise as mediation. Its primary objective, as stated in Rule 1.3, is to achieve a settlement to the dispute by agreement as quickly as possible.

As stated in the preface to the Procedure:

'The Conciliator's job is to explore with the Parties their interests, strengths and weaknesses, and perceived needs; to identify possible areas of accommodation or compromise and to search for possible alternative solutions. Anything can be explored which could lead the Parties to an agreed settlement. Where the conciliation follows an Engineer's decision, the Parties are equally free to explore options that were not available to the Engineer.'

As to procedure, the conciliator is permitted to communicate privately and separately with each party without subsequently revealing to the other what he has been told. And as stated in Rule 4.7:

'The Conciliator may consider and discuss such solutions to the dispute as he thinks appropriate or as suggested by any Party. He shall observe and maintain the confidentiality of particular information which he is given by any Party privately, and may only disclose it with the express permission of that Party. He will try to assist the Parties to resolve the dispute in any way which is acceptable to them.'

For the conciliator there can be a problem in endeavouring to assist the parties to reach what is often described as 'an agreement the parties can live with', but then in the absence of agreement being obliged to go on to give recommendations. Rule 5.2 of the Procedure deals with that as follows:

'The Conciliator's recommendation shall state his solution to the dispute which has been referred for conciliation. The recommendation shall not disclose any information which any Party has provided in confidence. It shall be based on his opinion as to how the Parties can best dispose of the dispute between them and need not necessarily be based on any principles of the contract, law or equity.'

28.8 Adjudication

Clause 66(6)(a) provides that either party has the right to refer a dispute as to a matter under the contract for adjudication and either party may give notice in writing to the other at any time of his intention so to do. The adjudication shall be conducted under The Institution of Civil Engineers' Adjudication Procedure 1997 or any amendment or modification in force at the time of the notice of adjudication.

Clauses 66(6)(b)–(f), 66(7) and 66(8) which follow are all intended to make the adjudication scheme within the Conditions compliant with the essential requirements of Section 108 of the Housing Grants, Construction and Regeneration Act 1996. They state:

'66(6)(b) Unless the adjudicator has already been appointed he is to be appointed by a timetable with the object of securing his appointment and referral of the dispute to him within 7 days of such notice

(c) The adjudicator shall reach a decision within 28 days of referral or such longer period as is agreed by the parties after the dispute has been referred

(d) The adjudicator may extend the period of 28 days by up to 14 days with the consent of the party by whom the dispute was referred

(e) The adjudicator shall act impartially

(f) The adjudicator may take the initiative in ascertaining the facts and the law.

66(7) The decision of the adjudicator shall be binding until the dispute is finally determined by legal proceedings or by arbitration (if the contract provides for arbitration or the parties otherwise agree to arbitration) or by agreement.

66(8) The adjudicator is not liable for anything done or omitted in the discharge or purported discharge of his functions as adjudicator unless the act or omission is in bad faith and any employee or agent of the adjudicator is similarly not liable.'

'at any time'

The point is made earlier in this chapter that the entitlement to adjudication 'at any time' is restricted under the Conditions by the matter of dissatisfaction procedure in clause 66(2) and for that reason the adjudication provisions in clause 66 may not be compliant with the Housing Grants, Construction and Regeneration Act 1996.

'a matter under the contract'

The right to adjudication under clause 66, as under the Act, is confined to a dispute as to 'a matter under the contract'. This will not always cover all the matters which are referable to the engineer under clause 66(2), which include matters 'arising under or in connection with the Contract'.

Decision of the adjudicator

Clause 66(9(b) states that where an adjudicator has given a decision under Clause 66(6) in respect of the particular dispute, the notice to refer to arbitration must be served within three months of the giving of the decision otherwise it shall be final as well as binding.

Adjudication generally

Adjudication differs considerably from conciliation in that it is a merit-based process whereby the adjudicator is required to decide the entitlements of the parties in accordance with the contract and the law.

The adjudicator is required to be impartial but not necessarily independent. The extent to which he is obliged to follow the rules of natural justice is not yet clearly defined. In the first court case on adjudication under the Act, *Macob Civil Engineering* v. *Morrison Construction* (1998), the judgment appeared to suggest that the adjudicator was not bound by the restraints and conventions which govern arbitration, but in *Discain* v. *Opecprime* (2000) an adjudicator's decision was not enforced because of breach of the rules of natural justice. The judge in that case distinguishing his decision from *Macob*, suggested there were two types of natural justice – procedural (with no effect on the outcome) and serious (with possible effect on the outcome).

The majority of cases on adjudication which have come before the courts, and they number 100 or so within three years of the Act taking effect, have been for enforcement of decisions on payment given under the Act. For the most part adjudicators' decisions have been upheld and the courts seem determined to uphold the will of Parliament notwithstanding the injustice that that may cause in particular cases.

It is not obvious why the courts should follow the same policy in respect of contractual adjudications, nor is it certain that they will.

ICE Adjudication Procedure 1997

In itself the ICE Adjudication Procedure 1997 is compliant with the Act. It has a number of differences from the Scheme for Construction Contracts but differences are permissible providing the essential requirements of the Act are met.

Under the Procedure the adjudicator has discretion as to how to conduct the adjudication. Rule 5.5 states:

'The Adjudicator shall have complete discretion as to how to conduct the adjudication, and shall establish the procedure and timetable, subject to any limitation that there may be in the Contract or the Act. He shall not be required to observe any rule of evidence, procedure or otherwise, of any court. Without prejudice to the generality of these powers, he may:

(a)   ask for further written information;
(b)   meet and question the Parties and their representatives;
(c)   visit the site;
(d)   request the production of documents or the attendance of people whom he considers could assist;
(e)   set times for (a)–(d) and similar activities;
(f)   proceed with the adjudication and reach a decision even if a Party fails:
      (i)   to provide information;
      (ii)   to attend a meeting;
      (iii) to take any other action requested by the Adjudicator;
(g)   issue such further directions as he considers to be appropriate.'

Under rule 6.1 the adjudicator cannot be required to give reasons for his decisions but it would be unusual for an experienced adjudicator to refuse to do so if requested by both parties.

28.9 Arbitration

Clauses 66(9) to 66(11) of the Conditions deal with the referral of disputes to arbitration.

Clause 66(9)(a) states that all disputes arising under or in connection with the contract or the carrying out of the works other than failure to give effect to a decision of an adjudicator shall be finally determined by reference to arbitration. The party seeking arbitration shall serve on the other party a notice in writing (called the notice to refer) to refer the dispute to arbitration.

Clause 66(9)(b) states that an adjudicator's decision becomes final and binding if not referred to arbitration within 3 months.

Clause 66(10) deals with the procedure for appointment of the arbitrator.

Clause 66(11)(a) states that any reference to arbitration shall be deemed to be a reference under the Arbitration Act 1996 or any later enactment and that the arbitration shall be conducted under the procedure set out in the appendix to the form of tender. It goes on to state that the arbitrator shall have full power to open up, review and revise, any decision, opinion, instruction, direction, certificate or valuation of the engineer or adjudicator.

Clause 11(b) states a point frequently overlooked by the parties that neither party shall be limited in the arbitration to the evidence or arguments put to the engineer.

Clause 11(c) states that the award of the arbitrator shall be binding on all parties.

Clause 11(d) states that unless the parties otherwise agree in writing, any reference to arbitration may proceed notwithstanding that the works are not complete.

Time limits for arbitration

Clause 66(5)(b) sets a one month time limit on referring a dispute to arbitration after a conciliator's recommendation. Clause 66(9)(b) sets a three month time limit after an adjudicator's decision.

Missing from the Seventh edition, although a prominent feature of earlier editions of Conditions, is a time limit following the engineer's decision or the date on which it should have been given. This is a consequence of trying to make clause 66 compliant with the Housing Grants, Construction and Regeneration Act 1996.

Failure to refer within the time limits has the effect of making a conciliator's recommendation or an adjudicator's decision final and binding but it is not clear from the Conditions whether, or how, an engineer's decision can itself become final and binding.

Some indication of the approach by the courts to the application of time limits for arbitration under the Arbitration Act 1996 can be found from the case of *Harbour & General* v. *Environment Agency* (1999) where the court ruled that a mistake by the contractor as to the time for commencing proceedings under the ICE Sixth edition was not sufficient to prevent operation of the time limits. The court held that the position under the 1996 Arbitration Act was different from the position under the Arbitration Act 1950 which gave a discretionary power under section 27 to extend time for the commencement of arbitration proceedings. See for example in *McLaughlin & Harvey plc* v. *P&O Developments* (1991) where a contractor who gave notice to refer to arbitration a day late was granted relief because there was no possible prejudice to the employer, and the hardship that the contractor would suffer would be out of proportion to his fault.

Arbitration generally

Section 1 of the Arbitration Act 1996 sets out the governing principles for arbitration under English law. Section 1 reads as follows:

> 'The provisions of this Part are founded on the following principles, and shall be construed accordingly –
> (a) the object of arbitration is to obtain the fair resolution of disputes by an impartial tribunal without unnecessary delay or expense;
> (b) the parties should be free to agree how their disputes are resolved, subject only to such safeguards as are necessary in the public interest;
> (c) in matters governed by this Part the court should not intervene except as provided by this Part.'

The 1996 Act goes well beyond previous statutes on arbitration in detailing the law. It says enough on the conduct of arbitration proceedings to make special procedures for arbitration largely redundant. Nevertheless the Act did lead to publication of a variety of new arbitration procedures, two of which attract mention in the Seventh edition.

Arbitration procedures

The Sixth edition originally required arbitrations arising from disputes under or in connection with ICE Contracts to be conducted under the ICE Arbitration Procedure 1983 – later replaced by the ICE Arbitration Procedure 1997.

The Seventh edition allows a choice between the ICE Arbitration Procedure 1997 and the Construction Industry Model Arbitration Rules (CIMAR). The choice is exercised by the employer in completion of part one of the appendix to the form of tender. There is not a great deal between the two sets of rules in procedural terms but whereas the ICE Procedure is drafted principally for use with the ICE Conditions, the CIMAR procedure is intended for general application in the construction industry.

Whichever procedure is used, there is always the likelihood that it will contain rules which go beyond details set out in the contract. The question then is how the rules should be applied in determining the parties' rights.

In *Christiani & Nielsen* v. *Birmingham City Council* (1994) the court held that a contractor who had served a notice under the ICE Arbitration Procedure 1983 which did not comply in full with the stated requirements in the Procedure for the notice had nevertheless served an effective notice.

In *Mid Glamorgan County Council* v. *The Land Authority for Wales* (1990) it was held that Rule 4.1 of the ICE Arbitration Procedure 1983 only entitled a party to refer disputes to arbitration which had previously been referred to the engineer for a clause 66 decision and that for Rule 4.2 to apply, entitling the arbitrator to deal with connected issues, the arbitrator had to make a finding of fact that such issues were connected and necessary to determination of matters already referred.

28.10 *Witnesses*

Clause 66(12)(a) states that no decision, opinion, instruction, direction, certificate or valuation given by the engineer shall disqualify him from being called as a witness and giving evidence before a conciliator, adjudicator or arbitrator on any matter relevant to the dispute.

Clause 66(12)(b) states that all matters and information placed before a conciliator shall be deemed to be submitted to him without prejudice and the conciliator shall not be called as witness in connection with any adjudication, arbitration or other legal proceedings arising out of, or connected with, any matters referred to him.

The engineer

The position of the engineer in formal dispute resolution proceedings is frequently difficult. He or his firm may have a liability to the employer if the contractor's claims are upheld. Professional indemnity insurers may be involved or may wish to be involved. Nevertheless it is better that the engineer appears in such proceedings as a witness rather than as an advocate or member of the employer's support team if he is to preserve the requirement for impartiality imposed by clause 2(7) of the Conditions.

The conciliator

Although information may be put before a conciliator on a 'without prejudice' basis, that is not to say such information is privileged for the purpose of other proceedings. If that were the case it would take out of circulation much essential material. As to whether the conciliator can be called as a witness, clause 66(12)(b) is probably effective only so far as the confidential parts of the conciliation are concerned. If there are disputes as to any agreements made between the parties in joint session it is questionable whether the conciliator is protected or precluded from being called as a witness.

Chapter 29
Miscellaneous provisions

29.1 Introduction

This concluding chapter covers clauses in the Conditions not considered in earlier chapters and which for the most part are clauses of miscellaneous nature found towards the end of the Conditions.

Clauses considered here are:

- clause 63 – frustration/war clause
- clause 67 – application to Scotland/Northern Ireland
- clause 69 – tax matters
- clause 70 – value added tax
- clause 71 – CDM regulations
- clause 72 – special conditions.

Clause 68 (notices) is considered in Chapter 2 on definitions and notices.

Frustration/war clause

In the Sixth edition frustration was covered in clause 64 and provisions on the outbreak of war in the lengthy clause 65. In the Seventh edition frustration and war are covered together in clause 63.

Application to Scotland/Northern Ireland

ICE Conditions have long contained a clause to the effect that if the works are situated in Scotland the contract shall be construed and shall operate as a Scottish contract in accordance with Scots Law. The 1993 corrigenda introduced a similar clause into the Sixth edition in respect of Northern Ireland.

Clause 67 of the Seventh edition has provisions relating to application to Scotland, application to Northern Ireland and application 'elsewhere'.

Tax matters

ICE Conditions have never contained, as have some other standard forms, clauses stating generally that risks of additional cost and time arising from changes in statute are to be carried by a particular party – usually the employer. In the absence of such clauses the risks lie where they fall – which is usually on the contractor.

However, ICE Conditions have traditionally put the risk of fluctuations arising from labour taxes (national insurance contributions and the like) on the employer. In the Fifth and Sixth editions that was done in clause 69.

In the Seventh edition clause 69 is expanded to cover both labour tax fluctuations and landfill tax fluctuations.

CDM Regulations 1994

A clause relating to the Construction (Design and Management) Regulations 1994 was first included in the Sixth edition by an amendment sheet issued in 1995. That clause now appears in the Seventh edition as clause 71.

29.2 *Frustration/war clause*

As stated above, clause 63 of the Seventh edition has provisions relating to both frustration and war. Clause 63(1) is specific to frustration, clause 63(2) is specific to war, and clauses 63(3) to 63(6) cover both situations.

Frustration

Clause 63(1) of the Seventh edition has the marginal note 'Frustration'. The clause reads as follows:

> 'If any circumstance outside the control of both parties arises during the currency of the Contract which renders it impossible or illegal for either party to fulfil his contractual obligations the Works shall be deemed to be abandoned upon the service by one party upon the other of written notice to that effect.'

In the Sixth edition, clause 64 is headed 'Frustration' but it carries the marginal note 'Payment in the event of frustration'. That clause reads:

> 'In the event of the Contract being frustrated whether by war or by any other supervening event which may occur independently of the will of the parties the sum payable by the Employer to the Contractor in respect of the work executed shall be the same as that which would have been payable under Clause 65(5) if the Contract has been determined by the Employer under Clause 65.'

It can be seen that there is a significant difference between these two clauses. The clause in the Seventh edition is, in effect, defining frustration, whereas the clause in the Sixth edition is more concerned with stating what payment should be made in the event of frustration.

Frustration generally

At common law a contract is discharged and further performance excused if supervening events make the contract illegal or impossible or render its performance commercially sterile. Such discharge is known as frustration. A plea of frustration acts as a defence to a charge of breach of contract.

In order to be relied on, the events said to have caused frustration must be:

- unforeseen
- unprovided for in the contract
- outside the control of the parties
- beyond the fault of the party claiming frustration as a defence.

In the case of *Davis Contractors* v. *Fareham UDC* (1956) Lord Radcliffe said:

> 'Frustration occurs whenever the law recognises that without default of either party a contractual obligation has become incapable of being performed because the circumstances in which performance is called for would render it a thing radically different from that which was undertaken by the contract. *Non haec in foedera veni*. It was not this that I promised to do.'

In that case a contract to build 78 houses in eight months took 22 months to complete due to labour shortages. The contractor claimed the contract had been frustrated and he was entitled to be reimbursed on a *quantum meruit* basis for the cost incurred. The House of Lords held the contract had not been frustrated but was merely more onerous than had been expected.

Frustration in construction contracts

Frustration, in the true legal sense of a radical change of obligation, is uncommon in construction contracts.

One of the few recorded cases in the UK is *Metropolitan Water Board* v. *Dick Kerr & Co Ltd* (1918) where the onset of World War I led to a two year interruption of progress. It was held that the event was beyond the contemplation of the parties at the time they made the contract and the contractor was entitled to treat the contract as at an end.

More recently in a Hong Kong case, *Wong Lai Ying* v. *Chinachem Investment Co Ltd* (1979), a landslip which obliterated the site of building works was held by the Privy Council to be a frustrating event.

A brief flurry of excitement was caused in the minds of contractors and claims consultants in 1990 when the judge at first instance in the case of *McAlpine Humberoak Ltd* v. *McDermott International Inc* held that a contract had been frustrated by the increased number of drawings issued. But the Court of Appeal in 1992 firmly rejected this. Lord Justice Lloyd said:

'If we were to uphold the judge's finding of frustration, this would be the first contract to have been frustrated by reason of matters which had not only occurred before the contract was signed, and were not only well known to the parties, but had also been provided for in the contract itself.'

Payment in the event of frustration/abandonment

The rules governing the transfer of monies from one party to the other in the event of frustration are exceedingly complex unless the contract itself provides how losses are to fall.

Hence the purpose of clause 64 in the Sixth edition was not so much to regulate a code of conduct or define frustrating circumstances. It was more to ensure that the contractor was paid for the work he had done and was not forced by the Law Reform (Frustrated Contracts) Act 1943 into proving that the employer had received valuable benefit.

In the Seventh edition the scheme for payment on abandonment is set out in clause 63(4).

Termination in the event of frustration

In the Seventh edition the purpose of clause 63(1) is to define those circumstances which, should they arise, would lead to the contract being deemed to be abandoned.

Under clause 63(1) any such circumstance must:

- be outside the control of both parties
- arise during the currency of the contract (that is, be a supervening event)
- render impossible or illegal the fulfilment of contractual obligations by either party.

Clause 63(1) further requires written notice to be served by one party on the other that abandonment is deemed to have occurred.

The question of whether the scope of clause 63(1) is wider or narrower than common law frustration is best left to lawyers, but two points may be worth brief comment:

- the phrase 'fulfil his contractual obligations' in clause 63(1) looks potentially wider than the common law test of performance. Perhaps it could be argued that if any, rather than major, contractual obligations are impeded the contract is deemed to be abandoned
- the contractor's general obligations as derived from clauses 8(1) and 13(1) are to construct and complete the works 'save in so far as it is legally or

physically impossible'. These would appear to apply to pre-existing as well as to supervening events. The Conditions may, therefore, have made physical or legal impossibility a defence to non-performance on a much broader scale than would be implied by the doctrine of frustration.

Termination in the event of war

The provisions in the Seventh edition for termination in the event of war are shorter than those in earlier editions.

Clause 63(2)(a) provides that if there is an outbreak of war the contractor is to continue to execute the works as far as possible for a period of 28 days.

Clause 63(2)(b) provides that if substantial completion is not achieved within the period of 28 days the works are deemed to have been abandoned unless the parties otherwise agree.

Effects of abandonment

Clauses 63(3) to 63(6) of the Conditions deal with the effects of abandonment whether arising from frustration or war.

Removal of contractor's equipment

Clause 63(3) requires the contractor to remove all contractor's equipment from the site with all reasonable dispatch.

In the event of the contractor failing to do so the employer is given powers to dispose of the contractor's equipment and to retain his costs before accounting for the proceeds.

Payment on abandonment

Clause 63(4) provides that the contractor is to be paid on abandonment:

- the value of all work carried out
- proportionate amounts in respect of preliminaries
- costs of goods and materials for which the contractor is legally obliged to accept delivery
- any expenditure reasonably incurred in expectation of completing the whole of the works
- the reasonable cost of removing contractor's equipment.

Clause 63(4) further provides that the timescale for finalisation of accounts in clause 60(4) applies as if the date of abandonment was the date of 'issue' of the defects correction certificate.

The word 'issue' may have been inserted in the clause unintentionally because clause 60(4) itself uses the phrase 'the date of the Defects Correction Certificate'.

Works substantially complete

Clause 63(5) deals with allowances for remedying defects and also the release of retention money.

The contractor may at his discretion allow a sum to be set off against the amount due to him on abandonment in lieu of his obligation in respect of making good defects.

The employer is not entitled to withhold payment of retention money except as part of any sum agreed as above.

Contract to continue in force

Clause 63(6) states that, save as 'aforesaid', by which it presumably means the earlier provisions of clause 63, the contract shall continue to have full force and effect.

This would seem to apply principally to the disputes resolution procedures of clause 66. It is difficult to see how much else of the contract can continue to have full force and effect after abandonment.

29.3 Application to Scotland, Northern Ireland and elsewhere

Clause 67 is in four parts:

- 67(1) – application to Scotland generally
- 67(2) – particulars as to application to Scotland
- 67(3) – application to Northern Ireland
- 67(4) – application elsewhere.

Application to Scotland

Clause 67(1) provides that if the works are situated in Scotland the contract shall:

- in all respects be construed as a Scottish contract
- operate as a Scottish contract
- be interpreted in accordance with Scots law
- be subject to the provisions of clause 67(2).

Clause 67(2) details the changes necessary to clause 66 in respect of arbitration procedures to bring the Conditions into line with Scots Law. These are:

- 67(2)(a) – 'arbiter' is substituted for 'arbitrator'
- 67(2)(b) – any reference to the 'Arbitration Act 1996' shall be substituted by reference to the law of Scotland and/or Section 66 and Schedule 7 of the 'Law Reform (Miscellaneous Provisions) (Scotland) Act 1990' as appropriate
- 67(2)(c) – any reference to 'the Institution of Civil Engineers' Arbitration Procedure (1997)' or 'the Construction Industry Model Arbitration Rules' shall be substituted by a reference to 'the Institution of Civil Engineers Arbitration Procedure (Scotland) (1983)' or any amendment or modification in force at the time of the appointment of the arbiter
- 67(2)(d) – nothing in any other provisions of the Conditions shall be construed as excluding or affecting the right of a party to call for the arbiter to state a case under Section 3 of the Administration of Justice (Scotland) Act 1972
- 67(2)(e) – where either party wishes to register the decision of an adjudicator in the Books of Council and Session for preservation and execution the other party shall consent to such registration.

Application to Northern Ireland

Clause 67(3) provides, in like manner to clause 67(1), that if the works are situated in Northern Ireland the contract shall:

- in all respects be construed as a Northern Irish Contract
- operate as a Northern Irish contract
- be interpreted in accordance with the law of Northern Ireland.

Application elsewhere

Clause 67(4) provides that where the works are situated outside a country or jurisdiction other than England and Wales, Scotland or Northern Ireland the contract, and the provisions for the settlement of disputes, shall be construed, shall operate, and shall be interpreted in accordance with the law of that country or jurisdiction.

Thus for works in the Channel Islands the contract is subject to the laws of the particular Channel Island.

29.4 Tax matters

Standard forms vary considerably in the manner in which risks arising from statutory changes are allocated. In the IChemE model forms the risks are placed on the purchaser (employer); in the New Engineering Contract the risks are on the contractor unless a secondary option clause is selected to transfer the risk to the employer; in ICE Conditions the risks generally are left with the contractor except for certain specific risks, which in the Seventh edition are those arising from labour tax and landfill tax fluctuations.

Clause 69(1) states that the rates and prices in the bill of quantities are deemed to take account only of the levels and incidence of labour taxes and landfill taxes in force at the date for return of tenders. Such rates and prices are not deemed to take account of known or foreseeable changes taking place at some later date.

Clause 69(1)(a) applies to tax levies, national insurance contributions etc. payable under statute (but excluding income tax) either by the contractor or his subcontractors in respect of workpeople engaged on the contract.

Clause 69(1)(b) applies to any landfill tax payable by the contractor or his subcontractors.

Clause 69(2) provides that:

- if after the date for return of tenders
- there is a change of level and/or incidence in such taxes
- the contractor shall inform the engineer, and
- the net increase or decrease shall be taken into account in arriving at the contract price.

The contractor is to provide information necessary to support any adjustment of the contract price and the amount is to be reflected in certificates of payment issued after submission of such information.

Usage of clause 69

The wording of clause 69 leaves much to be desired by way of certainty. It is not clear, for instance:

- whether it applies to labour only subcontractors
- whether it applies to sub-subcontractors
- whether it applies to workpeople off-site
- whether it applies to staff
- how it operates if taxes reduce but the contractor fails to notify the engineer or provide information necessary for changing the contract price.

In the Sixth edition where clause 69 related only to labour taxes the clause was frequently deleted by the employer and even if not so deleted, was frequently not operated by the contractor because of the complexities of its administration and calculations.

The landfill tax part of the clause is much simpler to operate and the probability is that under the Seventh edition this part will be put into general use.

29.5 *Value added tax*

The general rule seems to be that prices for services or goods are deemed to be inclusive of value added tax unless it is expressly excluded. However, to the extent that it is custom to give prices excluding value added tax the general rule may be modified. To put the matter beyond doubt most contracts state whether rates and prices are inclusive or exclusive of value added tax.

Clause 70(1) of the Conditions does that by stating that the contractor shall be deemed not to have allowed in his tender for value added tax.

Certificates

Clause 70(2) provides that all certificates issued by the engineer under clause 60 shall be net of value added tax and that in addition to payments due under certificates, the employer shall separately identify and pay to the contractor any value added tax properly chargeable under the contract.

Various schemes are operated to overcome the difficulty of contractors becoming liable for value added tax before its receipt from the employer.

Disputes on value added tax

Application of the value added tax regime to construction contracts can be exceedingly complex. There can be uncertainty on whether all or parts of the works are fully rated, zero rated or excluded. There can be uncertainty on whether certain payments, for instance damages paid in respect of breach of contract or other amounts paid in settlement of claims, attract value added tax.

Clause 70(3) provides that if disputes arise between either the employer or the contractor and Customs and Excise in relation to value added tax in connection with the works or the contract, each shall render to the other such support and assistance as necessary to resolve the disputes.

Clause 70(4) states that clause 66 shall not apply to any disputes or differences relating to value added tax.

This confirms that neither the engineer, an adjudicator nor an arbitrator has the power to make decisions on value added tax which bind the parties or Customs and Excise.

29.6 *CDM Regulations 1994*

The Construction (Design and Management) Regulations 1994 apply to all construction contracts except site surveys, certain quarrying activities, certain small projects and construction of a house for a private client.

Under the Regulations the employer is obliged to appoint as soon as practicable a planning supervisor to prepare and monitor a health and safety plan for the contract. In contracts under ICE Conditions the name of the planning supervisor is stated by the employer in the appendix to the form of tender.

The employer is also obliged to appoint as soon as practicable a principal contractor with responsibility for overall safety on the project.

It is common, but not essential, in contracts under ICE Conditions for the engineer to be appointed planning supervisor and for the contractor to be appointed principal contractor. Clause 71 of the Conditions deals with the consequences of the engineer giving instructions to the contractor when they occupy those roles.

Definition

Clause 71(1) defines what is meant by the Regulations, planning supervisor, principal contractor and health and safety plan.

Actions to be stated in writing

Clause 71(2) states that where the Regulations apply to the works and the engineer is planning supervisor and/or the contractor is the principal contractor, then in taking any action under the Regulations they are to state in writing that action is being taken under the Regulations.

Actions deemed to be engineer's instructions

Clause 71(3)(a) states that any action under Regulations taken by either the planning supervisor or the principal contractor, and in particular any alteration or amendment to the health and safety plan, shall be deemed to be an engineer's instruction pursuant to clause 13. It goes on to say that the contractor shall not be entitled to any additional payment and/or extension of time in respect of any such action to the extent that it results from any action, lack of action or default on his part.

Clause 71(3)(b) states that if any action by either the planning supervisor or the principal contractor could not in the contractor's opinion reasonably have been foreseen by an experienced contractor, the contractor shall as early as practicable give written notice thereof to the engineer.

The intention and effect of clause 71(3) is to enable the contractor to recover delay and extra cost arising from instructions given by the planning supervisor, whether or not the planning supervisor is the engineer, as though such instructions are instructions given by the engineer under clause 13 of the Conditions.

29.7 *Special clauses*

Clause 72 in the printed version of the Seventh edition is not a condition of contract as such but simply a suggested starting point for the inclusion of special conditions.

However, the printed version does include a specimen set of clauses for contract price fluctuations (i.e. inflationary or deflationary changes) and the Guidance Notes for the Seventh edition contain a useful appendix on the interpretation and application of the price fluctuation clauses. These clauses can be included in the contract as clause 72 (in which case other special clauses follow on as clause 73 etc.) but since the mid 1990s, with inflation under control, it has no longer been common practice to include price fluctuation clauses in contracts of less than 2 years duration.

Table of Cases

Note: references are to chapter sections.
The following abbreviations of law reports are used:

AC   Law Reports Appeal Cases Series
ALJR   Australian Law Journal Reports
All ER   All England Law Reports
BLISS   Building Law Information Subscriber Services
BLR   Building Law Reports
CILL   Construction Industry Law Letter
CLY   Construction Law Yearbook
Con LR   Construction Law Reports
Const LJ   Construction Law Journal
DLR   Dominion Law Reports
Exch   Exchequer Reports
Giff   Gifford's Reports
Greens WD   Greens Weekly Digest
HBC   Hudson's Building Contracts
KB   Law Reports, King's Bench Division
LGR   Local Government Reports
Lloyd's Rep   Lloyd's List Law Reports
NZLR   New Zealand Law Reports
ORB   Official Referees Business
SALR   South African Law Reports
SCLR   Scottish Civil Law Reports
SCR   Canada Supreme Court Reports
TR   Term Reports
WLR   Weekly Law Reports

Index

Note: references are to chapter sections.